WÖHL

OF

ORGANIC CHEMISTRY.

BY

RUDOLPH FITTIG, PH.D., NAT. SC. D.,

PROFESSOR OF CHEMISTRY IN THE UNIVERSITY OF TÜBINGEN.

TRANSLATED

FROM THE

EIGHTH GERMAN EDITION, WITH ADDITIONS,

BY

IRA REMSEN, M.D., PH.D.,

PROFESSOR OF CHEMISTRY AND PHYSICS IN WILLIAMS COLLEGE, MASSACHUSETTS.

PHILADELPHIA:

HENRY C. LEA.

1873

PHILADELPHIA:

COLLINS, PRINTER, 705 JAYNE STREET.

PREFACE

TO THE

AMERICAN EDITION.

In presenting this book to the American scientific public, I need only, as an excuse, refer to the success which it has met in Germany, as indicated by the appearance of eight editions in rapid sequence.

The grounds of its success may, in part, be looked for in the fact that it is adapted as well to the use of beginners as to that of those advanced in the science. The beginner will find a simple principle of classification, carefully carried out, eminently fitted to his first object of obtaining a general view of the subject; the advanced will find it exceedingly rich in statements of facts with which he has constantly to deal.

The year that has elapsed since the appearance of the last German edition, with its quota of investigations in this branch of science, has caused the necessity of a revision in order that the work might be equal to its avowed object. The additions and corrections have been made as nearly as possible in the spirit of the original, with the view merely of rendering the book a representative of the science at the date of publication.

An introductory chapter on the "Constitution ot Chemical Compounds" has been prefixed in order to aid the beginner in his attempt to comprehend certain terms, upon which he would otherwise, perhaps, stumble at the very outset of his study, and to render his entrance into the apparently labyrinthic structure somewhat less dark and indefinite.

The time, during which the strict division of Chemistry into Inorganic and Organic was held upright, has long passed away, and we now recognize that this division is merely conventional, intended to aid the work of classification. There is but one chemistry, but one set of laws govering the formation, existence, and decomposition of chemical compounds. The compounds of carbon, owing, in general terms, to their comparative instability and other properties, are, however, particularly susceptible to the action of reagents, and are, hence, particularly adapted to the uses of the investigator, who is endeavoring to discover the secrets of the science. Hence, further, most of the great advances of chemistry of late years have been due to the results of the study of chemical phenomena in connection with so-called organic compounds, and the subsequent application of the results obtained to the whole field.

It is, therefore, natural that of late the attention of Americans should have been attracted towards this field; and there begin to be slight indications of a desire on their part to aid in clearing up its many mysteries, a work in which for some years the Germans have been engaged almost to the exclusion of the chemists of other countries. Should the publication of this work tend in the slightest degree to increase this desire, I shall feel that my labor has not been in vain.

IRA REMSEN.

WILLIAMSTOWN, Mass., October, 1872

AUTHOR'S PREFACE

TO THE

SEVENTH EDITION.

At the desire of Professor Wöhler, I have with pleasure again undertaken the preparation of the present edition of the "Outlines," required by the publishers. Since the appearance of the last edition, however, such great advances have been made in the field of Organic Chemistry, that the book demanded a material transformation to place it in concordance with the later theoretical views; and the entire rewriting of several sections was necessitated. The principle upon which it is based remains, however, the same as before. It is not intended to be a text-book of Organic Chemistry in the usual acceptation of the term, but a guide in connection with instruction. Hence, facts have been placed in the foreground, and particular attention has been given to the occurrence, the formation, and the characteristic properties of individual compounds. The development of theoretical relations, the demonstration of the connections between the various groups of bodies, and of the general laws which govern them, must be left to the teacher in his oral exercises.

In the treatment of the individual groups, the guiding principle throughout has been this: Of every homologous series, that compound, which is most thoroughly investigated, and which may be considered as a type of the whole series (as, for instance, ethyl alcohol in the series of saturated, monatomic alcohols, acetic acid in the fatty-acid series), is, with its derivatives, considered very exhaustively; while, for the other members of the same series, only the physical properties of the more important ones are briefly stated, and their characteristic derivatives mentioned.

Although the book is intended as a guide in first instruction in Organic Chemistry, it still contains much more than is required for this purpose. A practical object was here kept in view. The teacher is, of course, always obliged to confine himself in first instruction to the drawing of a sketch of the science in rough outlines, as it were; for the student, however, a course of lectures of this character undoubtedly becomes much more comprehensible, if the material is at hand, by the aid of which he can, in private study, follow the general rules and laws more in detail.

Further, the book is designed for reference in connection with laboratory work, and, in order to make it comply with this object, it was necessary to embody in it a great deal that could otherwise have been omitted.

R. FITTIG.

GÖTTINGEN, 1868.

PREFACE TO THE EIGHTH EDITION.

THE rapid advances in the field of Organic Chemistry have again necessitated the entire rewriting of some parts of the "Outlines." The same principle has been followed as in previous editions, but the system of selecting the hydrocarbons as the starting-points in the consideration of all the other groups of bodies, has been more rigidly carried out than formerly; and more attention has been paid to those isomeric relations which are theoretically possible, and those which have been really observed in the individual groups.

RUD. FITTIG.

TÜBINGEN, 1871.

TABLE OF CONTENTS.

II. BENZENE DERIVATIVES (AROMATIC COMPOUNDS).

FIRST GROUP.

CONSTITUTION OF CHEMICAL COMPOUNDS.

It is noticed that certain elements combine with each other in only one proportion, forming thus but one kind of compounds. If we take, for instance, hydrogen and chlorine, and allow them to combine under the most varied conditions, the result is always hydrochloric acid, and this always contains 35.5 parts by weight of chlorine to 1 part by weight of hydrogen. The same is true of a number of other elements, as bromine, iodine, potassium, sodium, etc. Further, we notice that in the case of other elements, as oxygen and sulphur, nitrogen and phosphorus, carbon and silicium, a greater variety presents itself in their combinations, not only with each other, but with the elements of the first class referred to. Oxygen combines with hydrogen in two proportions, forming water and hydrogen peroxide; nitrogen combines with oxygen in five proportions, forming nitrous oxide, hyponitric acid, nitrogen binoxide, nitrous anhydride, and nitric anhydride. This distinction, between elements that combine with each other only in one proportion, and those which combine with each other and all other elements in more than one proportion, is fundamental and characteristic. The recognition of this distinction led to the acceptation of the hypothesis of the valence of elements. This hypothesis may be stated as follows: Every atom of an element has an inherent power of holding in combination a certain number of other atoms of known combining power. The simplest examples of this principle, we find in the first class of elements mentioned above; they combine with each other in only one proportion, *i. e.*, each atom can retain in combination only one other atom of any kind, and its combining power, as well as that of the atom with which it is united, represents the unit of this power. The atoms of such elements are said to possess one *affinity;* and the elements are called *monovalent.*

In order to determine which elements are monovalent,

we have to subject the formulæ and nature of their compounds, as far as they are known, to the most careful study. We thus find, in the first place, that hydrogen, chlorine, bromine, iodine, etc., are monovalent. Having once established this fact, knowing which elements are monovalent, we have a basis upon which we can work to determine the valence of other elements. Here again the determination of the empirical formulæ of the compounds, of the elements to be investigated, with monovalent elements must be the first step in the inquiry.

If we take oxygen, for example, we find that its simplest compound with hydrogen is water, and by the aid of familiar means we determine its formula to be H^2O, *i. e.*, it consists of two atoms of hydrogen united with one atom of oxygen. Hence we see that in this case, the atom of oxygen exhibits a combining power, twice as great as that of hydrogen, and, not finding any fact to conflict with this, we say that oxygen is a *bivalent* element—its free atom possesses two free affinities. In a similar manner we find that sulphur, selenium, tellurium, etc., are also bivalent.

Proceeding further, nitrogen, phosphorus, arsenic, and other elements are found to possess three times the combining power of the monovalent elements; their simplest compounds with hydrogen are NH^3, PH^3, AsH^3, etc. Elements of this class are called *trivalent*.

Carbon, silicium, etc., are *tetravalent*, or the uncombined atoms of these elements possess four free affinities. Their simplest hydrogen compounds are CH^4, SiH^4, etc.

In this way all the elements have been classified into groups, the individual members of which are said to be *monovalent*, *bivalent*, *trivalent*, *tetravalent*, or *pentavalent*. The elements are designated by the names *monads*, *dyads*, *triads*, *tetrads*, *pentads*, etc. This subdivision is dependent merely upon the combining powers of the elements, and tells us merely that the atoms of the elements of each group can unite with, or hold in combination, a certain number (indicated by the name) of monovalent atoms, such as hydrogen, chlorine, etc. When we say an element is monovalent, bivalent, trivalent, etc., we intend merely to say that each one of its atoms possesses the power of combining with one, two, three, etc., monovalent atoms or atomic units; that each one of its atoms in the free state possesses one, two, or three free affinities.

Now compounds are formed by virtue of the mutual action of these free affinities upon each other; and, the compounds once formed, the affinities are no longer free. Upon this mutual neutralization or *saturation* of free affinities are based our fundamental ideas in regard to the *constitution* of chemical compounds, or chemical *structure*. The compounds of the elements with hydrogen alone are very simple. We have hydrochloric acid, for instance, consisting of one atom of hydrogen united with one atom of chlorine; and the molecule of the compound is represented by the formula, H.Cl; and so also for hydrobromic acid, H.Br., etc. For water we have H.O.H or $O \begin{cases} H \\ H, \end{cases}$ which signifies that each of the free affinities of the bivalent oxygen atom is *saturated* by a hydrogen atom; for ammonia we have $N \begin{matrix} .H \\ .H \\ \cdot H \end{matrix}$ or $N \begin{cases} H \\ H \\ H \end{cases}$; for marsh gas, CH^4, we have $\begin{matrix} H. & & .H \\ & C & \\ H\cdot & & \cdot H \end{matrix}$ or $C \begin{cases} H \\ H \\ H \\ H. \end{cases}$ These formulæ indicate the *constitution* of the compounds, *i. e.*, *the arrangement of the atoms in the molecule*. By the expression "arrangement of the atoms in the molecule," however, we do not intend to go so far as to refer to the actual relative position of the atoms in space, as our present knowledge will not permit conclusions of any value in regard to this point. We only mean to give an account of the employment of the affinities of the atoms, which are the essential causes of the formation of the molecules. In the case of hydrochloric acid, for instance, we mean that the one free affinity originally possessed by the hydrogen-atom, and that possessed by the chlorine-atom, as inherent, characteristic powers, have been mutually satisfied, and, ceasing to be *free affinities*, now perform a function in holding together the two atoms, in order to form the molecule of the compound. It may be here mentioned that the so-called free affinities are in almost all cases never free except for an infinitesimally short space of time. An atom of hydrogen or of chlorine does not exist in a free condition, but, if nothing else be present with which it can combine, it combines with another atom of the same kind, forming a molecule of the element instead of a molecule of a compound. The molecule of hydrogen, or of chlorine, has the same chemical constitution as hydrochloric acid, *i. e.*, it consists of one monovalent atom united with another monovalent atom, thus H.H and Cl.Cl; and, in order that the compound H.Cl may be formed, it is necessary that the union of atoms already existing be broken up; which is accomplished by virtue of the stronger affinity of the hydrogen-atom for the chlorine-atom, than of the hydrogen-atom for hydrogen-atom, or the chlorine-atom for chlorine-atom. Thus the affinities, as stated, are not free, but the instant they become free they are taken up, neutralized, *saturated* by those of other atoms present.

In the simple cases which we have considered, viz., hydrochloric acid H.Cl, water H.O.H, ammonia $N\begin{matrix}\cdot H\\ \cdot H\\ \cdot H,\end{matrix}$ and marsh gas $\begin{matrix}H\cdot & & \cdot H\\ & C & \\ H\cdot & & \cdot H,\end{matrix}$ the constitution of the compounds is plainly indicated by the formulæ. By replacing one or all of the elements in the above formulæ by other elements of equal valence, we have the formulæ of a number of known compounds:—

H.Cl	similar to	H.Br, Na.Cl, K.Br, etc.
H.O.H	"	K.O.H, Cl.O.H, Cl.O.Cl, H.S.H, etc.
$N\begin{matrix}\cdot H\\ \cdot H\\ \cdot H\end{matrix}$	"	$N\begin{matrix}\cdot K\\ \cdot H\\ \cdot H,\end{matrix}$ $P\begin{matrix}\cdot H\\ \cdot H\\ \cdot H,\end{matrix}$ $P\begin{matrix}\cdot Cl\\ \cdot Cl\\ \cdot Cl,\end{matrix}$ $As\begin{matrix}\cdot H\\ \cdot H\\ \cdot H,\end{matrix}$ etc.
$\begin{matrix}H\cdot & & \cdot H\\ & C & \\ H\cdot & & \cdot H\end{matrix}$	"	$\begin{matrix}Cl\cdot & & \cdot H\\ & C & \\ H\cdot & & \cdot H,\end{matrix}$ $\begin{matrix}Cl\cdot & & \cdot Cl\\ & C & \\ H\cdot & & \cdot H,\end{matrix}$ $\begin{matrix}Cl\cdot & & \cdot Cl\\ & C & \\ Cl\cdot & & \cdot H,\end{matrix}$ $\begin{matrix}Cl\cdot & & \cdot Cl\\ & Si & \\ Cl\cdot & & \cdot H,\end{matrix}$ etc.

But by far the greater number of chemical compounds are more complicated in constitution, and may be looked upon as formed by the replacement of one or all of the elements in the above formulæ by atomic groups, which have the same valence as the replaced atoms. The four fundamental formulæ, inasmuch as they illustrate the functions of the elements of different valence, may be conveniently employed for the purpose of comparison with more complicated formulæ with the object of rendering the explanation of the latter more simple.

If we take any of the fundamental formulæ, and divide them at any part, we obtain two *residues* of equal valence. For instance, if we divide H.Cl, we obtain H and Cl, both monovalent; if we divide H.O.H, we obtain H and OH, and these are both monovalent, for, as can be readily seen, the group OH requires a monovalent atom or group in order that it may become saturated, and this is what we understand by a monovalent group. If we divide $N\begin{smallmatrix}\cdot H\\ \cdot H\\ \cdot H\end{smallmatrix}$, we obtain H and NH^2, or H^2 and NH; by the former division there are left two monovalent, by the latter two bivalent factors. And so in the case of $\begin{smallmatrix}H\cdot\\ H\cdot\end{smallmatrix}C\begin{smallmatrix}\cdot H\\ \cdot H\end{smallmatrix}$; if we divide this formula, the following cases are possible: H and CH^3, H^2 and CH^2, H^3 and CH, leaving in the first case two monovalent, in the second, two bivalent, and in the third, two trivalent factors. This principle may be carried out further in connection with other and more complicated formulæ, and so are obtained the formulæ of a great variety of these so-called *residues;* in most cases, however, the division made, and the residues resulting, may be compared to the simpler forms described. We speak of a water-residue, OH, which, on account of the exceedingly important part it plays in the constitution of chemical compounds, has received a distinct name, *hydroxyl;* the ammonia-residue, NH^2, is called *amide;* the residue, NH, is called *imide;* the residue, CH^3, of marsh gas, is called *methyl;* the residue, CH^2, methylene, etc. etc.

If we now operate with the groups mentioned instead of with atoms alone, we shall find that we are able to build up a larger number of formulæ representing compounds, as follows:—

$$N\begin{smallmatrix}\cdot H\\ \cdot H\\ \cdot H\end{smallmatrix} \text{ similar to } N\begin{smallmatrix}\cdot OH\\ \cdot H,\\ \cdot H,\end{smallmatrix}\quad N\begin{smallmatrix}:O\\ \cdot OH,\end{smallmatrix}\quad N\begin{smallmatrix}\cdot CH^3\\ \cdot I\\ \cdot H,\end{smallmatrix}\quad P\begin{smallmatrix}\cdot OH\\ \cdot OH\\ \cdot OH,\end{smallmatrix}\quad P\begin{smallmatrix}\cdot CH^3\\ \cdot CH^3\\ \cdot CH^3,\end{smallmatrix}$$

etc.

$$\begin{smallmatrix}H\cdot\\ H\cdot\end{smallmatrix}C\begin{smallmatrix}\cdot H\\ \cdot H\end{smallmatrix} \text{ similar to } \begin{smallmatrix}H\cdot\\ H\cdot\end{smallmatrix}C\begin{smallmatrix}\cdot NH^2\\ \cdot H,\end{smallmatrix}\quad \begin{smallmatrix}H\cdot\\ H\cdot\end{smallmatrix}C\begin{smallmatrix}\cdot OH\\ \cdot H,\end{smallmatrix}\quad O:C\begin{smallmatrix}\cdot OH\\ \cdot H,\end{smallmatrix}\quad N\vdots C.OH,$$

$$O:C\begin{smallmatrix}\cdot OH\\ \cdot OH,\end{smallmatrix} \text{ etc.}$$

Still further complications are introduced when, instead of compounds consisting of atoms of different valence, we have atoms of the same element, or of different elements of the same valence, united together, forming chains. Examples of the first kind are met with particularly in the case of carbon. If two carbon atoms unite in the simplest manner possible, we have a group $\cdot\dot{\underset{\cdot}{C}}\cdot\dot{\underset{\cdot}{C}}\cdot$, which must have six free affinities; if three atoms unite, $\cdot\dot{\underset{\cdot}{C}}\cdot\dot{\underset{\cdot}{C}}\cdot\dot{\underset{\cdot}{C}}\cdot$, the resulting group must have eight free affinities, etc.; and, as this chain combination may be continued indefinitely, and the free affinities may be saturated by the greatest variety of groups, it is evident that the number of compounds, the possibility of whose existence is thus indicated, is unlimited. The atoms of oxygen also possess this property of uniting with each other to form chains, as we see in the compounds:—

H.O.O.H, Cl.O.O.H, Cl.O.O.O.H, Cl.O.O.O.O.H, Br.O.O.O.H, etc.

In these cases the oxygen-atoms, which with one of their affinities are united with hydrogen, impart to this hydrogen-atom characteristic properties; and, whenever this kind of combination is found, we say, for convenience sake, the compound contains hydroxyl.

Examples of compounds formed by the chain-combination of different elements of the same valence are the following: H.O.S.O.O.H, H.O.O.S.O.O.H, in which we have sulphur and oxygen, forming a continuous chain.

Having thus seen the various methods of combination of atoms, let us briefly illustrate the applications of these forms to the characteristic classes of compounds.

In the first place, chemists have long recognized the existence of two classes of compounds, *bases* and *acids*, the representative members of which have, in certain respects, opposite or complementary properties. The larger number of acids, as well as bases, contain hydrogen and oxygen; and either all or a part of the hydrogen-atoms contained in them are united with oxygen. To the presence of these hydrogen-atoms or of the hydroxyl groups (see above), of which they form a part, are due the properties which distinguish the compounds as acids or bases. Examples of acids and bases are:—

$$\mathrm{Cl.O.H = Cl(OH),\quad K.O.H = K(OH).}$$

$$\mathrm{H.O.O.S.O.O.H = SO^2(OH)^2,\quad H.O.Ca.O.H = Ca(OH)^2.}$$

$$\mathrm{H.O.P}\begin{smallmatrix}\cdot\mathrm{O.H}\\ \cdot\mathrm{O.H}\end{smallmatrix} = \mathrm{P(OH)^3},\quad \mathrm{H.O.Al}\begin{smallmatrix}\cdot\mathrm{O.H}\\ \cdot\mathrm{O.H}\end{smallmatrix} = \mathrm{Al(OH)^3},$$

$$\mathrm{O{:}C}\begin{smallmatrix}\cdot\mathrm{H}\\ \cdot\mathrm{O.H}\end{smallmatrix} = \mathrm{HCO(OH)},\quad \mathrm{H\cdot\overset{H}{\underset{H}{C}}\cdot\overset{H}{\underset{H}{C}}\cdot O.H} = \mathrm{C^2H^5(OH)}.$$

Acids are *usually* derived from the so-called metalloids; bases from the metals. When acids act upon bases, *salts* are formed, water being given off, thus:—

H.O.O.S.O.O.H and K.O.H give H.O.O.S.O.O.K and H.O.H.
H.O.O.S.O.O.H and 2(K.O.H) give K.O.O.S.O.O.K and 2(H.O.H).

Now, if we examine these formulæ, we see that the salt may be considered either as the base, in which the original hydrogen is replaced by the acid-residue, or as the acid, in which the original hydrogen is replaced by the base residue or metal. As the latter is the simpler view, it is the one usually held, though be it remembered, that it is immaterial, for the constitution of the salt, which of the two views is held.

Compounds similar to salts in their constitution are anhydrides and metallic oxides. In the former two acid-residues, in the latter two base-residues, are employed in the formation.

$$\text{O:N.O.H} + \text{O:N.O.H} = \text{O:N.O.(N:O)} + \text{H.O.H},$$
$$\text{K.O.H} + \text{K.O.H} = \text{K.O.K} + \text{H.O.H},$$
$$\text{H.O.Ca.O.H} + \text{H.O.Ca.O.H} = \text{Ca}\begin{matrix}\cdot\text{O}\cdot\\ \cdot\text{O}\cdot\end{matrix}\text{Ca} + 2(\text{H.O.H}),$$
$$\text{H.O.O.S.O.O.H} + \text{H.O.O.S.O.O.H} = \text{S}\begin{matrix}\cdot\text{O.O.O}\cdot\\ \cdot\text{O.O.O}\cdot\end{matrix}\text{S} + 2(\text{H.O.H}).$$

It is, however, probable that in such cases as the two latter, the product splits up into two molecules, and the union of the atoms takes place in a different manner in consequence:—

$$\text{Ca}\begin{matrix}\cdot\text{O}\cdot\\ \cdot\text{O}\cdot\end{matrix}\text{Ca} = 2(\text{Ca:O}),\quad \text{S}\begin{matrix}\cdot\text{O.O.O}\cdot\\ \cdot\text{O:O.O}\cdot\end{matrix}\text{S} = 2\left(\text{S}\begin{matrix}\cdot\text{O}\cdot\\ \cdot\text{O}\cdot\end{matrix}\text{O}\right).$$

The compounds of carbon resemble those of other elements, but, owing to certain properties of the element, variations are met with in this connection that require special notice.

For one group of carbon-compounds, marsh gas, CH^4, is the mother substance. From this, other substances containing only carbon and hydrogen can be obtained, thus:—

$$H.\overset{H}{\underset{H}{C}}.H + Cl.Cl = H.\overset{H}{\underset{H}{C}}.Cl + H.Cl.$$

$$2\left(H.\overset{H}{\underset{H}{C}}.Cl\right) + 2Na = H.\overset{H}{\underset{H}{C}}.\overset{H}{\underset{H}{C}}.H + 2Na.Cl, \text{ etc.}$$

Each one of the compounds obtained in this way, as well as marsh gas itself, may be looked upon as a compound of one or more hydrogen atoms, with a residue or residues of corresponding valence; and each one of these residues can and does play the part of an element. Marsh gas, when divided as above, leaves, as we have seen, the residues $H.\overset{H}{\underset{\cdot}{C}}.H$ (methyl), $.\overset{H}{\underset{\cdot}{C}}.H$ (methylene), $.\underset{\cdot}{\dot{C}}.H$, and $.\underset{\cdot}{\dot{C}}.$, which are respectively mono-, bi-, tri- and tetravalent. Now, in the formation of the *hydrocarbons* (substances containing only hydrogen and carbon) from marsh gas, these four residues are the "*elements*," which are employed, and, by careful examination, we see that here an infinite variety presents itself. If we take the hydrocarbon $C^4H^{10} = H.\overset{H}{\underset{H}{C}}.\overset{H}{\underset{H}{C}}.\overset{H}{\underset{H}{C}}.\overset{H}{\underset{H}{C}}.H$, we see that it consists of $2CH^3$ and $2CH^2$; but these atoms can be arranged in another way, and the composition C^4H^{10} still be retained:—

$$\begin{array}{c} HHH \\ \dot{C} \\ H\cdot\qquad\cdot H \\ H\cdot C\cdot C\cdot C\cdot H; \\ H\cdot\qquad\cdot H \\ \dot{H} \end{array}$$

in this case we have $3CH^3$ and $1CH$. This principle can be carried out further, showing the possibility of a very large number of compounds of the same composition, but different constitution. This difference in constitution gives rise to a difference in the properties of the compounds.

In the hydrocarbons we can replace hydrogen by other elements or groups, and thus obtain the other possible compounds. The replacement by monovalent elements requires no explanation, as the constitution of the resulting compounds is exactly the same as that of the hydrocarbon itself.

Just as marsh gas may be considered as the mother-substance of a whole group of carbon-compounds, so, by replacing its hydrogen-atoms by various groups, we obtain compounds, each of which may, in turn, be looked upon as the mother-substance of a subordinate group. It has already been shown that the *water-residue*, hydroxyl, OH, plays an important part in the structure of the two classes of compounds known as acids and bases. If, for a hydrogen-atom of marsh gas, we substitute OH, we obtain a compound $\mathrm{H.\overset{H}{\underset{H}{C}}.OH} = CH^3(OH)$. This possesses the properties of the bases in general, corresponding to the simpler base K.O.H. We have in this case only CH^3, instead of the element K. Here, too, the hydrogen-atom, which is in combination with oxygen, imparts to the compound its characteristic properties, whereas the other hydrogen-atoms present exhibit only those other general properties which are met with in connection with the hydrogen-atoms of other groups of carbon-compounds. One or all of these latter can be replaced by other elements or groups, and the compound still retains the properties originally imparted to it by the hydroxyl group. We can, for instance, replace one of these atoms by CH^3, thus obtaining a compound, $\mathrm{H{\cdot}\overset{H}{\underset{H}{C}}{\cdot}\overset{H}{\underset{H}{C}}{\cdot}O.H} = C^2H^5(OH)$. This in every way resembles the body from which it is derived. We can, further, in this compound replace one or more hydrogen-atoms by elements or groups, without disturbing the hydroxyl-group. Let us again employ the group CH^3. We find that two products are formed, dependent upon the hydrogen-atoms replaced:—

$$1.\quad \mathrm{H.\overset{H}{\underset{H}{C}}{\cdot}\overset{H}{\underset{H}{C}}{\cdot}\overset{H}{\underset{H}{C}}{\cdot}O.H}, \text{ and } 2.\quad \mathrm{H{\cdot}\overset{H}{\underset{H}{C}}\cdot\overset{H\overset{H}{C}H}{\underset{H}{C}}{\cdot}O{\cdot}H},$$

and, just as in the case of the hydrocarbon C^4H^{10}, the two products differ from each other in properties as well as in constitution.

According to this method we can build up an indefinite number of compounds, all containing hydroxyl, and all exhibiting certain common properties owing to the presence of this group. Although these compounds strictly belong to the general heading *bases*, they have, as a class, received a distinct name to designate some properties which the bases do not possess in common with them. They are called *alcohols.*

The acids of carbon-compounds have an equally simple constitution. Let us start again with CH^4. Replacing one atom by OH and two by O, we obtain a compound $O{:}C{\cdot H \atop \cdot O.H.}$ This is formic acid. Here we have an oxidized carbon-atom, and in combination wtih it a water-residue. Again the hydrogen of the hydroxyl is the characteristic ingredient of the compound, but its characterizing powers have been imparted to it not alone by the fact that it is in combination with oxygen, but that this group is in its turn in combination with an *oxidized carbon-atom.* Organic acids all contain the group O:C.O.H, which may be looked upon as a residue of formic acid. It is monovalent, and can take the place of hydrogen in the most varied compounds. It has received the name *carboxyl.* The consideration of organic acids may be still further simplified by comparing them with certain derivatives of sulphuric acid. When sulphuric acid, and a number of other acids containing one hydroxyl-group, are allowed to act upon a compound containing replaceable hydrogen-atoms, one of the hydroxyl groups of the acid is given off in company with one of the hydrogen-atoms of the other compound in the form of water, and the two residues unite, thus:—

$$\underset{\text{Benzene.}}{C^6H^6} + \underset{\text{Sulphuric acid.}}{SO^2\begin{cases}OH\\OH\end{cases}} = C^6H^5.SO^2.OH + H^2O.$$

The resulting compound may be called a substituted sulphuric acid, one of its hydroxyls having been replaced by a monovalent group. Now carbonic acid resembles sulphuric acid in the fact that it contains 2(OH), and, although the acid itself is unknown to us, we can, under certain circumstances, induce a substitution similar to that noticed in connection with sulphuric acid, and thus obtain *substituted carbonic acids*, which are nothing but the so-called organic acids:—

$$C^6H^6 + CO{\cdot OH \atop \cdot OH} = C^6H^5.CO.OH + H^2O.$$

$$\text{or,}\quad C^6H^5.H + O{:}C{\cdot OH \atop \cdot OH} = O{:}C{\cdot OH \atop \cdot C^6H^5} + H^2O.$$

When acids and bases, in general terms, act upon each other, salts are formed, water being eliminated. Just so when alcohols and organic acids act upon each other, bodies, similar to salts, are formed, water being eliminated:—

$$H.CO.OH + C^2H^5.OH = H.CO.O.C^2H^5 + H^2O.$$

Salts were defined as acids, in which the hydrogen of the hydroxyl-group is replaced by a base-residue. In this case we have the hydrogen of the hydroxyl-group of the carboxyl replaced by an alcohol residue, and the resulting compound has received the name *ether*. The name ether is applied to all similar compounds, it being, as is clear, but a special form of the salt.

In regard to anhydrides the remarks made above are here equally applicable.

The carbon-compounds, which are formed like metallic oxides, and which correspond to them, have also been called *ethers*, though the same differences between them and the ethers mentioned may be found, that are met with between metallic oxides and salts.

$$\left\{ \begin{array}{c} K.O.H + K.O.H = K.O.K + H^2O, \\ C^2H^5.O.H + C^2H^5.O.H = C^2H^5.O.C^2H^5 + H^2O. \end{array} \right\}$$

Among carbon-compounds there are other series, which do not occur among inorganic compounds, the character of which is dependent upon the peculiar properties or carbon. If in marsh gas, CH^4, we replace two hydrogen-atoms by one O, we obtain the body $O{:}C{\cdot H \atop \cdot H}$, which occupies a position intermediate between that of ${H \atop H}{\cdot}C{\cdot OH \atop \cdot H}$ and $O{:}C{\cdot H \atop \cdot OH}$. These bodies consist of two hydrogen-atoms united by the bivalent group C:O, which has been called *carbonyl*. Now either one or both of these hydrogen-atoms may be replaced by alcohol-residues, as CH^3, C^2H^5, etc. If only one be replaced, we obtain the compounds known as *aldehydes*, as $O{:}C{\cdot CH^3 \atop \cdot H}$, $O{:}C{\cdot C^2H^5 \atop \cdot H}$, etc. If, however, both be replaced, we obtain compounds of a somewhat different character. These have been designated as *acetones* or *ketones*, as $O{:}C\begin{smallmatrix}\cdot CH^3\\ \cdot CH^3\end{smallmatrix}$, $O{:}C\begin{smallmatrix}\cdot C^2H^5\\ \cdot CH^3\end{smallmatrix}$, $O{:}C\begin{smallmatrix}\cdot C^2H^5\\ \cdot C^2H^5\end{smallmatrix}$, etc.

Aldehydes are, hence, compounds which consist of an alcohol-residue and a hydrogen-atom united by means of the group CO: and acetones consist of two alcohol-residues united by the group CO. It will thus be seen that the body $O{:}C\begin{smallmatrix}\cdot H\\ \cdot H\end{smallmatrix}$ may be considered as the simplest representative of both classes of compounds.

In addition to the various classes of compounds, there are others, but they are all variations on these principal classes, and demand here no special explanation. We have, for instance, compounds which partake of the properties of both acids and bases, acids and alcohols, aldehydes and acids, aldehydes and alcohols, etc. etc., but with the aid of the few principles laid down these will be readily understood. We have, further, instances, especially among carbon-compounds, in which atoms of the same kind are united with each other by means of more than one affinity, and also those in which each carbon-atom of the compound is united with two other carbon-atoms, on the one hand with one, on the other with two affinities, etc. etc.; the constitution of such compounds can, however, be easily comprehended by the application of the fundamental principles.

The main question which now presents itself is: What grounds have we for the acceptation of these fundamental principles? It can only be answered, they have been proposed and accepted as affording the simplest explanation of innumerable investigations concerning the properties of chemical compounds. At present no facts are known that conflict with their acceptation. They are by no means established beyond a doubt, but, as they simplify known facts, and have been exceedingly fruitful in widening the field of observation, they are worthy of our most careful study.

It would lead too far in this place to recall the individual investigations and the methods of reasoning which have led to the acceptation of our present ideas concerning the constitution of chemical compounds. In order to draw our conclusions we must know the methods of formation, the decompositions, and all the varied changes which individual compounds or groups of compounds undergo.

A simple example may suffice to illustrate the *rationale* of the process. Let us take ordinary alcohol. We can first establish the formula by means of analysis and the determination of the specific gravity of its vapor. This we find to be C^2H^6O. This formula is the expression of a fact and a hypothesis. The fact expressed is that alcohol consists of 52.17 per cent. carbon, 13.04 per cent. hydrogen, and 34.78 per cent. oxygen. The hypothesis of which it is an expression is that the molecules of all chemical compounds in the form of vapor have the same volume as a molecule of hydrogen. This hypothesis, when applied, tells us the weights of the atoms contained in the molecule of alcohol and the weight of the molecule of alcohol, and hence, further, the number of atoms of carbon, hydrogen, and oxygen contained in the molecule under consideration. We know that hydrogen is monovalent, oxygen bivalent, and carbon tetravalent. It now remains to decide how those atoms are united—what the constitution of alcohol is? If we take marsh gas, CH^4, which, according to our ideas, as we have seen, can only have the constitution $\begin{matrix}H.&&.H\\ H.&C&.H,\end{matrix}$ we can produce from it (see above) the hydrocarbon $\begin{matrix}H&H\\ H.\dot{C}.&\dot{C}.H\\ \dot{H}&\dot{H}\end{matrix} = C^2H^6$; if we now replace one of the hydrogen-atoms of this compound by chlorine, we have the compound $\begin{matrix}H&H\\ H.\dot{C}.&\dot{C}.Cl\\ \dot{H}&\dot{H}\end{matrix} = C^2H^5Cl$, and experience shows us that only one compound of this composition can result, it being immaterial which one of the hydrogen-atoms is replaced. If we, further, allow the substance K.O.H, in regard to the constitution of which, according to the principles already laid down, there can be no question, to act upon this compound, two products are formed, thus:—

$$\begin{matrix}H&H\\ H.\dot{C}.&\dot{C}.Cl\\ \dot{H}&\dot{H}\end{matrix} + K.O.H = \begin{matrix}H&H\\ H.\dot{C}.&\dot{C}.OH\\ \dot{H}&\dot{H}\end{matrix} + KCl.$$

The water-residue, the hydroxyl group, before in combination with K, has changed places with Cl. The resulting compound, $C^2H^5(OH)$, is ordinary alcohol, and we have thus from one point of view determined its constitution. Again, by the action of certain reagents we find that an atom of oxygen *and* an atom of hydrogen are given off, and their place is taken by *one atom of chlorine*, thus showing that the hydrogen and oxygen were present in the compound in the form of a monovalent group, or as hydroxyl, which is the only form that satisfies these conditions. These and other similar facts are looked upon as proofs of the constitution of alcohol.

It is in work of this kind that chemists are at present largely engaged, and the results achieved are already of great magnitude. The constitution of a large number of substances occurring in nature has been discovered, and the discovery of their constitution has in many cases led directly to the artificial preparation (*synthesis*) of the substances. Although this cannot be considered the highest aim of the science of chemistry, yet the cultivation of this field promises rich reward, direct and indirect, and its development will place us a step nearer that state in which all chemical phenomena can be dealt with as other physical phenomena are now dealt with, viz., as subject to mathematical laws.

CHEMISTRY OF ORGANIC COMPOUNDS.

INTRODUCTION.

ORGANIC CHEMISTRY is the chemistry of the compounds of carbon. It includes those compounds of carbon which have had their origin in the organs of plants and animals, as well as those which have been produced exterior to the living organism.

Most organic compounds are solid, partially crystalline, partially amorphous bodies; many are liquids; only a few are gaseous at ordinary temperatures. All of them are destroyed when heated above their melting or boiling point without access of air; a very large number cannot even be melted nor volatilized without undergoing decomposition. The melting point and boiling point are very characteristic properties for those bodies, which are not readily decomposed at higher temperatures. The difference in the boiling points of organic compounds is very frequently made use of for the purpose of separating them from each other, and preparing them in a pure condition from a mixture (partial distillation).

Another very important property of those organic bodies, which are volatile without decomposition, is their specific gravity in the form of gas or vapor (vapor density). Experience has shown, that the molecules (the smallest quantity that can exist in a free condition) of the various chemical compounds in the form of gas or vapor possess the same volume, and that this volume is the same as that of two atoms (one molecule) of hydrogen.

2

The proportion of the molecular weight to the specific gravity of the vapor (molecular weight divided by the specific gravity) is, therefore, for all compounds the same, and is represented by the constant number 28.9.

This conformity yields an important and frequently the only means of determining the molecular weight of an organic compound. This is obtained by multiplying the specific gravity found by the number 28.9.

Carbon is the characterizing element of all organic compounds. In most of these compounds it is in combination with hydrogen and oxygen, in very many together with nitrogen, sulphur, etc. Furthermore, nearly all other elements can be made constituents of carbon compounds.

The fact that so great a number and variety of carbon compounds exist is principally due to the tendency, possessed by the atoms of carbon more than by the atoms of any other element, to unite with each other in chains.

Carbon is tetravalent in all its compounds. When two or more carbon atoms unite with each other, a portion of the affinities of each atom is used in holding the atoms together, so that two atoms of carbon have always less than eight free affinities, three always less than twelve. In most cases the union of several atoms of carbon takes place in such a manner that each of them loses one of its four affinities. Hence, two atoms of carbon have six, three atoms of carbon eight, or in general terms x carbon atoms have $2x+2$ free affinities.

The valence of any group of atoms containing carbon may be found by subtracting the sum of the affinities of the other atoms present from the affinities of the carbon atoms. The group methyl CH^3 must be monovalent, inasmuch as three of the four affinities of the carbon atom are saturated by the affinities of the three monovalent hydrogen atoms. Carbonic oxide CO must be bivalent, because but two of the four affinities of the carbon atom are saturated by the bivalent oxygen atom. A similar reflection shows us that ethylene C^2H^4 must be bivalent, that acetylene C^2H^2 must be tetravalent, for in the first case four of the six affinities of the two carbon atoms are saturated by hydrogen, in the latter only two.

Compounds, in which all the affinities of the carbon atoms thus united are saturated by other elements, are called *saturated;* compounds which possess free affinities, *non-saturated.*

Experience has shown that only such non-saturated carbon compounds can exist in an isolated condition, in which two, four, or in general terms an even number of affinities, are unsaturated. Atomic groups, in which an odd number of affinities are unsaturated, cannot be isolated.

It is, however, questionable whether, with the exception of carbonic oxide CO, non-saturated compounds of carbon are really capable of existence. In the so-called non-saturated compounds the carbon atoms are probably united with each other with more than one affinity. Ethylene C^2H^4 may be regarded as the saturated compound of a group of two carbon atoms, which are united with each other by means of two affinities, as $\left\{\begin{array}{l} CH^2 \\ \| \\ CH^2. \end{array}\right.$ It is not positive proof of the contrary that this body conducts itself in most reactions as a non-saturated compound, and, for instance, unites with the greatest ease with two atoms of a monovalent element or monovalent group, as in this reaction the double union of the carbon atoms can be broken up and the simple union, quite sufficing for the sustenance of the atoms in their position, re-established. According to this view the carbon atoms in acetylene C^2H^2 must be united with each other by means of three of their free affinities each.

In a large number of compounds, especially the so-called aromatic bodies, we are compelled to admit that the union of the carbon atoms takes place in a manner different from that mentioned above; that, for the purpose of holding together the single atoms of carbon, more than one of the affinities of each of the carbon atoms is employed. Benzol C^6H^6, for instance, according to the above method of consideration, should have eight free affinities; in most reactions, however, it conducts itself as a saturated compound.

It is frequently the case that two organic bodies contain the same elements in the same proportion by weight, and still have entirely different physical and chemical properties. In general such bodies are called *isomeric*. For this relation there may be two different causes, viz.:—

1. A dissimilarity of constitution, *i. e.*, a dissimilarity in the method of grouping or joining of the atoms in the two bodies, as, for instance, in ethyl formate, methyl acetate, and in propionic acid. All three of these compounds have the formula $C^3H^6O^2$. In ethyl formate, however, the atoms are grouped together according to the formula $CHO.O.C^2H^5$, or further reduced $CHO.O.\ CH^2.CH^3$, whereas in methyl acetate the method of grouping is $C^2H^3O.O.CH^3$ or $CH^3.CO.O.CH^3$, and in propionic acid $C^3H^5.O.OH$ or $CH^3.CH^2.CO.OH$. Ammonium cyanate $CN.O.NH^4$ and urea $CO\begin{cases}NH^2\\ \|\\ NH^2\end{cases}$ bear a similar relation to each other. In the latter case the change in the arrangement of the atoms from the first manner of grouping to the second takes place spontaneously at the ordinary temperature.

Such bodies are called *metameric* or *isomeric* in the narrower application of the word.

Or, 2. A different molecular weight, as, for instance, acetic aldehyde and butyric acid, which both contain the same percentages of their constituents. The molecular weight of aldehyde C^2H^4O is, however, only half as large as that of butyric acid $C^4H^8O^2$. According to the same principle acetic acid $C^2H^4O^2$, and grape-sugar $C^6H^{12}O^6$, acetylene C^2H^2, and benzol C^6H^6, and many other compounds, are isomeric. Such bodies are called *polymeric*.

Compounds which contain the same elements in the same proportions by weight, have the same molecular weight, and show no essentially different chemical properties, and yet conduct themselves somewhat differently in connection with certain physical properties, especially in the action on polarized light, are said to be *physically isomeric.*

By the expression *homologous* bodies, are understood such bodies as conduct themselves in their chemical properties in a similar manner, and differ in their composition by the group CH^2, or a multiple of it. We are, for instance, acquainted with a series of compounds which, in their conduct, show the greatest similarity to ordinary alcohol, and of which each succeeding member differs in its composition from the preceding by the group CH^2, as may be seen in the following schedule:—

CH^4O . . . . Wood-spirits, methyl alcohol,
C^2H^6O . . . . Spirits of wine, ethyl alcohol,
C^3H^8O . . . . Propyl alcohol,
$C^4H^{10}O$ . . . . Butyl alcohol,
$C^5H^{12}O$ . . . . Fusel-oil, amyl alcohol, etc.

Another series, of which acetic acid is the principal representative, runs parallel to this:—

CH^2O^2 . . . . Formic acid,
$C^2H^4O^2$ . . . . Acetic acid,
$C^3H^6O^2$ . . . . Propionic acid,
$C^4H^8O^2$ . . . . Butyric acid,
$C^5H^{10}O^2$ . . . . Valeric acid, etc.

For several of these homologous series, as, for instance, for the two mentioned, experience has shown that the following law exists: The boiling point of a compound is 19–20° higher if it contains CH^2 more than another member of the series. The boiling point of ethyl alcohol is, for instance, 78°; that of normal amyl alcohol, which differs from ethyl alcohol by $3 \times CH^2$, should, according to this law, be $3 \times 19°$ higher. The observed boiling point is 137°.

In connection with other homologous series, a similar conformity is observed, but the difference in the boiling points effected by the addition of every CH^2 is not the same. With the hydrocarbons, which are homologous with benzol C^6H^6, viz.: toluol C^7H^8, xylol C^8H^{10}, and cumol C^9H^{12}, for instance, the difference is 28–29°. This conformity is, however, only observed in the case of those bodies which, being homologous according to their empirical composition, are also of an analogous constitution. Conditions dependent upon isomerism can at times entirely withdraw it from observation. While with ethyl alcohol C^2H^6O, boiling point 78°, normal propyl alcohol, boiling point 97–98°, normal butyl alcohol, boiling point 115–116°, complete regularity takes place, we observe no regularity in comparing the three following dissimilarly constituted alcohols with each other:—

Ethyl alcohol . . . . C^2H^6O boiling point 78°,
Isopropyl alcohol . . . C^3H^8O boiling point 85°,
Tertiary butyl alcohol . $C^4H^{10}O$ boiling point 82°.

Organic bodies undergo the most varied changes when subjected to the action of high heat. Frequently the action is such that hydrogen and oxygen are removed from the body in the form of water, or carbon and oxygen in the form of carbonic acid, and the other elements of the compound remain united as a new organic body; for instance:—

$$\underbrace{C^4H^6O^4}_{\text{Succinic acid.}} = \underbrace{C^4H^4O^3}_{\text{Succinic anhydride.}} + H^2O$$

$$\underbrace{(C^2H^3O^2)^2Ca}_{\text{Calcium acetate.}} = \underbrace{C^3H^6O}_{\text{Acetone.}} + \underbrace{CO^3Ca}_{\text{Calcium carbonate.}}$$

Or one organic body is separated into two new ones under the influence of heat, or there is formed at the same time a larger number of new organic compounds, which, in their turn, are often destroyed at the moment of their formation, thus giving rise to a complicated mixture of products, generally ending in leaving behind a residue of carbon. These products are different, according as the heat is more or less strong, slower or more rapid. The products of decomposition of bodies free of nitrogen are frequently acid, from the formation of acetic acid; of those which contain nitrogen, they are mostly alkaline, from the formation of ammonia and other bases.

Many organic substances, especially such as contain nitrogen, are decomposed when exposed to the influence of air and water at ordinary temperatures, their elements being rearranged during the process to form simpler substances. This kind of decomposition is called *putrefaction.* Putrefaction occurs only under certain conditions. It can only take place in the presence of water, and access of atmospheric air is necessary to its commencement. Once begun, however, it continues without access of air. Everywhere in the air are present microscopical germs of vegetable and animal organisms. When these fall on a soil favorable to their growth they are developed quickly, they multiply with great rapidity, and in consequence of the vital process and the dying off of these organisms, that species of decomposition of organic compounds takes place which is called putrefaction. The air loses its power to start the process of putrefaction when previously passed through a strongly heated tube, or a dense cotton stopper, or even only through a tube which has a large number of curves, as by these means the germs, which are present in the atmosphere, are either destroyed or held back. Further, putrefaction occurs only within certain limits of temperature, most readily between 20° and 30°. Below 0° and above 100° it does not take place.

If the oxygen of the air takes part in the decomposition, and thus a simultaneous oxidation takes place, the decomposition is called *decay.* The last products of decaying organic substances are water, carbonic acid, and ammonia.

A phenomenon very similar to putrefaction is *fermentation.* This will be treated of more in detail in connection with alcohol.

Organic compounds can be changed in a variety of ways under the influence of many inorganic bodies.

Free *oxygen* acts on but a very few organic bodies at the ordinary temperature; it acts, however, more energetically in *statu nascendi*, or in the presence of certain substances, particularly of spongy platinum. When it acts at all, it is either added directly to the compound, or the hydrogen contained in the compound is oxidized to form water, or both of these changes take place together. At times a more material decomposition takes place. At a red heat, all organic substances burn in oxygen, forming carbonic acid, water, and nitrogen.

Hydrogen, especially in *statu nascendi*, likewise transforms very many organic compounds, either a direct addition of hydrogen, or an elimination of oxygen, or both at the same time taking place. In most cases in which hydrogen acts upon compounds containing chlorine, bromine, or iodine, these elements are eliminated and replaced by hydrogen. Hydriodic acid and sulphuretted hydrogen act similarly to free hydrogen. At times, bodies containing iodine or sulphur result, but generally the iodine or sulphur is set free, and merely the hydrogen acts.

Chlorine and *bromine* act very energetically upon organic bodies. Non-saturated organic compounds (those in which the carbon atoms are united by means of more than one of each of their four affinities), usually combine directly with these elements, and take up as many atoms as are sufficient to produce saturated bodies, the simple union of the carbon atoms being re-established. In this way are formed from ethylene, $C^2H^4 = \begin{matrix} CH^2 \\ \| \\ CH^2 \end{matrix}$, the compounds $C^2H^4Cl^2 = \begin{matrix} CH^2Cl \\ | \\ CH^2Cl \end{matrix}$ and $C^2H^4Br^2 = \begin{matrix} CH^2Br \\ | \\ CH^2Br \end{matrix}$. With saturated compounds, however, the action generally takes place in such a manner, that a certain number of atoms of hydrogen are eliminated, and replaced in the compound by an equal number of atoms of chlorine or bromine; for instance:—

$$\underbrace{C^2H^4O^2}_{\text{Acetic acid.}} + 2Cl = \underbrace{C^2H^3ClO^2}_{\text{Chloracetic acid.}} + ClH$$

This kind of action is called *substitution*, and the newly-formed body a substitution-product of the original body. Iodine, in other respects so similar to chlorine and bromine, when alone acts never, or at least only exceptionally, in the manner described, as hydriodic acid is formed at the same time, and this has the tendency to cause a reverse substitution, *i.e.* a displacement of the iodine in organic compounds, containing iodine, by hydrogen. If, however, a body be added with the iodine which has the property of removing the hydriodic acid as soon as formed, for instance iodic acid, substitution-products containing iodine can in many cases be obtained. An addition of small quantities of iodine aids materially the substituting action of chlorine upon organic compounds. In the presence of water, chlorine sometimes acts as an oxidizing agent. Organic compounds containing chlorine likewise result, as a rule, by the action of hydrochloric acid or the chlorine compounds of phosphorus.

Concentrated *nitric acid* acts in most cases in a similar manner to chlorine. A certain number of hydrogen atoms is eliminated, and for each of them the monovalent group NO^2 (hyponitric acid) enters the compound; for example:—

$$\underbrace{C^6H^6}_{\text{Benzol.}} + NO^2.OH = \underbrace{C^6H^5.NO^2}_{\text{Nitrobenzol.}} + H^2O.$$

Compounds resulting in this way are called *nitro-compounds*, or *nitro-substitution-products*. The formation of these bodies is very much aided by mixing the concentrated nitric acid with twice its volume of concentrated sulphuric acid. Nitric acid acts frequently, especially by continued boiling, only as an oxidizing agent.

Concentrated *sulphuric acid* acts upon a great many organic bodies similarly to nitric acid. One or more hydrogen atoms of the compound are displaced by the monovalent group $SO^2.OH$; for example:—

$$\underbrace{C^6H^6}_{\text{Benzol.}} + SO^2\left\{\begin{matrix}OH\\OH\end{matrix}\right. = \underbrace{C^6H^5.SO^2.OH}_{\text{Sulphobenzolic acid.}} + H^2O.$$

Bodies formed in this way are called *sulpho-compounds*, or, as they all possess the character of acids, *sulpho-acids*.

Frequently, however, the action of concentrated sulphuric acid consists in the elimination of the elements of water from organic compounds, the latter being completely destroyed (carbonized), or converted into others containing less hydrogen and oxygen; for example:—

$$\underbrace{C^2H^5.OH}_{\text{Alcohol.}} = \underbrace{C^2H^4}_{\text{Ethylene.}} + H^2O.$$

When *ammonia* acts upon organic compounds, especially those which contain chlorine, bromine, or iodine, bodies containing nitrogen are formed as a rule, the halogens being eliminated and replaced by the group NH^2; for example:—

$$\underbrace{C^2H^3ClO^2}_{\text{Chloracetic acid.}} + 2NH^3 = \underbrace{C^2H^3(NH^2)O^2}_{\text{Amidoacetic acid.}} + NH^4Cl.$$

The new compounds which result in this way are called *amides*. They are also formed by the action of hydrogen in *statu nascendi* (from tin and hydrochloric acid), or of sulphuretted hydrogen upon nitro-compounds, the group NO^2 contained in the latter being transformed into NH^2; for example:—

$$\underbrace{C^6H^5.NO^2}_{\text{Nitrobenzol.}} + 6H = \underbrace{C^6H^5.NH^2}_{\text{Anilin.}} + 2H^2O.$$

The elementary composition of organic bodies can be determined with the greatest exactitude. The analysis of the ordinary ones consists in the oxidation of the carbon to carbonic acid, and of the hydrogen to water, and the calculation, from the quantity of these products of combustion, of the quantity of carbon and hydrogen in the compound. Nitrogen is separated as nitrogen and measured, or it is transformed into ammonia. Oxygen is calculated indirectly by loss.

The most common method of estimating carbon and hydrogen consists in submitting an accurately weighed quantity of the substance to be analyzed, with a large excess of perfectly dry copper oxide or lead chromate, to a red heat, finally employing a current of pure oxygen. The water formed during the process is taken up by a tube filled with calcium chloride, the carbonic acid by a small apparatus which is filled with a solution of potassium hydroxide, and, to secure absolute safety, is joined to a small tube containing pieces of solid potassium hydroxide. The gain in weight of these three pieces shows the quantity of water and carbonic acid.

The conversion of nitrogen into ammonia is accomplished by heating the body strongly with a large excess of a dry mixture of sodium hydroxide and calcium hydroxide. The ammonia formed is either taken up by hydrochloric acid and weighed as ammonium chloroplatinate, or by dilute sulphuric acid of a known strength, and the quantity of acid which remains free afterwards estimated by means of a standard test-solution of sodium hydroxide. This subtracted from the quantity of acid employed shows how much of the acid has been neutralized by ammonia, from which the quantity formed and the nitrogen contained therein may be easily calculated.

Nitrogen is not, however, given off in all cases by heating with soda-lime. This applies especially to such cases in which the nitrogen is in close combination with the oxygen, as for instance in nitro-compounds. In analyzing such substances, the nitrogen is set free by heating the substance with an excess of finely powdered copper oxide, and passing the escaping gases over metallic copper for the purpose of destroying the oxides of nitrogen. This operation is carried out in a long tube, from which the air has been previously completely removed by means of carbonic acid. The mixture of carbonic acid and nitrogen is collected in a graduated tube over mercury, the carbonic acid absorbed by caustic potassa, the volume of nitrogen which has remained unabsorbed, measured and its weight calculated according to the formula

$$G = \frac{V}{1+0.00367t} \cdot \frac{B-f}{760} \cdot 0.001256,$$

in which V represents the volume of gas, t its temperature, B the pressure under which the gas stands (height of barometer), expressed in millimetres, f the tension of water vapor at the temperature t, and 0.001256 the weight of 1 cc. of nitrogen at 0°, and 760 mm. pressure.

When an organic compound contains chlorine, bromine, iodine, or sulphur, it must in most cases be thoroughly decomposed, before these can be detected by ordinary reagents and estimated. The estimation of the halogens is accomplished by igniting the substance with pure lime, free from water; the estimation of sulphur by heating with nitric acid in sealed tubes, or igniting with a mixture of sodium carbonate and potassium nitrate. Chlorine, bromine, and iodine can in many cases be detected in the ordinary manner, and estimated by previously treating the substance with hydrogen in *statu nascendi* (from sodium-amalgam and water).

As an example of the method of calculating an elementary analysis, that of acetic acid may be taken.

0.234 grm. of acetic acid were ignited with copper oxide. The gain in weight of the calcium chloride tube amounted to 0.1405 grm.; that of the potassa bulbs and tube 0.3432 grm. From 0.234 grm. acetic acid were hence produced 0.1405 grm. of water, and 0.3432 grm. carbonic acid, and these contain 0.0156 grm. hydrogen, and 0.0936 grm. carbon. These numbers show 40.00 per cent. of carbon, and 6.67 per cent. of hydrogen. As acetic acid only contains carbon, hydrogen, and oxygen, its composition expressed in percentages is—

$$\begin{aligned} C &= 40.00 \text{ per cent.} \\ H &= 6.67 \quad \text{“} \\ O &= 53.33 \quad \text{“} \end{aligned}$$

In order to find the atomic proportion from these numbers, we must divide them by the respective atomic weights of the elements.

$$\begin{aligned} C &= 40.00 \div 12 = 3.33 \\ H &= 6.67 \div 1 = 6.67 \\ O &= 53.33 \div 16 = 3.33 \end{aligned}$$

The elements in acetic acid hence stand to each other in the atomic proportion of 3.33 : 6.67 : 3.33, or, 1 : 2 : 1. The chemical formula of the acid could hence be expressed by CH^2O, but, of course, with exactly the same right, we might express it by $C^2H^4O^2$, or $C^6H^{12}O^6$, for all these formulæ show the same percentages of the elements. Simply the elementary analysis is not sufficient to determine which of these formulæ is the correct one; an estimation of the molecular weight must be united with it. With an acid this is simple if we know its basicity. We know that, for the purpose of forming a neutral salt, one molecule of acetic acid gives up one atom of hydrogen, and takes up in its place one atom of a monovalent metal. Hence, in order to find the molecular weight of acetic acid, we need only determine the amount of metal contained in one of its salts.

0.412 grm. silver acetate on being ignited leave a residue of 0.2665 grm. metallic silver. This represents 64.7 per cent.

In 100 parts of silver acetate are hence contained

Organic substance	35.3
Silver	64.7

The molecular weight of the organic substance in silver acetate can now be found by means of the following proportion:—

$$64.7 : 35.3 :: 108^* : x$$

$$\text{Result} = 59$$

Free acetic acid contains one atom more of hydrogen, therefore the molecular weight of the free acid is 60.

The simplest formula, agreeing with the results of the analysis, has the molecular weight 30. This must hence be doubled, and the composition must be expressed by the formula $C^2H^4O^2$.

When basic bodies are under investigation, a neutral

* Atomic weight of silver.

salt is also prepared for the estimation of the molecular weight, and from the quantity of acid contained in this salt the molecular weight of the base is calculated in a similar manner.

The molecular weight cannot, however, in all cases be determined by this method—only when experiments have shown how many atoms of a monovalent element an acid, or how many molecules of a monobasic acid a base, needs to form a neutral salt. If the substance is volatile without decomposition, the molecular weight can be found more simply by an estimation of the specific gravity of its vapor.

The specific gravity of acetic acid vapor, for instance, was found to be 2.08 at 300°. This number, multiplied by the constant number 28.9 (see *ante*, p. 14), gives as a result for the molecular weight of acetic acid the number 60.1, hence, taken together with the results of the analysis, the formula $C^2H^4O^2$.

The processes, more intimately connected with the formation of the primitive organic compounds in the living organism of plants and animals, are almost entirely unknown to us. We only know with certainty that all organic material is originally formed in plants, that for this purpose plants make use of the elements of existing compounds particularly of carbonic acid, water, ammonia, and the inorganic acids of nitrogen, and that this process of formation takes place only under the influence of sunlight and of certain inorganic salts, which are absorbed from the soil; the manner in which this takes place is, however, up to the present, inexplicable. The animal organism, on the other hand, receives its constituents in the food in the form of organic compounds already existing.

A great many of the organic compounds occurring in nature can be produced artificially from the elements, but in by far the most cases the conditions and the chemical processes are entirely different from those through the instrumentality of which the formation occurs in nature.

I. MARSH GAS DERIVATIVES (FATTY BODIES).

FIRST GROUP.

A. HYDROCARBONS, $C^n H^{2n+2}$ (MARSH GAS SERIES).

COMPOUNDS consisting merely of carbon and hydrogen are called hydrocarbons. The simplest compound of this kind is marsh gas CH^4, in which the four affinities of the carbon are saturated with four hydrogen atoms. Marsh gas is the first member of an homologous series of compounds, which have for their general formula C^nH^{2n+2}. All facts as yet known justify the conclusion that each of the four hydrogen atoms in marsh gas has exactly the same value, and that, as far as the properties of compounds are concerned, which are produced by the displacement of hydrogen atoms in marsh gas by means of other elements or groups of atoms, it is immaterial which of the four hydrogen atoms are displaced. Assuming this to be the case, we see that for the first three members of this series, CH^4, C^2H^6, and C^3H^8, but one manner of constitution is possible, viz.: CH^4,—$CH^3.CH^3$ and $CH^3.CH^2.CH^3$. There can hence exist only one modification of each of these three hydrocarbons. Isomeric compounds are not possible. Of the fourth member, C^4H^{10}, there are two different modifications possible: $CH^3.CH^2.CH^2.CH^3$ and $CH^3.CH\left\{\begin{matrix}CH^3\\CH^3\end{matrix}\right.$; of the fifth member, C^5H^{12}, there are three modifications possible; of the sixth member, five, etc.

1. *Marsh Gas* (*Fire Damp, Methyl Hydride*), CH^4.

Occurrence and Formation. Together with the homologous hydrocarbons, and mixed with carbonic acid and nitrogen, it issues in many localities from the earth; frequently collects in mines and coal-beds. Is formed in the process of putrefaction under water and in the destructive distillation of a great many organic bodies, and is hence contained in ordinary coal gas. It is further formed when a mixture of the vapor of carbon bisulphide with sulphydric acid is conducted over ignited metals; from ethylene at a red heat; and is most readily obtained in a pure state by heating 2 parts of crystallized sodium acetate with 2 parts of potassium hydroxide and 3 parts of lime.

Properties. Inodorous, inflammable gas, insoluble in water, of specific gravity 0.559. Mixed with oxygen or air it explodes with great violence when ignited; also when mixed with chlorine it forms a gas, which explodes violently when exposed to direct sunlight. In dispersed sunlight chlorine acts upon it in another manner, displacing its hydrogen and forming the compounds, CH^3Cl, CH^2Cl^2, $CHCl^3$ and CCl^4 (treated of in connection with methyl alcohol).

2. *Ethyl Hydride*, C^2H^6.

Is contained in a state of solution in crude petroleum. Is produced from the first substitution-products of marsh gas (methyl chloride CH^3Cl, methyl iodide CH^3I) by the action of sodium or zinc; by the decomposition of a concentrated solution of sodium acetate by means of an electrical current; by the action of water on zinc ethyl, or by heating ethyl iodide with water and zinc in sealed tubes to 180°. Colorless, almost inodorous gas. Burns with a slightly luminous flame. Is but slightly absorbed by water, more by alcohol. Chlorine displaces its hydrogen, forming successively the compounds C^2H^5Cl, $C^2H^4Cl^2$, $C^2H^3Cl^3$, $C^2H^2Cl^4$, C^2HCl^5, and C^2Cl^6 (see Ethyl Chloride).

3. *Propyl Hydride* (*Trityl Hydride*), C^3H^8.

In petroleum. Is formed like ethyl hydride, and can be obtained most readily, though not free from hydrogen, by the action of hydrogen in *statu nascendi* (from zinc and hydrochloric acid) on propyl iodide or isopropyl iodide. Colorless gas; liquid below —17°.

4. *Butyl Hydride* (*Tetryl Hydride*), C^4H^{10}.

In petroleum. The normal hydrocarbon *diethyl* $CH^3.CH^2.CH^2.CH^3$ is produced by the action of zinc or sodium on ethyl iodide. Colorless gas; liquid at +1°.

Pseudobutyl hydride (Trimethylformene), $CH^3.CH\left\{\begin{matrix}CH^3\\CH^3\end{matrix}\right.$ is isomeric with diethyl. It is obtained from the corresponding iodide (see tertiary butyl alcohol) by the action of zinc and water. Colorless gas; condensable at —17°.

5. *Amyl Hydride*, C^5H^{12}.

The normal hydrocarbon $CH^3.CH^2.CH^2.CH^2.CH^3$ is contained in petroleum, together with the following compound; also in products of distillation of cannel and boghead coal.—Mobile liquid; boils at 37°—39°.

The hydrocarbon $CH^3.CH^2.CH\left\{\begin{matrix}CH^3\\CH^3\end{matrix}\right.$ is contained in large quantity in American petroleum. It is formed by heating the iodide $C^5H^{11}I$ from ordinary amyl alcohol, with zinc and water to 142°; by distilling ordinary amyl alcohol over zinc chloride. (In both reactions other hydrocarbons are formed at the same time, particularly amylene C^5H^{10}.) Colorless liquid; boils at 30°; does not solidify at —24°; specific gravity 0.626.

The third hydrocarbon (*tetramethylformene*) $\left.\begin{matrix}CH^3\\CH^3\end{matrix}\right\}C\left\{\begin{matrix}CH^3\\CH^3\end{matrix}\right.$ is produced by the action of zinc methyl on the iodide obtained from tertiary butyl alcohol.—Colorless, mobile liquid. Boiling point 9°.5. Solidifies at —20°, forming crystals, which resemble sublimed sal-ammoniac.

The higher members of this series form the principal ingredients of American petroleum and of the oils (solar oil, photogene) obtained by the distillation of peat, bituminous slates, lignite and certain varieties of anthracite. The hydrocarbons, which are obtained from these sources by means of partial distillation, are mostly mixtures of isomeric compounds. By means of transforming these mixtures into the corresponding alcohols and oxidizing the latter, the chemical constitution of the principal ingredients has been discovered. The accompanying hydrocarbons, however, which occur in but very small quantity, are not well investigated. Others have been prepared artificially by means of reactions, that permit of a conclusion in regard to their constitution.

6. *Hexyl Hydride* (*Hexan*), C^6H^{14}.

There are three methods known for the preparation of this hydrocarbon; by partial distillation of American petroleum; by the action of tin and hydrochloric acid on the iodide of secondary butyl alcohol; by the action of sodium on an ethereal solution of propyl iodide. The first product boils at 70°; the second and third at 71.5°. The two latter have the specific gravity 0.663. These products are probably all identical and represent the normal hydrocarbon $CH^3.CH^2.CH^2.CH^2.CH^2.CH^3$.

Ethyl-isobutyl, $C^6H^{14} = CH^3.CH^2.CH^2.CH \begin{cases} CH^3. \\ CH^3. \end{cases}$ By the action of sodium in a mixture of ethyl iodide and isobutyl iodide. Boiling point, 62°; specific gravity, 0.7011.

Di-isopropyl, $C^6H^{14} = \left.\begin{matrix} CH^3 \\ CH^3 \end{matrix}\right\} CH.CH \begin{cases} CH^3. \\ CH^3. \end{cases}$ By the action of sodium on an ethereal solution of isopropyl iodide. Boiling point, 58°; specific gravity, 0.67.

7. *Normal Heptyl Hydride (Heptan)*, $C^7H^{16} = CH^3.CH^2.CH^2.CH^2.CH^2.CH^2.CH^3$.

Is contained in the light oil of cannel coal-tar and in large quantity in petroleum. Can be obtained from these sources by partial distillation. Boiling point, 99°; specific gravity, 0.699.

Ethyl-amyl, $C^7H^{16} = CH^3.CH^2.CH^2.CH^2.CH.\left\{\begin{matrix} CH^3. \\ CH^3. \end{matrix}\right.$ By the decomposition of a mixture of ethyl and amyl iodides (the latter from ordinary amyl alcohol) with sodium. Boiling point, 90.5; specific gravity, 0.6819 at 17°.

Dimethyldiethylformene, $C^7H^{16} = \left.\begin{matrix} CH^3 \\ CH^3 \end{matrix}\right\} C \left\{\begin{matrix} CH^2.CH^3. \\ CH^2.CH^3. \end{matrix}\right.$ By the action of zinc ethyl on acetone-chloride. Boiling point, 86–87°; specific gravity, 0.711 at 0°.

8. *Normal Octyl Hydride (Octan)*, $C^8H^{18} = CH^3.CH^2.CH^2.CH^2.CH^2.CH^2.CH^2.CH^3$.

The hydrocarbons obtained by the action of sodium on butyl iodide, from methylhexyl carbinol by reduction, from sebasic acid and from octyl alcohol, all appear to be normal octyl hydride. Boiling point, 123–125°; specific gravity at 17°, 0.7032.

In regard to the constitution of the remaining discovered hydrocarbons nothing is as yet known.

	Boiling point.	Specific Gravity.
Nonyl hydride, C^9H^{20} . .	136–138°	0.741
Decatyl hydride, $C^{10}H^{22}$ . .	158–162°	0.757
Undecyl hydride, $C^{11}H^{24}$ . .	180–182°	0.766
Lauryl hydride, $C^{12}H^{26}$ . .	198–200°	0.778
Cocinyl hydride, $C^{13}H^{28}$ . .	218–220°	0.796
Myristyl hydride, $C^{14}H^{30}$ . .	236–240°	0.809
Benyl hydride, $C^{15}H^{32}$ . .	258–262°	0.825
Palmityl hydride, $C^{16}H^{34}$ . .	280°	?

Paraffin. The portions of petroleum or of the oils obtained by the distillation of peat, bitumen, etc., which boil above 300°, solidify wholly or partially on cooling, forming, when purified, a colorless, translucent mass, called paraffin. Paraffin is not a distinct chemical body, but a mixture of several solid hydrocarbons, homologous with marsh gas, which, up to the present, have not been separated. The melting point of commercial paraffin varies from 45° to 65°.

B. Monatomic Alcohols, $C^nH^{2n+2}O$.

A large class of organic compounds has been designated by the name alcohols. These are formed by the displacement of one or more atoms of hydrogen in the hydrocarbons by the same number of hydroxyl atoms (OH). These bodies possess the common property of readily taking up acid radicles in the place of the hydrogen of the hydroxyl group, thus forming compounds, analogous to inorganic salts, called *ethers*.

According to the number of hydroxyl atoms contained in them, alcohols are divided into *monatomic*, *diatomic*, *triatomic*, etc.

The monatomic alcohols, which are derived from the hydrocarbons of the marsh-gas series, have the general formula $C^nH^{2n+2}O$ or $C^nH^{2n+1}.OH$. Only one monatomic alcohol can be derived from marsh gas and ethyl hydride each. These two alcohols have the constitutional formulæ $CH^3.OH$, and $CH^3.CH^2.OH$. With the third member C^3H^8O, however, the case is different. Here, according as in the hydrocarbon $CH^3.CH^2.CH^3$ an atom of hydrogen of one of the terminal carbon atoms, or of the central one is displaced by OH, two isomeric alcohols must result, which have respectively the constitutional formulæ $CH^3.CH^2.CH^2.OH$, and $CH^3.CH.OH.CH^3$.

A similar method of consideration shows that four isomeric modifications of the fourth member $C^4H^{10}O$ are possible, of the fifth, eight, etc.

The conduct of the alcohols in a chemical point of

view, especially under the influence of oxidizing agents, is dependent upon their constitution. They are divided into primary, secondary, and tertiary alcohols.

Primary alcohols contain the group $CH^2.OH$. Under the influence of oxidizing agents they are at first converted into aldehydes by the transformation of the group $CH^2.OH$ into CHO, and then, by further oxidation of the group CHO to COOH, into acids containing the same number of carbon atoms.

Secondary alcohols contain the group CH.OH. When oxidized, they are at first converted into acetones, the group CH.OH being changed to CO. These acetones, when further oxidized, are resolved into simpler compounds, yielding acids with a smaller number of carbon atoms.

Tertiary alcohols contain the group C.OH. They are decomposed by oxidation without previous formation of aldehydes or acetones, and yield acids with a smaller number of carbon atoms.

Normal alcohols are the primary alcohols of normal hydrocarbons.

1. *Methyl Alcohol* (*Wood Spirit*), $CH^4O = CH^3.OH$.

Formation and Occurrence. By the destructive distillation of cellulose, hence contained in wood vinegar obtained by distilling wood. The volatile oil of *Gaultheria procumbens* is the methyl ether of salicylic acid. Pure methyl alcohol may be obtained by distilling this oil with a solution of potassa.

Preparation. From wood vinegar by distilling with calcium hydroxide; only practicable on a large scale. The volatile distillate which at first goes over (wood spirit) contains the methyl alcohol, still, however, containing impurities in the form of other volatile products. After distilling again over quicklime, it is placed in contact with calcium chloride, and the whole distilled on a water bath, by which process the volatile impurities distil over, and the methyl alcohol remains behind in combination with calcium chloride. By mixing with water and distilling, these are separated, and by means of repeated distillations over quicklime, the alcohol is purified.

Or, volatile methyl oxalate is prepared from commercial wood spirit by mixing the wood spirit gradually with its own weight of concentrated sulphuric acid and distilling the brown mixture over two parts by weight of finely powdered acid potassium oxalate. At first a combustible liquid passes over, which, on being evaporated gently, leaves the oxalic ether behind, then the principal part of the ether passes over and congeals in a crystalline form. By pressing and allowing it to stand over sulphuric acid, or by continued fusing, it is obtained pure. By boiling with water or caustic potassa, the alcohol is obtained from the ether.

Properties. A limpid, colorless liquid, of a peculiar odor, similar to that of spirits of wine, and a pungent taste; specific gravity, 0.798; boiling point, 60–65°; combustible; miscible with water, alcohol, and ether. Combines with anhydrous baryta, and with calcium chloride, forming crystalline compounds which are easily decomposed by water. Potassium and sodium are dissolved by it, the action being accompanied by an evolution of hydrogen, and the formation of potassium and sodium methylate, CH^3KO, readily crystallizing compounds.

DERIVATIVES OF METHYL ALCOHOL.

These are perfectly analogous to the derivatives of ethyl alcohol, and are formed from methyl alcohol in the same manner as those from ethyl alcohol. As the corresponding ethyl compounds are of greater importance and generally better investigated, they will be treated of more in detail in the following section, and only a few of the more important methyl compounds will be here described.

Methyl chloride, CH^3Cl. Is formed by the action of chlorine on marsh gas, and of hydrochloric acid on methyl alcohol. Colorless gas, with an ethereal odor; condensable at —22°.

Methylene chloride, CH^2Cl^2. Is produced by the action of chlorine on methyl chloride or methylene iodide, and by treating chloroform with zinc and ammonia.—Colorless liquid of specific gravity 1.36 at 0°; boiling point, 40°; insoluble in water.

Chloroform, $CHCl^3$. Produced by the action of chlorine on the preceding compounds, and in many other ways, particularly by the action of calcium hypochlorite on alcohol, wood spirit, acetone, and several other organic bodies. It is prepared most expediently by distilling 3 parts of alcohol, 100 parts of water, and 50 parts of calcium hypochlorite. It is purified by shaking successively with water and sulphuric acid and subsequent distillation.

Colorless liquid, not miscible with water, with a sweetish ethereal taste and odor; specific gravity, 1.48. Boiling point, 62°; not inflammable; dissolves iodine, the solution taking a bluish-purple color. Its vapor on being inhaled causes unconsciousness and insensibility. With an alcoholic solution of potassa it forms potassium chloride and potassium formate; with sodium ethylate, a colorless ether, *orthoformic ether* $CH(O.C^2H^5)^3$, which boils at 146°. Heated with aqueous or alcoholic ammonia to 180° it yields ammonium cyanide and chloride. If potassa is present this decomposition takes place at 100°.

Carbon tetrachloride, CCl^4. Is obtained most readily by the action of chlorine on chloroform in direct sunlight.—Colorless liquid, of a pleasant odor, boiling at 77°; specific gravity, 1.6; below —25°, solid and crystalline; acts upon the organism analogously to chloroform; yields potassium carbonate and chloride when heated with an alcoholic solution of potassa.

Methyl bromide, CH^3Br. Is obtained by saturating methyl alcohol with hydrobromic acid, or better, by mixing 6 parts of methyl alcohol with 1 part of amorphous phosphorus, carefully adding 6 parts of bromine, at the same time cooling the mixture, and afterward gently heating the whole.—Liquid, of a leeky odor, boiling at 13°; specific gravity, 1.66.

Bromoform, $CHBr^3$. Is produced by the action of bromine on a solution of potassa in wood spirit. —Colorless liquid, boiling at 150–152°; congealing at —9°; of specific gravity 2.9.

Carbon tetrabromide, CBr^4. Is obtained by heating carbon bisulphide or bromoform with bromine in the presence of iodine or antimony bromide ($SbBr^3$) in sealed tubes to 150–160°.—Colorless, lustrous plates. Fusing point, 91°. Insoluble in water, easily soluble in alcohol and ether. It is decomposed when heated in an alcoholic solution.

Methyl iodide, CH^3I. Is prepared in the same manner as the bromide.—Colorless liquid of an ethereal odor. Boils at 43°; specific gravity, 2.2.

Methylene iodide, CH^2I^2. Is produced by the action of sodium ethylate on iodoform, by heating iodoform alone or with iodine, and can be prepared most readily by heating chloroform or iodoform for several hours with very concentrated hydriodic acid to 125–130°.—Yellow liquid, of specific gravity 3.34. Congeals at a low temperature, forming lustrous plates, which fuse at +4°. Boils at 180°, undergoing partial decomposition.

Iodoform, CHI^3. Is formed, when iodine, together with caustic alkalies, acts on alcohol, aldehyde, acetone, and a great many other organic bodies.—Yellow scales, which fuse at 119°. Can be distilled with the vapors of water without undergoing decomposition. Readily soluble in alcohol and ether.

Nitroform, $CH(NO^2)^3$. The ammonium compound of this body $C(NH^4)(NO^2)^3$, a yellow, crystalline substance, soluble in water and alcohol, is produced when trinitroacetonitrile (see fulminuric acid) is treated with water or alcohol. By agitating with sulphuric acid, free nitroform is obtained from this.—Colorless, cubical crystals. Fusing point, 15°; easily soluble in water. Strong acid. When rapidly heated it is decomposed with explosion.

Nitrocarbon, $C(NO^2)^4$. Is produced from nitroform by heating with a mixture of concentrated sulphuric acid and fuming nitric acid.—White crystalline mass, fusing at about 13°, and boiling at 126°. Not inflammable. Insoluble in water; soluble in alcohol and ether.

Nitrochloroform (Chloropicrin), $C(NO^2)Cl^3$. Is formed when alcohol or wood-spirit is distilled with sodium chloride, saltpetre and sulphuric acid, by the distillation of a number of nitro-compounds with calcium hypochlorite or hydrochloric acid and potassium chlorate. Further, by heating chloroform with nitric acid (containing hyponitric acid) in sealed tubes to 90–100° for 12 hours. Is most readily prepared by adding 45 parts of calcium hypochlorite, mixed with water so as to form a thick pasty mass, to a saturated aqueous solution of 4½ parts of picric acid at 30°. The reaction begins immediately and spontaneously, and the greater part of the chloropicrin distils over.—Colorless oil, not combustible; boiling at 112°; specific gravity, 1.66. When heated with acetic acid and iron filings, it yields methylamine; heated with sodium ethylate, it yields *orthocarbonic ether* $C(O.C^2H^5)^4$, a liquid which boils at 158–159°.

A compound very similar to chloroform, *Marignac's oil* $C(NO^2)^2Cl^2$, is produced by the distillation of naphthalene chloride with nitric acid.—Colorless liquid; explodes when heated alone; can be distilled, however, with vapors of water.

Nitrobromoform (Bromopicrin), $C(NO^2)Br^3$. Is prepared, like chloropicrin, by distilling picric acid with calcium hypobromite (lime-water containing bromine).—Colorless, prismatic crystals, which melt at

10°, forming a liquid of specific gravity 2.8. Can only be distilled in a vacuum without decomposition.

Acetonitrile (Methyl cyanide), $C^2H^3N = CH^3.CN$. Is obtained by gently heating acetamide with phosphoric anhydride or phosphorus pentasulphide; and by distilling a mixture of potassium methylsulphate with potassium cyanide.—Colorless liquid, boiling at 82°. Combines with two atoms of bromine, with hydrobromic and hydriodic acids, and with several metallic chlorides. Is decomposed by boiling with potassa, yielding ammonia and potassium acetate, and gives, with hydrogen in *statu nascendi*, ethylamine. For the substitution-products of acetonitrile, see fulminic acid.

Methyl carbylamine, $C^2H^3N = CH^3.NC$ (isomeric with acetonitrile). Is formed by the action of methylamine on chloroform in the presence of potassa; by heating one molecule of methyl iodide with two molecules of silver cyanide to 130–140°, and distilling the resulting crystalline compound $C^2H^3N + AgCN$ with half its weight of potassium cyanide and a little water. Is formed in small quantity, together with acetonitrile, by the distillation of a mixture of potassium methylsulphate with potassium cyanide.—Colorless liquid, possessing an exceedingly strong odor. Soluble in ten parts of water. Boiling point, 58–59°. Combines with thoroughly dried hydrochloric acid gas; is decomposed by dilute hydrochloric acid, however, and by being heated with water to 180°, yielding methylamine and formic acid.

Methylether, $(CH^3)^2O$, is formed, but with difficulty, by distilling methyl alcohol with four times its weight of concentrated sulphuric acid.—Colorless gas, of ethereal odor, congealing at —21°; combustible, exploding violently with chlorine; specific gravity, 1.617. Water absorbs thirty-seven times its volume of the gas. Combines with sulphuric anhydride, forming methyl sulphate.

Methyl nitrate, $CH^3.O.NO^2$. Results in small quantity when a mixture of wood-spirit with saltpetre and sulphuric acid is subjected to distillation.—Colorless liquid, boiling at 66°.

Methylsulphate, $(CH^3.O)^2SO^2$, is formed by distilling wood-spirit with eight to ten times its weight of concentrated sulphuric acid.—Colorless liquid, possessing the odor of garlic, of specific gravity 1.324; boiling point, 188°. Is decomposed by heating with water, yielding methyl alcohol and methylsulphuric acid.

Methylsulphuric acid, $CH^3.O.SO^2.OH$, is formed by mixing one part of methyl alcohol with two parts of concentrated sulphuric acid. Crystallizes in colorless needles, when carefully evaporated; forms easily soluble salts with bases. The potassium salt, crystallizing in deliquescent plates, yields by distillation methyl sulphate.

Methylsulphurous acid (sulphomethylic acid), $CH^3.SO^2.OH$. The potassium salt, $CH^3.SO^2.OK$, is produced by heating methyl iodide with neutral potassium sulphite to 100–120°. The free acid is a syrupy liquid.

Trichlormethylsulphurous acid, $CCl^3.SO^2.OH$. The barium salt, $(CCl^3.SO^3)^2Ba$, is obtained by digesting trichlomethyl sulphochloride with baryta water. The acid, set free from this salt by means of sulphuric acid, crystallizes in small, colorless, very deliquescent prisms. Very strong acid.

Trichlormethyl sulphochloride, $CCl^3.SO^2Cl$. Is formed by the action of hydrochloric acid and black oxide of manganese, or of hydrochloric acid and potassium bichromate on carbon bisulphide. An addition of nitric acid aids the reaction.—Colorless, crystalline mass; insoluble in water; easily soluble in alcohol and ether. Melting-point, 135°; boiling-point, 170°; also volatile with the vapor of water without decomposition.

Methylamine, $CH^3.NH^2$. Gas, of ammoniacal odor; liquid below 0°; water absorbs more than 1000 times its volume of the gas. The solution is strongly alkaline, smells like ammonia, and acts on solutions of metallic salts like ammonia, but does not, however, redissolve the precipitated hydroxides of nickel, cobalt, and cadmium, when added in excess. It forms neutral, easily soluble salts with acids.

Dimethylamine, $(CH^3)^2HN$. Inflammable gas; liquid below +8°; strongly alkaline.

Trimethylamine, $(CH^3)^3N$. Is formed in *Chenopodium vulvaria*, in the blossoms of *Cratægus oxyacantha*, and several other plants; is contained in herring brine, in liver oil, coal-tar oil, and bone oil. At ordinary temperatures it is gaseous; below +9°, a clear liquid, of a peculiar odor somewhat resembling that of ammonia; in water and alcohol very easily soluble. Strong base.

The compounds of methyl with phosphorus and the metals bear the strongest resemblance to the corresponding ethyl compounds, which will be treated of later; hence, only the methyl compounds of arsenic, which are better investigated than the ethyl compounds, will be here treated of.

Arsendimethyl (Cacodyl), $\left.\begin{matrix}(CH^3)^2As\\(CH^3)^2As\end{matrix}\right\}$. By distilling dry potassium acetate with arsenious acid is obtained a liquid (alkarsin), which contains cacodyl together with the products of its oxidation. Treated with concentrated hydrochloric acid, this liquid yields cacodyl chloride, and this chloride treated with zinc filings in an atmosphere of carbonic anhydride at 100° yields pure cacodyl, the zinc chloride having been dissolved out with water.—Clear liquid, of a disgusting odor; congeals at —6°; boils at 170°; but slightly soluble in water, easily soluble in alcohol and ether. In contact with the air it gives off fumes and takes fire; its vapor is very poisonous; it combines directly with oxygen, sulphur, and chlorine.

Cacodyl chloride, $(CH^3)^2AsCl$. Liquid, boiling at 100°; heavier than water; unites with metallic chlorides. The iodide and bromide are similar to the chloride. The cyanide forms large prisms, fusing at 30°, boiling at 140°. Exceedingly poisonous.

Cacodyl oxide, $[(CH^3)^2As]^2O$. Is formed by slow oxidation of cacodyl, simultaneously with cacodylic acid, and can be separated from the latter by distillation. Liquid, boiling at 150°, of disagreeable odor. It does not give off fumes in contact with the air, and does not take fire; is oxidized slowly, however, forming cacodylic acid. It combines with $2HgCl^2$, yielding a crystalline compound.

Cacodyl sulphide, $[(CH^3)^2As]^2S$. By distilling cacodyl chloride with potassium or barium sulphhydrate.—Colorless liquid, of a disagreeable odor; insoluble in water, easily soluble in alcohol and ether. Yields cacodyl chloride and hydrosulphuric acid when treated with hydrochloric acid.

Cacodyl disulphide, $(CH^3)^4As^2S^2$, is formed by dissolving sulphur in cacodyl or cacodyl sulphide.—Large colorless crystals, fusing at 50°; not volatile without decomposition.

Cacodylic acid, $(CH^3)^2As.OH$. Is produced by slow oxidation of cacodyl, and by the action of mercury oxide on cacodyl under water (or on the crude liquid alkarsin).—Large, colorless, deliquescent prisms, which fuse at 200°; are inodorous and not poisonous. Phosphorous acid reduces it, forming cacodyl.

Cacodyl trichloride, $(CH^3)^2AsCl^3$. Is formed by the action of phosphorus pentachloride (under ether) on cacodylic acid, or when chlorine is conducted upon the surface of a solution of cacodyl in carbon bisulphide.—Crystallizes in transparent prisms, or large laminæ. Heated up to 40–50° it is resolved into methyl chloride and

Arsen-monomethyl dichloride, $(CH^3)AsCl^2$. This is also formed by the action of dry hydrochloric acid gas on cacodylic acid.—Colorless, heavy liquid, boiling at 133°; easily soluble in water; does not give off fumes in contact with the air. It takes up two atoms of chlorine, but the resulting crystalline compound is decomposed even below 0° into methyl chloride and arsenic trichloride. On being treated with hydrosulphuric acid, it yields crystals of *arsen-monomethyl sulphide* $(CH^3)AsS$, which fuse at 110°.

Arsen-monomethyl oxide, $(CH^3)AsO$. Is formed by the action of potassium carbonate on the dichloride under water.—Crystals fusing at 95°; not volatile alone without decomposition, readily with vapors of water; soluble in water, alcohol, and ether.

Arsen-monomethylic acid, $(CH^3)As(OH)^2$. Is formed when the dichloride is treated under water with silver oxide.—Large crystalline laminæ, soluble in water and alcohol; bibasic acid; forms crystalline salts.

2. *Ethyl Alcohol (Spirits of Wine).*

$$C^2H^6O = CH^3.CH^2.OH.$$

Formation. By the fermentation of sugar.

When the clear juice of a plant containing sugar is left to itself at the ordinary summer temperature, it soon begins to grow turbid, and small bubbles of carbonic anhydride appear in it, which gradually increase in number, at the same rate that the liquid, accompanied by a simultaneous and spontaneous increase in warmth, shows signs of a more or less marked internal motion (fermentation). After a time this phenomenon ceases, the juice is then no longer sweet, its sugar has disappeared, and the liquid now contains alcohol instead of sugar. The turbidness has settled in the form of an ill-looking, grayish mass, which is called yeast.

A solution of pure sugar in water does not undergo this change alone. If, however, a small quantity of yeast be added to it, the phenomena observed in connection with the plant-juice make their appearance, though more slowly than in the former case. Cane-sugar, grape-sugar, and fruit-sugar, according to all appearances, conduct themselves in a similar manner. Grape-sugar and fruit-sugar are in reality, however, the only varieties capable of fermentation; cane-sugar only undergoes fermentation after having been previously converted into these varieties. From one molecule of grape-sugar result two molecules of alcohol, and two molecules of carbonic anhydride; but in addition to these are always formed small quantities of succinic acid and glycerin.

Yeast consists of microscopical vesicles (yeast-cells), the walls of which are formed by an elastic membrane consisting of cellulose.—Their contents are, in the young cells, a liquid, but in the older ones, a thick, granular, nitrogenous mass.

The germs of the yeast-cells come from the air. Hence, contact of the plant-juice with the air is essential to the beginning of fermentation; once begun, however, fermentation continues regularly even though the air be excluded. The germs, which have fallen from the air into the solution, develop when they meet with the substances necessary to their growth. But, in addition to the saccharine solution, nitrogenous substances and inorganic salts are essential. For this reason albuminous substances aid fermentation materially, but they are not, as was supposed for a long time, the real ferment which causes fermentation. As these substances are not present in a pure solution of sugar, the germs cannot develop in it. They are contained in plant-juices, however, and hence in these the development and rapid multiplication of the cells by means of the formation of buds begin immediately. The splitting up of sugar into alcohol and carbonic anhydride stands in the closest relation to the growth of these vegetable organisms in the saccharine solution. It has been proven with certainty, that the formation of alcohol and carbonic anhydride only takes place in the interior of the plant cells, but, regarding the details of this process and the character of the chemical reaction, nothing is positively known.

Fermentation only takes place between 3–35°, it progresses most rapidly at 25–30°. The character of the ferment (the variety of vegetable organism) that is undergoing development in the saccharine solution, exerts the most marked influence upon the products of the fermentation. Under certain circumstances, which appear to be unfavorable to the development of yeast-cells, the germs of another ferment are developed, and now entirely different products result (see Lactic Acid).

Yeast loses its efficacy by being thoroughly dried, by being heated up to 60°, by being immersed in alcohol, and by being acted upon by acids and alkalies. Various substances, particularly the volatile oil of mustard, sulphurous, nitrous, and arsenious acids, mercury chloride, prevent the beginning of fermentation, when added in exceedingly small quantity to a fermentable liquid.

Starch is not fermentable, but, as it can be readily converted into sugar, alcohol can also be obtained from substances containing starch, such as potatoes, grain, etc.

Preparation. By partial distillation of a fermented liquid, the alcohol goes over still mixed with more or less water. Such a mixture containing between 30 and 40 per cent. of alcohol is *brandy*. Subjected again to distillation, it is separated into water, which remains behind, and an alcohol containing less water (spirits of wine), which distils over. The last portions of water cannot be removed from this by means of distillation, but only by means of desiccating agents, such as fused calcium chloride, fused potassa, quicklime, etc., most efficiently, however, by means of anhydrous baryta, which is brought in contact with the alcohol, and the latter afterward distilled off from it.

Properties. Colorless, thin liquid; in a perfectly

pure condition and free from water, almost inodorous. Specific gravity, 0.78945 at +20°, 0.80625 at 0°. Does not solidify even at 100°. Boiling point, 78°. Easily inflammable, burning with a flame, which has a weak light and does not soot. Attracts moisture from the air, and is miscible with water in all proportions with the accompaniment of heat and contraction of the volume of the mixture. The greatest contraction takes place when one molecule of alcohol is mixed with three molecules of water. 100 volumes of this mixture contain 53.939 volumes of alcohol and 49.836 volumes of water, hence the contraction amounts to 3.775 volumes. With an increase of the amount of water contained in it the boiling point is elevated and the specific gravity increased.

Like water, it is a solvent for a great many substances; it combines, also, with salts, forming crystalline compounds.

Decompositions. By means of oxidizing agents (black oxide of manganese and sulphuric acid, chromic acid, etc.) and oxygen in the presence of spongy platinum or certain organic substances, it is converted into aldehyde and acetic acid. When heated with nitric acid a violent reaction takes place, and a large number of products result. Mixed with sulphuric acid there result, according to the proportions of the two and the temperature, either ethylsulphuric acid, ether, or ethylene (C^2H^4). Potassium and sodium are dissolved by it, hydrogen being evolved, and potassium and sodium ethylate $C^2H^5.OK$ being formed.

DERIVATIVES OF ETHYL ALCOHOL.

Ethyl chloride, C^2H^5Cl. Absolute alcohol is saturated with dried hydrochloric acid gas, the liquid heated to boiling after standing for twenty-four hours, the evolved ethyl chloride passed through water of the temperature of 25° for the purpose of cleansing it, and then condensed in a vessel which is cooled at least down to 0°. It is formed by the action of chlorine on ethyl hydride.—Colorless, very mobile liquid, of a pleasant odor; specific gravity, 0.874; boiling point, 12°, hence at the ordinary temperature gaseous. Burns with a green-bordered flame. But slightly soluble in water. It is converted into alcohol, with formation of hydrochloric acid or potassium chloride, when heated for a long time with water at 100°; more rapidly with an alcoholic solution of potassa.

With chlorine, ethyl chloride yields a series of substitution-products.

Ethylidene chloride, $C^2H^4Cl^2 = CH^3.CHCl^2$. Is the first product of the action of chlorine on ethyl chloride. Is also produced by the action of phosphorus pentachloride on aldehyde.—Colorless liquid, boiling at 58–59°, of specific gravity 1.198.

Further action of chlorine, finally with the aid of heat and direct sunlight, produces the liquid compounds $C^2H^3Cl^3$, boiling point, 75°; $C^2H^2Cl^4$, boiling point, 102°; C^2HCl^5, boiling point, 158°; and the final product

Carbon trichloride, C^2Cl^6. Colorless crystals of a camphorous odor. Fusing point, 160°; boiling point, 182°. But slightly soluble in water, readily in alcohol and ether.

Ethyl bromide, C^2H^5Br. 1 part of red phosphorus is put into 6 parts of alcohol and 6 parts of bromine added, the vessel being kept cool. After a time the mixture is distilled. The distillate is shaken with caustic soda, the oil which separates is freed of water and rectified.—Colorless, heavy liquid, boiling at 40°, of specific gravity 1.47. Bromine acts upon this compound, displacing its hydrogen, forming thus *ethylidene bromide* (ethyl bromobromide) $C^2H^4Br^2 = CH^3.CHBr^2$ (colorless liquid, boiling at 110°) and higher substitution-products.

Ethyl iodide, C^2H^5I, is prepared in the same manner as bromine from 1 part of red phosphorus, 5 parts of alcohol, and 10 parts of iodine.—Very similar to the bromide. Boiling point, 72°; specific gravity, 1.975.

Propionitrile (Ethyl cyanide), $C^3H^5N = C^2H^5.CN$, is prepared by distilling a mixture of potassium cyanide and potassium ethylsulphate, or of ammonium propionate and phosphoric andydride.—Colorless liquid, specific gravity, 0.787; boiling point, 98°; in a pure condition possessing a pleasant odor; does not mix with water. Combines directly with bromine, with hydrochloric, hydrobromic, and hydriodic acids, with phosphorus terchloride, and several metallic subchlorides. Heated with caustic potassa it is transformed into ammonia and potassium propionate; hydrogen in *statu nascendi* converts it into propylamine. When allowed to drop gradually on potassium a violent reaction and the formation of potassium cyanide and volatile products ensue, and it is transformed into *cyanethine*, $C^9H^{15}N^3$, which is polymeric with ethyl cyanide. This substance crystallizes in colorless and inodorous laminæ, is difficultly soluble in water, and possèsses strong basic properties.

Ethylcarbylamine, $C^3H^5N = C^2H^5.NC$ (isomeric with propionitrile), is produced with a violent reaction when an alcoholic solution of ethylamine is poured upon caustic potassa, or when silver cyanide is heated with ethyl iodide. It is also formed in small quantity, as a secondary product, in the preparation of propionitrile from potassium ethylsulphate.—Oily liquid, lighter than water, of an unendurable, garlic-like odor. Boiling point, 79°. Unites with silver cyanide, forming a crystalline compound; is with great difficulty decomposed by means of potassa, easily by hydrochloric acid, yielding ethylamine and formic acid, the elements of water being assimilated for the purpose.

Ethylether, $(C^2H^5)^2O$. Is formed by the action of sulphuric acid, phosphoric acid, or anhydrous zinc subchloride and a few similar metallic chlorides on alcohol at a temperature of 140°; by means of the double decomposition of sodium ethylate $C^2H^5.ONa$ and ethyl iodide C^2H^5I. For its preparation a mixture of 9 parts of concentrated sulphuric acid and 5 parts of 85–90 per cent. alcohol is heated to boiling, *i.e.* up to 140°, in a retort connected with a good condensing apparatus. During the operation just as much alcohol is allowed to flow into the retort, through a tube passing to the bottom of the retort, as liquid distils off. The distillate consists of ether and water. The formation of the ether in this reaction takes place in two phases. At first, from one molecule of alcohol and one molecule of sulphuric acid, water and ethyl-sulphuric acid are formed; the latter then acts on a second molecule of alcohol, the result being ether and sulphuric acid. Hence a small quantity of sulphuric acid can transform a large (theoretically an unlimited) amount of alcohol into ether.

Ether prepared in this way contains alcohol, which has distilled over unchanged, especially when the too rapid addition of alcohol to the mixture caused the temperature to sink much below 140°; it also often contains sulphurous acid, when, the addition of the alcohol having been too slow, the temperature in the retort has risen much above 140°. Both impurities may be removed by shaking the distillate with water containing an alkali and rectifying the ether, after separating from the water, over calcium chloride or quicklime. Ether can be obtained perfectly anhydrous and free from alcohol by being allowed to stand for some time in contact with metallic sodium.

Colorless, limpid liquid, strongly refracting, of a peculiar penetrating odor and taste. Specific gravity at + 20° = 0.713, at 0° = 0.736. Very volatile, boiling at 35°.5. At —31° congeals, forming a crystalline mass. Easily inflammable, burning with a luminous, sooty flame. Mixed with air in the form of vapor it is exceedingly explosive. Inhaled as vapor it causes unconsciousness and insensibility. Does not mix with water; ether, however, does dissolve some water $\frac{1}{36}$

and water some ether $\frac{1}{10}$. Mixes with alcohol in all proportions.

Chlorine acts very energetically on ether, yielding substitution-products: $C^4H^9ClO = CH^3.CHCl.O.C^2H^5$, boiling point 97–98°; $C^4H^8Cl^2O = CH^2Cl.CHCl.O.C^2H^5$, colorless liquid, boiling at 140–147°; $C^4H^6Cl^4O$, heavy, yellow liquid with a fennel-like odor; $C^4Cl^{10}O$, colorless crystals, fusing at 69°. Concentrated sulphuric acid forms ethylsulphuric acid; sulphuric anhydride forms ethyl sulphate together with other products. Heated with water and a little sulphuric acid to 150–180°, it is reconverted into alcohol.

Ethyl-methylether (ethyl-methyl oxide), $C^2H^5.O.CH^3$, is formed by the double decomposition of sodium ethylate and methyl iodide.—Liquid, boiling at + 11°.

Compound ethers. Alcohol combines with acids to form ethers, water being eliminated. These may be considered as salts, in which the atomic group C^2H^5 (ethyl) takes the part of a metal. Monobasic acids can form only one kind of ethers, and this is a neutral substance; bibasic acids, as for instance sulphuric acid SO^4H^2, can take up one or two atoms of ethyl. In the first case there is formed an acid ether, a so-called *ether acid*, which conducts itself as a monobasic acid; in the latter case, however, a neutral ether is the result. Tribasic acids, finally, as for instance phosphoric acid PO^4H^3, yield three different ethers, of which one is a bibasic, the second a monobasic acid, and the third a neutral compound.

By boiling with alkalies the ethers are decomposed into alcohol and acids.

The ethers of most of the weaker acids can only be produced by the simultaneous action of sulphuric or hydrochloric acid.

Ethyl nitrate, $C^2H^5.O.NO^2$. 15 grm. of urea nitrate are added to a mixture of 80 grm. of nitric acid free of hydrochloric acid, of specific gravity 1.4, which has been previously heated with a little urea, and the liquid distilled off down to one-eighth of the original volume. The distillate is agitated with water; the ether, which is precipitated, is separated from the water, desiccated by means of calcium chloride, and rectified on a water bath. Without the presence of the urea, a violent reaction takes place and the acid and the alcohol are thoroughly decomposed, forming nitrous ether together with many other products.—Colorless liquid, of pleasant odor; of specific gravity, 1.132 at 0°; boiling point, 87.° Does not mix with water; burns with a white flame; its vapor explodes when heated above the boiling point.

Ethyl nitrite, $C^2H^5.NO^2$, is formed when nitrous anhydride is mixed with well-cooled aqueous alcohol, in which case the ether separates immediately; or by conducting the acid in a gaseous form into the alcohol and condensing the gaseous ether that passes over by cooling. Is prepared most easily by adding a solution of potassium nitrate to a mixture of alcohol and sulphuric acid, or by pouring this mixture upon the dry salt.—Pale yellow, very thin liquid, of an agreeable fruity odor; specific gravity, 0.947; boiling point, 16°.5; does not mix with water; decomposes when kept for any length of time.

Ethyl sulphate, $(C^2H^5.O)^2SO^2$, is formed when the vapor of sulphuric anhydride is conducted into well-cooled ether, or, better, when absolute alcohol or ether is added drop by drop to sulphuryl oxichloride ($SO^2.Cl.OH$).—Colorless, thick liquid, undergoes decomposition at 130–140°.

Ethylsulphuric acid (Sulphovinic acid), $C^2H^5.O.SO^2.OH$, is formed when 1 part of alcohol and 2 parts of sulphuric acid are mixed together. When the mixture has cooled, it is diluted with water, neutralized with barium carbonate, and the dissolved barium ethylsulphate filtered off. The solution is then carefully evaporated, and the ethylsulphuric acid obtained in a free state by precipitating the barium with the exact amount of sulphuric acid required. The acid can, however, only be concentrated in a vacuum at the ordinary temperature. It forms a thick, very strongly acid liquid. The watery solution is resolved, by heating, into alcohol and sulphuric acid. Its salts are soluble in water, the alkaline salts also in alcohol. The *barium salt* $(C^2H^5.SO^4)^2Ba + 2H^2O$ forms large laminated crystals.

Ethyl sulphite, $(C^2H^5)^2SO^3$, is formed by the action of sulphur chloride S^2Cl^2 or chlorothionyl $SOCl^2$ on alcohol.—Liquid, boiling at 160°; of specific gravity, 1.106; of a peppermint odor; is decomposed gradually by water.

An ether isomeric with this, *ethylsulphonic ether* $(C^2H^5)^2SO^3$, is produced by the action of sodium ethylate on ethylsulphonchloride.—Colorless liquid; boiling at 207–208°; of specific gravity, 1.1712.

Ethylsulphurous acid, $C^2H^5.SO^2.OH$, is formed by the oxidation of mercaptan, ethyl sulphide, and ethyl sulphocyanide by means of nitric acid; by the action of zinc ethyl on sulphurous acid or sulphuric anhydride. The potassium salt is formed by boiling ethyl iodide with a concentrated solution of potassium sulphite.—Crystalline, very deliquescent mass, much more stable than ethylsulphuric acid. Its solution can be evaporated on a water bath. By oxidation it is converted into ethylsulphuric acid. Its salts are all easily soluble and are decomposed only at a high temperature. The *lead salt* $(C^2H^5SO^3)^2Pb$ forms colorless laminæ, soluble in alcohol and water.

Ethylsulphonchloride, $C^2H^5SO^2Cl$. Is produced by the action of phosphorus pentachloride on potassium ethylsulphite.—Colorless liquid, boiling at 173°.5.

Ethyl phosphate, $(C^2H^5.O)^3PO$, is formed by the action of phosphoric anhydride on absolute alcohol in the presence of ether; by heating silver phosphate with ethyl iodide; and by heating lead diethylphosphate to 200°. Is prepared most readily by the action of phosphorus oxichloride on sodium ethylate.—Clear, transparent liquid; soluble in water, alcohol and ether. Boiling point, 211°; specific gravity, 1.072 (at 12°). Is decomposed slowly by water.

Diethylphosphoric acid, $(C^2H^5.O)^2PO.OH$, is formed when phosphoric anhydride is allowed to slowly absorb the vapor of alcohol. By neutralizing the liquid, diluted with water, with lead carbonate, the soluble *lead salt* $[(C^2H^5)^2PO^4]^2Pb$ is obtained, which crystallizes in needles. The free acid decomposes by evaporation. Monobasic acid.

Ethylphosphoric acid, $C^2H^5.O.PO(OH)^2$, is formed by heating a mixture of syrupy phosphoric acid and alcohol.—Strongly acid, thick liquid. Its aqueous solution does not undergo decomposition by boiling. Bibasic acid. The *barium salt* $C^2H^5.PO^4Ba$ crystallizes in prisms, and is soluble in water.

Ethylphosphoric chloride, $C^2H^5.O.POCl^2$. Is produced by conducting chlorine into a mixture of 1 molecule of PCl^3 and 2 molecules of alcohol.—Liquid, boiling at 167°.

Ethyl phosphite, $(C^2H^5.O)^3P$, is produced when sodium ethylate and phosphorus terchloride are brought together; and by the action of phosphorus cyanide on alcohol. Boiling point, 191°; specific gravity, 1.075. By the action of phosphorus terchloride on alcohol is produced *ethylphosphorous chloride* $C^2H^5.O.PCl^2$.—Colorless liquid; specific gravity, 1.316; boiling point, 117°. Is resolved rapidly by water into hydrochloric acid, phosphorous acid, and alcohol. Yields with bromine ethyl bromide and PCl^2BrO.

Ethyl arsenate, $(C^2H^5.O)^3AsO$, is formed when silver arsenate is heated with ethyl iodide to 120°.—Colorless liquid; boils with slight decomposition at 235–

238°; specific gravity, 1.3264 at 0°. Mixes with water and is decomposed by it.

Ethyl arsenite, $(C^2H^5.O)^3As$, is produced by the action of methyl iodide on silver arsenite; by heating silicic ether with arsenious acid to 220°.—Colorless liquid; boiling point, 166–168°; specific gravity, 1.224 at 0°. Decomposed immediately by water, arsenious acid being precipitated.

Ethyl borate, $(C^2H^5.O)^3B$, is formed when 2 parts of anhydrous borax are heated with 3 parts of potassium ethylsulphate; by the action of boron chloride on absolute alcohol; and by continued heating of boracic anhydride with absolute alcohol at 110–120°. Liquid; boiling point, 120°; specific gravity at 0° = 0.887. Decomposed rapidly by water.

Ethyl silicate, $(C^2H^5)^4Si$, is obtained by distilling a mixture of silicium chloride and absolute alcohol. —Colorless liquid; boiling point, 165–168°; specific gravity, 0.933 at 20°. Insoluble in water; is, however, slowly decomposed by it, silicic acid being thrown down. If the alcohol used in the preparation be not entirely free of water, a small quantity of an ether, $(C^2H^5)^6Si^2O^7$, is formed at the same time. This boils at 230–240°.—By heating silicic ether with silicium chloride, fluid ethyl-silicic chlorides are formed, as follows: $(C^2H^5.O)^3SiCl$, boiling point, 155–157°; $(C^2H^5.O)^2SiCl^2$, boiling point, 136–138°; $C^2H^5.OSiCl^3$, boiling point, 104°. When these chlorides are allowed to act upon different alcohols, compound silicic ethers are formed; for instance, *diethyldimethyl silicate*, $\begin{matrix}(C^2H^5.O)^2\\(CH^3.O)^2\end{matrix}Si$, boiling point, 143–147°; *triethylmethyl silicate*, $\begin{matrix}(C^2H^5.O)^3\\CH^3.O\end{matrix}Si$, boiling point, 155–157°; *ethyltrimethyl silicate*, $\begin{matrix}C^2H^5O\\(CH^3.O)^3\end{matrix}Si$, boiling point, 133–135°.

The ethers with organic acids will be treated of in connection with the latter.

Ethyl sulphhydrate (Mercaptan), $C^2H^6S = C^2H^5.SH$, is produced by distilling a mixture of concentrated solutions of potassium ethylsulphate and potassium sulphhydrate.—Very thin, colorless liquid, of an exceedingly nauseous smell; specific gravity, 0.831; boiling point, 36°. Does not mix with water; easily inflammable.

It dissolves potassium and sodium, hydrogen being evolved, and, on evaporating, granular compounds *potassium and sodium mercaptide*, $C^2H^5.SK$ and $C^2H^5.SNa$, are left behind. With a number of metallic oxides, it forms water and similar metallic compounds, the action being accompanied by an evolution of heat.

Mercury mercaptide, $(C^2H^5.S)^2Hg$, crystallizes from alcohol in colorless shining laminæ, fuses at 85–87°, and is decomposed by sulphuretted hydrogen, yielding mercury sulphide and mercaptan; hence used as a means of purification for crude mercaptan. When mercaptan is mixed with an alcoholic solution of corrosive sublimate, there results a difficultly soluble precipitate, $C^2H^5.S.HgCl$.

Ethyl sulphide, $(C^2H^5)^2S$, is best prepared by conducting ethyl chloride into an alcoholic solution of potassium sulphide and then distilling. It is precipitated from the distillate by means of water.—Colorless, thin liquid of an exceedingly disagreeable smell; specific gravity, 0.825; boiling point, 91°. Combines with several metallic chlorides. Mercury chloride causes a precipitate from an alcoholic solution $(C^2H^5)^2S.HgCl^2$; platinum chloride precipitates $2[(C^2H^5)^2S].PtCl^4$. On being oxidized with dilute nitric acid, it is converted into *sulphethyl oxide* $(C^2H^5)^2SO$. Thick liquid, not volatile without decomposition. Treated with fuming nitric acid *diethylsulphon* $(C^2H^5)^2SO^2$ is produced. Large, thin plates, which fuse at 70°, begin to sublime below 100°, and boil at 248° without decomposition. Easily soluble in alcohol and water. Hydrogen in *statu nascendi* (zinc and sulphuric acid) reconverts it into ethyl sulphide.

Ethyl sulphide, when heated, combines readily with ethyl iodide, forming *triethyl sulphiodide* $(C^2H^5)^3SI$, a crystalline substance, easily soluble in water and alcohol, which, when treated with silver oxide and water, yields *triethyl sulphhydroxide* $(C^2H^5)^3S.OH$. Indistinct deliquescent crystals. Strong base, combines with acids forming well characterized, easily soluble salts.

Ethyl bisulphide, $(C^2H^5)^2S^2$, is produced when ethyl chloride is conducted into an alcoholic solution of potassium bisulphide, and by the action of iodine on sodium mercaptide.—Liquid, boiling at 151°. When shaken with dilute nitric acid, it yields *ethyl disulphoxide* $(C^2H^5)S^2O^2$, a liquid, which cannot be distilled without decomposition.

The corresponding selenium and tellurium compounds are produced in a similar manner to the sulphur compounds, potassium selenide or telluride being employed instead of the sulphide.

Selenmercaptan, C^2H^6Se. Colorless, thin liquid, with an insupportable odor; with mercury oxide it also yields a mercaptide.—*Ethyl selenide* $(C^2H^5)^2Se$. Pale yellow liquid, with an exceedingly repulsive odor, heavier than water. Is oxidized by nitric acid, the action being accompanied by an evolution of nitrogen binoxide; from the resulting solution hydrochloric acid precipitates *ethyl chloroselenide* $(C^2H^5)^2SeCl^2$, a pale yellow, heavy oil.

Ethyl telluride, $(C^2H^5)^2Te$. Reddish-yellow liquid, heavier than water, of insupportable odor. Is dissolved by nitric acid as *tellurethyloxide nitrate*. From this solution hydrochloric acid precipitates an oily, colorless substance, *tellurethyl chloride* $(C^2H^5)^2TeCl^2$; hydriodic acid, an orange-yellow, powdery substance, *tellurethyl iodide* $(C^2H^5)^2TeI^2$. Aqueous ammonia decomposes the chloride, forming ammonium chloride and an oxichloride $(C^2H^5)^2TeCl^2+(C^2H^5)^2TeO$, which crystallizes in colorless and inodorous prisms. The iodide conducts itself in a similar manner.

Ethylamine, $C^2H^5.NH^2$. Ethyl bromide and aqueous ammonia combine gradually at the ordinary temperature, more rapidly when heated in sealed tubes to 100°, forming ethylamine hydrobromate (bromethyl-ammonium). Ethyl iodide and bromide act in the same way. By this reaction, however, small quantities of di- and triethylamine are formed at the same time.* It is obtained in a pure condition by distilling ethyl cyanate or cyanurate with potassa; the distillate, being neutralized by hydrochloric acid, yields, on evaporation, ethylamine hydrochlorate. Ethylamine nitrate is produced when ethyl nitrate is heated with an alcoholic solution of ammonia; ethylamine sulphate by treating acetonitrile (see p. 38) with zinc and sulphuric acid. By gently heating one of these salts with caustic potassa, ethylamine is set free; it evolves in gaseous form, is passed through a tube containing pieces of caustic potassa, for the purpose of drying it, and then conducted into a vessel cooled below 0°. This liquid (boiling point, 18°) smells almost exactly like ammonia; specific gravity, 0.696; inflammable; mixes with water; a caustic alkali; a more powerful base than ammonia. Its solution precipitates metallic salts the same as ammonia, redissolves precipitated alumina, however, when added in excess. Nitrous acid decomposes it, alcohol, nitrogen, and water being formed.

Ethylamine hydrochlorate, $C^2H^7N.HCl$, forms large, deliquescent, tabular crystals, soluble in alcohol. With platinum chloride it gives a yellow compound $(C^2H^7N.HCl)^2PtCl^4$.

Diethylamine, $(C^2H^5)^2NH$. Ethylamine in an aqueous solution combines in a short time with ethyl bromide, forming diethylamine hydrobromate, from which the free base can be obtained by means of potassa.—A liquid, easily inflammable, boiling at 57°,

* On the separation of these three bases from each other, see Diethyl Oxamid in connection with Oxalic Acid.

miscible with water. Strong base. The *hydrochlorate* $(C^2H^5)^2NH.HCl$, when distilled with a concentrated solution of potassium nitrite, yields *nitrosodiethyline* $(C^2H^5)^2NO.N$, a liquid boiling at 177°, which is decomposed by hydrochloric acid, forming nitrogen binoxide and diethylamine hydrochlorate.

Triethylamine, $(C^2H^5)^3N$. Is formed from diethylamine in the same way that this is formed from ethylamine.—Colorless, light, strongly alkaline liquid, but slightly soluble in water. Boiling point, 89°. The hydrochlorate, when heated in concentrated solution with potassium nitrite, yields nitrosodiethyline, the same as diethylamine.

Tetrethylammonium. Triethylamine and ethyl iodide combine slowly at the ordinary temperature, rapidly at 100°, forming *tetrethylammonium iodide* $(C^2H^5)^4NI$. Colorless crystals, easily soluble in water and alcohol. Is resolved into ethyl iodide and triethylamine by heating. Is converted into a triiodide $(C^2H^5)^4NI^3$, of a dark violet color, when treated with an alcoholic solution of iodine. Silver oxide precipitates silver iodide from the aqueous solution of the iodide, and the filtered solution, when carefully evaporated, leaves behind fine, deliquescent crystals of *tetrethylammonium hydroxide* $(C^2H^5)^4N.OH$. This is not volatile, but at 100° breaks up into triethylamine, ethylene, and water. Its watery solution conducts itself almost like caustic potassa, takes up carbonic anhydride from the air; has a very caustic action, saponifies fats, and causes the same precipitates as potassa in solutions of metallic salts.

Ethylphosphine, $C^2H^7P = C^2H^5.PH^2$. Is produced, together with some diethylphosphine, when iodophosphonium is allowed to act upon ethyl iodide in the presence of a metallic oxide. To prepare it 1 part of zinc white, 4 parts of iodophosphonium, and 4 parts of ethyl iodide are heated to 150° in sealed tubes. The product of the reaction is then brought into an apparatus filled with hydrogen, and water, which has been boiled and allowed to cool, slowly added. The ethylphosphine is by this means set free and then condensed in a spiral tube surrounded by ice. The distillate dried by means of caustic potassa is pure ethylphosphine.—Mobile, colorless, transparent liquid, insoluble in water; refracts light strongly; lighter than water; boils at 25°; entirely without action upon vegetable colors; exceedingly disagreeable odor. Takes fire when brought together with bromine, chlorine, and fuming nitric acid. Combines with sulphur and carbon bisulphide, forming liquid compounds.

It combines with hydrochloric, -bromic, and -iodic acids, forming salts. *Ethylphosphine hydriodate* $(C^2H^5)H^2P.HI$ forms white, four-sided plates, which can be sublimed in an atmosphere of hydrogen at the temperature of boiling water. Is soluble in water, undergoing complete decomposition; soluble in alcohol with partial decomposition; insoluble in ether; slightly soluble but without decomposition in concentrated hydriodic acid. The addition of ether to this solution causes the salt to separate in crystalline form. Oxidized by means of nitric acid it yields *ethylphosphinic acid* $(C^2H^5).PO.(OH)^2$. This is a solid body, that fuses at 44°. It is a bibasic acid.

Diethylphosphine, $C^4H^{11}P=(C^2H^5)^2.PH$. Is produced together with ethylphosphine in the preparation of the latter. To obtain it from the mixture, after having treated the product of the reaction with water in order to set the ethylphosphine free, a strong solution of caustic soda is added to the mixture in the flask, which still must be kept filled with hydrogen. The diethylphosphine is thus set free and condensed by means of an ordinary apparatus. The liquid dried with caustic potassa is diethylphosphine in a chemically pure condition.—Colorless, transparent, perfectly neutral liquid, insoluble in water, lighter than it, refracts light strongly. Boils at 85°. Penetrating odor, not at all similar to that of ethylphosphine.

Takes up oxygen very rapidly, and occasionally takes fire in contact with the air. Combines with sulphur and carbon bisulphide, forming liquid compounds. Dissolves readily in all acids. The salts crystallize with difficulty, with the exception of the hydriodate. The salts are not decomposed by water. Oxidized by means of nitric acid, it yields *diethylphosphinic acid* $(C^2H^5)^2PO.OH$, a liquid.

Triethylphosphine, $(C^2H^5)^3P$, is formed, when phosphorus terchloride is added drop by drop to an ethereal solution of zinc ethyl and the resulting viscid compound of zinc chloride with the phosphorus base distilled with potassa. Is most readily obtained by heating 1 molecule of iodophosphonium, PH^4I, with 3 molecules of absolute alcohol for eight hours in sealed tubes at 180°. On the addition of caustic soda to the solution, it is precipitated.—Colorless, strongly refracting liquid, which possesses an almost narcotic odor (in a dilute condition like hyacinthes), perfectly insoluble in water, mixes with alcohol and ether in every proportion; specific gravity, 0.812; boiling point, 127°.5. Combines slowly with acids forming very easily soluble salts, which crystallize badly. In contact with the air it forms *triethylphosphine oxide* $(C^2H^5)^3PO$, this being accompanied by an increase in temperature and an assimilation of oxygen. It crystallizes in needles, is exceedingly deliquescent, and boils at 240°. Sulphur is also dissolved by the free base, forming *triethylphosphine sulphide* $(C^2H^5)^3PS$. This crystallizes from water in long, brilliant, white needles, which fuse at 94°.

Phosphethylium iodide, $(C^2H^5)^4PI$. Is produced when an ethereal solution of triethylphosphine is mixed with ethyl iodide; is also formed in the preparation of triethylphosphine from iodophosphonium and alcohol, and crystallizes from the liquid after the addition of caustic soda and evaporation.—Crystals, easily soluble in water. Is not decomposed by caustic potassa; when treated with silver oxide, gives silver iodide and

Phosphethylium hydroxide, $(C^2H^5)^4P.OH$. Crystalline, very deliquescent, strong base; takes up carbonic anhydride from the air with avidity and forms very deliquescent salts with acids. Is decomposed at a high temperature into ethyl hydride and triethylphosphine oxide.

Triethylarsine, $(C^2H^5)^3As$, is formed, together with the following compound, when sodium arsenide, mixed with sand for the purpose of lessening the violence of the reaction, is distilled with ethyl iodide in a vessel filled with carbonic anhydride. By careful distillation of the oil which passes over, in an atmosphere of carbonic anhydride, triethylarsine distils over first.—Colorless liquid, strongly refracting, of exceedingly disagreeable odor; specific gravity, 1.151; begins to boil at 140°; gives off fumes in contact with the air, but takes fire only when heated. Combines with oxygen, forming *triethylarsine oxide* $(C^2H^5)^3AsO$, a colorless, oily liquid; with sulphur forming *triethylarsine sulphide* $(C^2H^5)^3AsS$, a beautifully crystallizing compound. It combines with ethyl iodide, forming crystals of *arsenethylium iodide* $(C^2H^5)^4AsI$, and this gives with silver oxide *arsenethylium hydroxide* $(C^2H^5)^4As.OH$, a white, alkaline, deliquescent mass.

Arsendiethyl (Ethylcacodyl) $\left.\begin{matrix}(C^2H^5)^2As\\(C^2H^5)^2As\end{matrix}\right\}$. Yellowish liquid, of a very disagreeable odor. Takes fire spontaneously in contact with the air; boils at 190°; is heavier than water. Combines with oxygen, sulphur, chlorine, etc., with evolution of heat. Conducts itself perfectly analogously to the methyl compound (p. 40).

Triethylstibine (Stibethyl), $(C^2H^5)^3Sb$, is produced when potassium antimonide is distilled with ethyl iodide in a current of carbonic anhydride. Colorless, very thin liquid, of a disagreeable odor like that of onions; specific gravity, 1.324; boiling point, 158°; gives off fumes in contact with the air, takes fire and burns with a white flame. When air is allowed entrance to it very slowly, it is oxidized, forming *triethylstibine oxide* $(C^2H^5)^3SbO$, a viscid, uncrystalline base, easily soluble in water; forms with acids crystallizing salts. From the solutions of these salts hydrochloric acid precipitates a chloride $(C^2H^5)^3SbCl^2$, in the form of a colorless, thick liquid. Triethylstibine combines with sulphur, forming *triethylstibine sulphide* $(C^2H^5)^3SbS$, crystals with a silvery lustre. Ethyl iodide combines with triethylstibine at 100°, forming

Stibethylium iodide, $(C^2H^5)^4SbI$.—Large transparent prisms, easily soluble in alcohol, but slightly soluble in ether. Silver oxide converts it into *stibethylium hydroxide* $(C^2H^5)^4Sb.OH$, a colorless, oleaginous liquid, which conducts itself like the analogous arsenic compound.

Triethylborine (Borethyl), $(C^2H^5)^3B$, is formed by the action of ethyl borate on zinc ethyl.—Colorless, very mobile liquid; specific gravity, 0.6961; boiling point, 95°; its vapor excites to tears. It combines with ammonia with great avidity. In contact with the air and in oxygen it is oxidized, forming *triethylborine oxide* $(C^2H^5)^3BO^2$, a colorless liquid, boiling at 125°, which breaks up into alcohol and $(C^2H^5)H^2BO^2$ in contact with water.

Triethylbismuthine, $(C^2H^5)^3Bi$, is formed from bismuth-potassium and ethyl iodide.—Heavy, unvolatile liquid, of a very disagreeable odor. It is extracted from the mass by means of ether. It fumes in the air and takes fire spontaneously. It conducts itself like triethylstibine; its compounds, however, are less stable.

Zincethyl, $(C^2H^5)^2Zn$, is formed by the action of zinc on an ethereal solution of ethyl iodide at 150° (if sieved zinc-filings or a small amount of zincethyl

6

be added, the reaction takes place at a lower temperature), or by gently heating equal parts of ethyl iodide and zinc-sodium in an atmosphere of carbonic anhydride. When the reaction ceases, the zincethyl iodide $C^2H^5.ZnI$ is decomposed by means of heat, and the zincethyl distilled off.—Colorless liquid; specific gravity, 1.18; boiling point, 118°. It takes fire in the air and burns with a white flame. When its solution in ether is slowly oxidized, it is transformed into zinc ethylate $(C^2H^5O)^2Zn$, a white, solid body. Sulphur converts it, in an ethereal solution, into zinc mercaptide $(C^2H^5S)^2Zn$. Water decomposes zincethyl instantaneously, forming zinc hydroxide and ethyl hydride. Sodium and potassium are dissolved by an excess of zincethyl, zinc being thrown down: when this solution is cooled or the excess of zincethyl evaporated in an atmosphere of hydrogen, a crystalline compound of zincethyl with sodium- or potassiumethyl separates. From these compounds the potassium or sodium compounds can be isolated.

Mercuryethyl, $(C^2H^5)^2Hg$, is produced by the distillation of mercury chloride or subchloride with an excess of zincethyl. Can be best prepared by bringing sodium-amalgam and ethyl iodide or bromine together and adding acetic ether ($\frac{1}{10}$ the weight of the bromide or iodide). The mixture is alternately shaken and cooled and finally subjected to distillation. The distillate is again treated with sodium-amalgam, water added, the oily liquid, which separates, shaken at first with an alcoholic solution of potassa for the purpose of decomposing the acetic ether, then with water, finally desiccated by means of calcium chloride and rectified.—Heavy, colorless liquid, boiling at 159°; specific gravity, 2.44.—Exceedingly poisonous. Insoluble in water, but slightly soluble in alcohol, easily in ether. When heated with zinc at 100° it is converted into zincethyl. By boiling its alcoholic solution with corrosive sublimate there is formed a crystalline precipitate of *mercuryethyl chloride* C^2H^5HgCl. The corresponding iodide C^2H^5HgI is formed slowly from mercury and ethyl iodide in dispersed light. Both compounds form iridescent scales; of an unpleasant odor, volatile without decomposition. The iodide is decomposed by silver oxide, forming silver iodide and *mercuryethyl hydroxide* $C^2H^5.Hg.OH$. An oleaginous, almost colorless, strongly akaline liquid, easily soluble in alcohol and water. The solution precipitates solutions of metallic salts the same as potassa, and expels ammonia from its salts. With acids it yields crystallizing salts.

Aluminiumethyl, $(C^2H^5)^3Al$, is produced by heating mercuryethyl with aluminium-filings at 100°. —Colorless liquid, boils at 194°, does not congeal at —18°. Gives off fumes in the air and in thin layers takes fire spontaneously. Is decomposed by water, with explosion.

Leadtetrethyl, $(C^2H^5)^4Pb$, is formed from zincethyl and lead chloride, metallic lead being thrown down. —Colorless, oily liquid, boiling at 198–202°, undergoing at the same time partial decomposition. Does not combine with oxygen, chlorine, or iodine.

Leadtriethyl (Methplumbethyl), $(C^2H^5)^6Pb^2$, is produced by bringing together ethyl iodide and an alloy of lead and sodium ($PbNa^3$) an evolution of heat accompanying the reaction. When the reaction is ended, the substance is extracted with ether.—Thin, yellowish oil, not volatile without decomposition; insoluble in water; specific gravity, 1.471. Is decomposed when exposed to the light or boiled for some time with water, lead being thrown down. With iodine it yields a very unstable iodide $(C^2H^5)^3PbI$. The corresponding chlorine compound $(C^2H^5)^3PbCl$ is formed in long needles of a silken lustre by the action of hydrochloric acid gas on leadtetrethyl, the reaction being accompanied by an escape of ethyl hydride. Both compounds give, with water and silver oxide, *leadtriethyl hydroxide* $(C^2H^5)^3Pb.OH$. Colorless, thick liquid, but slightly soluble in water, strong base, saponifies fats, expels ammonia from its salts, precipitates solutions of metallic salts, and forms with acids neutral crystalline salts.

Tindiethyl, $(C^2H^5)^2Sn$, is formed, together with tintriethyl, when ethyl iodide is brought together with an alloy of 1 part of sodium and 4 parts of tin.—Yellow, oily liquid, not volatile without decomposition; unites with oxygen, chlorine, bromine, and iodine. *Tindiethyl iodide* $(C^2H^5)^2SnI^2$ is formed when tin and ethyl iodide are heated together. It forms needly crystals, which fuse at 44°.5 and boil at 245°, are soluble in ether and hot alcohol, only with difficulty soluble in water. Zinc precipitates tindiethyl from its solutions. Alkalies precipitate *tindiethyl oxide* $(C^2H^5)^2SnO$; white, amorphous powder; insoluble in water, alcohol, and ether; soluble in an excess of caustic soda or potassa; combines with acids, forming crystalline salts.

Tintetrethyl, $(C^2H^5)^4Sn$, is formed by heating tindiethyl and distilling tindiethyl iodide with zincethyl. —Colorless liquid, of specific gravity 1.187; boiling point, 181°. Does not unite with oxygen, chlorine, or iodine.

Tintriethyl, $(C^2H^5)^6Sn^2$.—Thin liquid, boiling at 265–270°, but not entirely without decomposition; specific gravity, 1.4115. It absorbs oxygen and yields with it an oxide $(C^2H^5)^6Sn^2O$, volatile without decomposition, the hydrate of which is a strong base, consisting of colorless prisms fusing at 66° and boiling at 271°, forming with acids crystalline salts. The iodide $(C^2H^5)^3SnI$ is formed by direct union of tintriethyl with iodine, by continued heating of ethyl iodide with zinc-sodium (containing 2 per cent. sodium), and, together with ethyl iodide, by the action of iodine on tintetrethyl. A liquid boiling at 231°; specific gravity, 1.83. The further action of iodine resolves it into tindiethyl iodide and ethyl iodide.—The chloride $(C^2H^5)^3SnCl$ is formed, together with ethyl hydride, by the action of hydrochloric acid on tin tetrethyl. A liquid of pungent odor, congealing at 0°, boiling at 208–210°.

Siliciumethyl, $(C^2H^5)^4Si$, is formed by heating silicium chloride with zincethyl to 160°.—Colorless liquid, boiling at 153°, lighter than water and insoluble in it. Yields with chlorine a liquid $C^8H^{19}ClSi$, boiling at 180–190°.

3. *Propyl Alcohols.*
$C^3H^8O = C^3H^7.OH.$

Of the alcohols, which have the formula C^3H^8O, there are two isomeric modifications possible, as was shown at p. 32. Both are known.

1. Normal propyl alcohol, $CH^3.CH^2.CH^2.OH$. Is formed in the preparation of ethyl alcohol by fermentation, together with some of the other alcohols of this series, and is contained in the secondary products, which boil at a higher temperature (fusel-oil). It can be isolated from these by means of partial distillation, but only with difficulty can it thus be obtained in a pure condition. To prepare the pure alcohol, that portion of fusel-oil that boils between 85–110° is treated with amorphous phosphorus and bromine (see ethyl bromide, p. 46), and thus converted into bromides. These are then separated by partial distillation, the portion that boils at 71° decomposed with silver acetate or potassium acetate, and the ether thus formed decomposed by means of caustic potassa. It is also produced by the action of hydrogen in *statu nascendi* on propionic aldehyde, by the action of sodium-amalgam on propionic anhydride, and together with ethyl alcohol and other bodies by heating allyl alcohol with caustic potassa.

Colorless liquid, of a pleasant odor, of specific gravity 0.8205 at 0°; boiling point, 97°; mixes with water, but not with a concentrated solution of calcium chloride. Under the influence of oxydizing agents it yields propionic aldehyde and propionic acid.

The derivatives of propyl alcohol are prepared in the same manner as those of ethyl alcohol, and conduct themselves analogously.

Propyl chloride, C^3H^7Cl. Colorless liquid, boiling at 52°.

Propyl bromide, C^3H^7Br. Liquid; boiling point, 71°.

Propyl iodide, C^3H^7I. Liquid; boiling point, 102°.

Propylether, $(C^3H^7)^2O$. Very mobile liquid, boiling at 85–86°.

Propylamine, $C^3H^7.NH^2$. By the action of hydrogen in *statu nascendi* (zinc and hydrochloric acid) on propionitrile (p. 47), and by the distillation of propyl cyanate with caustic potassa.—Clear, strongly refracting liquid, possessing an ammoniacal odor. Boiling point, 49–50°. Mixes with water. Burns with a luminous flame. Strong base. The *hydrochlorate*, $C^3H^7.NH^2.HCl$, is deliquescent, also very easily soluble in alcohol. With platinum chloride it yields a double salt $(C^3H^7.NH^2.HCl)^2PtCl^4$, which is pretty easily soluble in hot water and in alcohol, and crystallizes in large, gold-colored, klinorhombic plates.

2. Secondary propyl alcohol (Pseudopropyl alcohol), $CH^3.CH.OH.CH^3$. Is formed by the action of hydrogen in *statu nascendi* (from water and sodium-amalgam) on acetone.—Colorless liquid, miscible with water in all proportions. Boiling point, 85°; specific gravity, 0.791 at 15°. Combines with calcium chloride, forming a solid compound. By oxidation it is at first reconverted into acetone and then yields acetic and formic acids.

Pseudopropyl iodide, C^3H^7I, is produced by the direct union of propylene with hydriodic acid, and by heating pseudopropyl alcohol, propylene alcohol, allyl iodide, or glycerin with the same acid.—Is prepared most readily by the simultaneous action of iodine and phosphorus on glycerin.—Colorless liquid, boiling at 89°; specific gravity, 1.7 at 15°. When heated with potassium cyanide, it is converted into *pseudopropyl cyanide* (*pseudobutyronitrile*), C^4H^7N. At the same time is formed a small quantity of the isomeric compound *pseudopropylcarbylamine*, $C^3H^7.NC$, which boils at 87°.

Pseudopropyl chloride, C^3H^7Cl, and **pseudopropyl bromide,** C^3H^7Br, are very similar to the iodide, and are obtained from the alcohol in the same way as the corresponding ethyl compounds. The former boils at 36–38°; the latter at 60–63°.

Pseudopropylether, $(C^3H^7)^2O$, is formed, together with pseudopropyl alcohol and propylene, by heating the iodide with silver oxide and water.—A liquid not miscible with water. Boiling point, 60–62°.

Pseudopropylamine, $C^3H^7.NH^2$. Colorless, very mobile liquid, of ammoniacal odor. Boiling point, 32°; specific gravity, 0.69.

4. *Butyl Alcohols.*

$$C^4H^{10}O = C^4H^9.OH.$$

The existence of four different alcohols of the formula $C^4H^{10}O$ is possible—two primary, one secondary, and one tertiary. These are all known.

1. Normal butyl alcohol, $CH^3.CH^2.CH^2.CH^2.OH$. Is obtained by the action of hydrogen in *statu nascendi* (sodium-amalgam and very dilute sulphuric acid) on butyric aldehyde, or by the action of sodium-amalgam on a mixture of butyric acid and butyryl chloride, and treatment of the product, chiefly consisting of butyl butyrate, with caustic potassa.—Colorless liquid of agreeable odor; specific gravity, 0.826; boiling point, 115–116°. But slightly soluble in water. Yields butyric acid by oxidation.

Butyl chloride, C^4H^9Cl. Clear liquid. Boiling point, 77.6°; specific gravity, 0.8874 at 20°.—The *bromide*, C^4H^9Br, boils at 100.4°; specific gravity, 1.2792 at 20°.—The *iodide*, C^4H^9I, boils at 129.6°; specific gravity, 1.6136 at 20°.

Butyl cyanide, $C^4H^9.CN$. Liquid, boiling at 140.4°, of exceedingly disagreeable odor. Specific gravity, 0.8164 at 0°.

Butyl-ethylether, $C^4H^9.O.C^2H^5$. Liquid, boiling at 91.7°; specific gravity at 20°, 0.7512.

Butylamine, $C^4H^9.NH^2$. Clear liquid, possessing a strongly ammoniacal odor, fumes in contact with the air, very hygroscopic. Mixes with water in all proportions. Boils at 75.5°; specific gravity, 0.7553 at 0°.—With hydrochloric acid and platinum chloride it yields a double salt, $(C^4H^9.NH^2.HCl)^2PtCl^4$, which crystallizes in gold-colored laminæ, but slightly soluble in cold water, more readily in hot water and in alcohol.

2. Isobutyl alcohol, $\left.\begin{matrix}CH^3\\CH^3\end{matrix}\right\}CH.CH^2.OH$. Is often contained in fusel-oil, and is obtained from this like propyl alcohol.—Colorless liquid of specific gravity 0.805. Boiling point, 108–109°. Soluble in 10 parts of water, and is precipitated from this solution by soluble salts. By oxidation it is converted into isobutyric acid.

Isobutyl chloride, C^4H^9Cl. Colorless liquid, boiling at 64–68°.—The *bromide*, C^4H^9Br, boils at 92°; the *iodide*, C^4H^9I, at 121°.

3. Secondary butyl alcohol (butylene hydrate), $CH^3.CH^2.CH.OH.CH^3$. The iodide corresponding to this alcohol is obtained by distilling erythrite with concentrated hydriodic acid. From this the alcohol is obtained by heating with silver oxide and water.—Colorless liquid, rather easily soluble in water, is precipitated from this solution by means of potassium carbonate. Of a strong, penetrating odor. Boiling point, 96–98°; specific gravity, 0.85 at 0°. When heated to 240–250° it is resolved into butylene and water. By oxidation it is at first converted into ethylmethylketone, and then into acetic acid. The *iodide*, C^4H^9I, boils at 117–118°.

4. Tertiary butyl alcohol (Pseudobutyl alcohol, trimethylcarbinol), $CH^3.C.OH \left\{ \begin{matrix} CH^3 \\ CH^3 \end{matrix} \right.$. Is contained in small quantity in commercial butyl alcohol of fermentation.—Can easily be prepared from isobutyl alcohol. Isobutyl iodide, when heated with an alcoholic solution of potassa, yields a hydrocarbon, C^4H^8 (isobutylene), which combines directly with hydriodic acid, forming pseudobutyl iodide. By means of silver oxide and water the alcohol is prepared from this. The alcohol can be obtained still more readily by conducting the isobutylene into concentrated sulphuric acid, and, after diluting with water, subjecting to distillation. When acetyl chloride (1 vol.) is poured very slowly into an excess (about 4 vols.) of zincethyl, kept at 0°, there separates from the mixture after a time, large, transparent prisms of $C^2H^3O.Cl + 2[(CH^3)^2Zn]$, which, in contact with water, are immediately decomposed, forming zinc oxide, zinc chloride, marsh gas, and pseudobutyl alcohol.—Colorless, thick liquid, which, when thoroughly free of water, congeals at 20–25° in a crystalline form, and boils at about 82°. Yields by oxidation carbonic, acetic, and propionic acids.—*Pseudobutyl chloride*, C^4H^9Cl, boils at 50–51°. The *iodide*, C^4H^9I, at 98–99°.

5. *Amyl Alcohols.*

$C^5H^{12}O = C^5H^{11}.OH.$

Eight isomeric alcohols of this composition can exist, viz.: 4 primary, 3 secondary, and 1 tertiary. Of these, five are known, as follows: 2 primary, 2 secondary, and the tertiary.

PRIMARY AMYL ALCOHOLS.

1. Normal amyl alcohol, $CH^3.CH^2.CH^2.CH^2.CH^2.OH$. Is obtained from the aldehyde of normal valeric acid by the action of hydrogen in *statu nascendi*, in the same manner as normal butyl alcohol.—Colorless liquid; insoluble in water; boiling point, 137°. By oxidation it yields normal valeric acid.

Amyl chloride, $C^5H^{11}Cl$, boils at 106.6°; specific gravity at 0° = 0.9013.—The *bromide* $C^5H^{11}Br$ boils at 128.7°, specific gravity at 0° = 1.246.—The *iodide* $C^5H^{11}I$ boils at 155.4°; specific gravity at 0° = 1.5435. —*Amyl acetate* $C^5H^{11}.O.C^2H^3O$ boils at 148.4°; specific gravity at 0° = 0.8963.

2. Amyl alcohol of fermentation, $\left.\begin{matrix}CH^3\\CH^3\end{matrix}\right\} CH.CH^2.CH^2.OH$. Is the principal constituent of fusel-oil, and is prepared from this by means of partial distillation.—Colorless liquid, boiling at 130–131°; specific gravity, 0.825; of an unpleasant odor and acrid taste, but slightly soluble in water. By oxidation it yields ordinary valeric acid. Its derivatives are prepared like those of ethyl alcohol, and thoroughly resemble them in their chemical conduct.

The *chloride* $C^5H^{11}Cl$ is a liquid, boiling at 102°. The *iodide* $C^5H^{11}I$ boils at 147°; the *bromide* $C^5H^{11}Br$, at 119°.

Amylether, $(C^5H^{11})^2O$, is a liquid, boiling at 170°.

SECONDARY AMYL ALCOHOLS.

3. Isoamyl alcohol, $CH^3.CH^2.CH^2.CH.OH.CH^3$. Is produced from methyl-propylketone by the action of hydrogen in *statu nascendi*. The iodide is formed by the direct combination of ethylallyl (see amylene) with hydriodic acid, and the alcohol obtained from this in the same way that normal propyl alcohol is obtained from propyl bromide (p. 65).—Colorless liquid, insoluble in water, of specific gravity, 0.8205; boiling point, 120°. By oxidation it yields first methyl-propylketone, then acetic and propionic acids.

The *iodide* $C^5H^{11}I$ is a liquid, boiling at 146°; of specific gravity, 1.537 at 0°.

4. Amylenehydrate, $\left.\begin{matrix}CH^3\\CH^3\end{matrix}\right\}CH.CH.OH.CH^3.$

Amylene C^5H^{10}, which results from the action of zinc chloride on amyl alcohol, combines with hydriodic acid, forming the iodide $C^5H^{11}I$, boiling at 128–130°, which, when treated with silver oxide yields the alcohol.—A liquid, boiling at 105–108°. Does not combine with sulphuric acid, but is decomposed by it, yielding water, amylene, and substances polymeric with it. Furnishes by oxidation carbonic and acetic acids; acetones are formed as intermediary products.

---

5. Tertiary amyl alcohol (Pseudoamyl alcohol, Ethyldimethylcarbinol), $\left.\begin{matrix}CH^3\\CH^3\end{matrix}\right\}C.OH.CH^2.CH^3.$ Is prepared, like pseudobutyl alcohol, from propionyl chloride and zincmethyl.—A liquid, boiling at about 100°. Yields acetic acid by oxidation. Congeals at —30°, forming a mass of small needles.—The *iodide* $C^5H^{11}I$ is a heavy liquid.

6. *Hexyl Alcohols (Caproyl Alcohols).*
$C^6H^{14}O = C^6H^{13}.OH.$

1. Primary hexyl alcohol. Is contained in fusel-oil obtained from grape skins. Hexyl hydride from petroleum (p. 30) yields the chloride $C^6H^{13}Cl$, from which hexyl acetate may be obtained by heating with potassium or silver acetate. This when boiled with potassa gives hexyl alcohol, a liquid boiling at 150–155°.—The *iodide* boils at 172–175°.

That portion of the volatile oil of *Heracleum giganteum*, which boils at 201–206°, consists partially of hexyl butyrate. The alcohol, prepared from this ether by means of saponification, boils at 156.6°.—The *iodide* boils at 179.5°. This alcohol, as well as the preceding one, yields an acid $C^6H^{12}O^2$ by oxidation, and is probably the normal alcohol. It is not decided whether these two alcohols are identical or not.

2. Secondary hexyl alcohol (β-Hexyl alcohol), $CH^3.CH^2.CH^2.CH^2.CH.OH.CH^3$. When mannite is distilled with concentrated hydriodic acid, there results an iodide, $C^6H^{13}I$, boiling at 167.5°. This yields the alcohol when heated with silver oxide and water. —A liquid, boiling at 137°; of specific gravity, 0.8327 at 0°. Its conduct towards sulphuric acid is similar to that of amylenehydrate. Yields by oxidation carbonic, acetic, and butyric acids; as an intermediary product, methyl-butylketone.

The chloride of the same alcohol ($C^6H^{13}Cl$, boiling point, 125–126°) appears to be formed together with the chloride of the primary alcohol by the action of chlorine on the hexyl hydride from petroleum.

In addition to these there are three tertiary hexyl alcohols known:—

3. Dimethylpropylcarbinol, $\left.\begin{matrix}CH^3\\CH^3\end{matrix}\right\}C.OH.CH^2.$ $CH^2.CH^3$. From butyryl chloride and zincmethyl like pseudobutyl alcohol.—Boiling point, 115°. By oxidation it yields acetic and propionic acids.

4. Diethylmethylcarbinol, $\left.\begin{matrix}CH^3.CH^2\\CH^3.CH^2\end{matrix}\right\}C.OH.CH^3.$ From acetyl chloride and zincethyl.—Boiling point, 120°. Yields by oxidation only acetic acid.

5. Dimethylpseudopropylcarbinol, $\left.\begin{matrix}CH^3\\CH^3\end{matrix}\right\}C.$ $OH.CH\left\{\begin{matrix}CH^3\\CH^3.\end{matrix}\right.$ Is obtained by the action of isobutyryl chloride on zincmethyl.—Colorless liquid, that congeals at —35°, forming a white crystalline mass. Boiling point, 112–113°; specific gravity at 0°, 0.8364. Yields by oxidation acetone, and by further oxidation of this, acetic acid.

7. *Heptyl Alcohols* (*Œnanthyl Alcohols*).
$C^7H^{16}O = C^7H^{15}.OH.$

1. Primary heptyl alcohol. Is contained in the fusel-oil from grape skins, and is prepared from heptyl hydride (obtained from petroleum) in the same way as hexyl alcohol. Is also formed by the action of hydrogen in *statu nascendi* on œnanthylic aldehyde.—Colorless liquid, insoluble in water, boiling at 164–165°.—The *chloride*, $C^7H^{15}Cl$, obtained from heptyl hydride by the action of chlorine, boils at 146–149°.

It is not positively known whether these alcohols, obtained from different materials, are identical.

2. Secondary heptyl alcohol, $\left.\begin{matrix} CH^3.CH^2.CH^2 \\ CH^3.CH^2.CH^2 \end{matrix}\right\}$ CH.OH. Is produced by the action of hydrogen on butyrone.—Liquid that boils at 149–150°; but slightly soluble in water; soluble in all proportions in alcohol; specific gravity at 25° = 0.814.—The *iodide*, $C^7H^{15}I$, boils at 180°, but not without undergoing partial decomposition.

3. Triethylcarbinol (Tertiary heptyl alcohol), $C^7H^{16}O = C^2H^5.C.OH \left\{\begin{matrix} C^2H^5 \\ C^2H^5 \end{matrix}\right.$. Is produced by the action of propionyl chloride on zincethyl.—Colorless liquid, of an odor similar to camphor; boiling point, 140–142°; specific gravity, 0.8593 at 0°. Yields by oxidation acetic and propionic (?) acids.

8. *Octyl Alcohols* (*Capryl Alcohols*).
$C^8H^{18}O = C^8H^{17}.OH.$

Primary octyl alcohol. That portion of the volatile oil of *Heracleum spondylium* which boils at 206–

208° is the acetic ether of this alcohol. By decomposing this with caustic potassa the alcohol is obtained.—Colorless liquid, insoluble in water; specific gravity, 0.83; boiling point, 190–192°.

The *chloride*, $C^8H^{17}Cl$, boils at 180°; the *bromide*, $C^8H^{17}Br$, at 198–200°; the *iodide*, $C^8H^{17}I$, at 220–222°.

Secondary octyl alcohol (Methylhexylcarbinol), $C^6H^{13}.CH.OH.CH^3$. Is formed by the distillation of castor oil with alkaline hydrates, and can be prepared from octyl hydride (obtained from petroleum) in the same way as hexyl alcohol.—Oil boiling at 181°. Yields by oxidation at first methyl-hexylketone and then acetic and caproic acids.

The *chloride*, $C^8H^{17}Cl$, boils at 175°.

Tertiary octyl alcohol (Propyldiethylcarbinol), $\left.\begin{matrix}C^2H^5\\C^2H^5\end{matrix}\right\}C.OH.C^3H^7$. Prepared from butyryl chloride and zincethyl in the same way as pseudobutyl alcohol.—A liquid, boiling between 145–155°.

---

9. *Nonyl alcohol*, $C^9H^{20}O$, a liquid, boiling at about 200°, and

10. *Decatyl alcohol*, $C^{10}H^{22}O$, a liquid, boiling at 210–215°, have been prepared from the corresponding hydrocarbons of petroleum in the same way as hexyl alcohol. They have not been subjected to closer study.

11. *Cetyl alcohol*, $C^{16}H^{34}O$. A compound ether of this alcohol is the principal constituent of spermaceti. By boiling this with an alcoholic solution of potassa the alcohol is obtained.—White crystalline mass, fusing at 50°, volatile without decomposition.

12. *Ceryl alcohol*, $C^{27}H^{56}O$. In Chinese wax and in opium wax in the form of ceryl cerotate and palmitate.

Prepared from this by boiling with an alcoholic solution of potassa.—A wax-like mass, fusing at 79°.

13. *Myricyl alcohol*, $C^{30}H^{62}O$. Is contained in Carnauba wax (from the leaves of *Copernica cerifera*) and as myricyl palmitate in beeswax. Separated by means of caustic potassa, it forms a crystalline mass, fusing at 85°.

C. Monobasic, Monatomic Acids, $C^nH^{2n}O^2$ (Fatty Acids).

The acids of this series are formed in general terms by oxidizing the primary alcohols, the group $CH^2.OH$ being hereby converted into CO.OH (carboxyl), and by heating the alcoholic cyanides (nitriles) with caustic potassa, the cyanogen group (CN) being transformed into COOH, and nitrogen in the form of ammonia being given off. The first member of the series is the hydrogen compound of carboxyl H.CO.OH; the homologous members, $C^2H^4O^2 = CH^3.COOH$, $C^3H^6O^2 = C^2H^5.CO.OH$, etc., must be considered as derivatives of the marsh gas hydrocarbons, formed by the displacement of an atom of hydrogen by the monovalent group, COOH. In regard to the isomeric compounds that are possible in connection with the individual members of the series, the remarks made under the head of alcohols are here equally applicable. Each hydrocarbon can yield just as many monobasic acids (carboxyl-derivatives) of different constitution, as it can form monatomic alcohols (hydroxyl-derivatives). Hence only one acid of the composition of each of the three first members of the series can exist. Of the fourth member, $C^4H^8O^2 = C^3H^7.COOH$, two differently constituted varieties are possible, $CH^3.CH^2.CH^2.COOH$ and $\left.\begin{matrix}CH^3\\CH^3\end{matrix}\right\} CH.CO.OH$; of the fifth member, $C^5H^{10}O^2 = C^4H^9.CO.OH$, four varieties are possible; of the sixth member, $C^6H^{12}O^2 = C^5H^{11}.CO.OH$, eight, etc.

1. *Formic Acid.*

$$CH^2O^2 = H.CO.OH.$$

Occurrence. In ants, in common nettles, in pine needles.

Formation. (*a*) From carbonic oxide; potassium hydroxide unites with it when heated for some time at 100°, forming potassium formate; (*b*) from carbonic anhydride; potassium spread out on a basin under a bell-jar inserted in lukewarm water and kept constantly filled with carbonic anhydride is converted into a mixture of potassium formate and bicarbonate; it, in fact, always results in small quantities whenever hydrogen in *statu nascendi* and carbonic anhydride in a state of transmission come together, as, for instance, by the action of sodium-amalgam on a concentrated solution of ammonium carbonate, by the addition of a mixture of zinc and zinc carbonate to hot caustic potassa; (*c*) from methyl alcohol by means of oxidation; (*d*) from prussic acid by treating with alkalies or dilute acids; (*e*) from oxalic acid, by heating, or by the action of sunlight upon an aqueous solution of the acid containing a salt of uranium; (*f*) from chloroform, iodoform, and bromoform by treatment with alcoholic potassa; (*g*) from a large number of organic substances, starch, sugar, tartaric acid, etc., by distillation with dilute sulphuric acid and black oxide of manganese or potassium chromate.

Preparation. By distilling ants with water.—Most practicably by treating crystallized oxalic acid with glycerin, from which the water has been separated as thoroughly as possible. The reaction commences at 70° and is in full progress at 90°. Carbonic anhydride escapes, and a very dilute formic acid distils over. When the evolution of carbonic anhydride begins to grow less active, a fresh quantity of oxalic acid is added and heat again applied. A more concentrated acid now goes over, and, by continued addition of oxalic acid, an acid containing 56 per cent. is finally obtained.—For the purpose of obtaining the acid in an anhydrous condition, the lead or copper salt is prepared, dried, and decomposed with sulphuretted hydrogen: the acid, which is by this means set free, is distilled off and rectified over dried lead or copper formate. Or anhydrous oxalic acid is dissolved in 70 per cent. formic acid (obtained by carefully heating glycerin with dried oxalic acid) by the aid of gentle heat, the solution allowed to cool, poured off from the oxalic acid that crystallizes out, and rectified.

Properties. Colorless liquid of a pungent odor, crystallizing below 0°; specific gravity, 1.223 at 0°; boiling point, 99°; fusing point, +1°. Acts as a vesicant.

Decompositions. Concentrated sulphuric acid resolves it into water and carbonic oxide. Heated with mercury or silver oxide, it is converted into water and carbonic acid, the oxides being reduced.

All formates are soluble in water.

The salts of the *alkalies* are deliquescent in the air.

Ammonium formate, $CHO^2.NH^4$, is decomposed when heated up to 110, forming prussic acid and water.

Barium formate, $(CHO^2)^2Ba$, crystallizes in prisms, which are not changed by contact with air.

Lead formate, $(CHO^2)^2Pb$. Lustrous, difficultly soluble needles.—*Copper formate* $(CHO^2)^2Cu + 4H^2O$. Large, blue, transparent crystals. When heated yields formic acid of 82 per cent.—*Silver formate* CHO^2Ag. White crystals, which are decomposed when heated, yielding carbonic anhydride, silver, and formic acid.—*Mercury formate* $(CHO^2)^2Hg$ conducts itself in a similar manner; when heated it is, however, at first converted into the difficultly soluble salt of the suboxide, carbonic anhydride being evolved.

Methyl formate, $HCO.O.CH^3$. By the distillation of sodium formate with methyl sulphate.—Colorless liquid of pleasant odor, boiling at 36°.

Ethyl formate, $HCO.O.C^2H^5$. By the distillation of 7 parts dried sodium formate with a mixture of 10 parts sulphuric acid and 6 parts 90 per cent. alcohol.

More readily by heating a mixture of glycerin with oxalic acid and alcohol, in an apparatus in which the vapors are condensed and returned to the flask. When the evolution of carbonic anhydride has ceased, the ether, which has been formed, is distilled off.—Colorless, spicy-smelling liquid, soluble in 10 parts of water; boils at 55°.

Amyl formate, $HCO.O.C^5H^{12}$, obtained like the ethyl ether.—A fluid, boiling at 112°, having a fruity odor.

Formylamide, $HCO.NH^2$, is formed when ethyl formate, which has been saturated with ammonia, is heated for several days at 100°; and by heating 2 parts dry ammonium formate with 1 part of urea up to 140°. Is formed also, together with other products, by the destructive distillation of ammonium formate, and by heating formates with ammonium chloride.—Colorless liquid, boiling at 192–195°. Can only be distilled in a vacuum without decomposition.

2. *Acetic Acid.*

$$C^2H^4O^2 = CH^3.CO.OH.$$

Formation and preparation. By the decay of a great many organic bodies; by the destructive distillation of wood, sugar, starch, tartaric acid, and numerous other substances.—From alcohol under the influence of oxidizing agents or such substances as cause its oxidation in contact with the air. Sodiummethyl combines with carbonic anhydride, forming sodium acetate.

Alcohol, in contact with black powdered platinum, is converted into concentrated acetic acid, an elevation of temperature and absorption of oxygen from the air accompanying the action. Certain organic substances act in a similar manner to platinum; through their agency dilute alcohol, at a temperature of 20–40°, is caused to absorb oxygen from the air and is tranformed into acetic acid. Hence the power of every fermented liquid, *i. e.* vegetable juice containing alcohol, to become acid when left in contact with the air. In this manner *vinegar* is formed, which is a mixture of acetic acid with a great deal of water and small quantities of accidental foreign substances.

It is obtained by allowing wine, beer, fermented fruit juices, particularly after the addition of a small quantity of vinegar, to acidify spontaneously, in vessels which permit the access of air and are kept warm. Or by a similar acidifying of fermented beer wort, or ot mixtures of brandy and water with honey and a ferment. This takes place most readily in the German process for the manufacture of vinegar (Schnellessigfabrikation); in which the liquid to be acidified is exposed to the air in such a manner that as much surface as possible may be presented to its action. This is effected by allowing the liquid to flow slowly through a high cask filled with beech shavings, the sides of the cask being furnished with air holes. The shavings must be previously steeped in vinegar.

By distilling vinegar the acetic acid can be freed from the foreign substances with which it is mixed, but the water cannot be removed by this means.

The anhydrous acid is obtained by distilling 5 parts anhydrous sodium acetate with 6 parts concentrated sulphuric acid, or also by distilling an intimate mixture of equal parts of anhydrous lead acetate and fused potassium bisulphate.

A large quantity of acetic acid is obtained by the destructive distillation of wood (wood vinegar). The watery distillate is saturated with sodium carbonate, evaporated, the dried sodium salt heated for a length of time at 230–250° for the purpose of destroying any organic impurities which may be present, dissolved in water, filtered, evaporated and the heating repeated if necessary.

Properties. Colorless liquid of a penetrating and pleasant acid odor, of a sharp acid taste, caustic; specific gravity, 1.056 at 15.5°; fumes slightly in the air; boils at +119°; its vapor is inflammable and burns with a blue flame. It crystallizes in lustrous, transparent tablets, which fuse at +17°. Miscible with water in all proportions. At first the specific gravity of this mixture increases. The acid containing 77–80 per cent. has the highest specific gravity, 1.0754 at 15.5°. Then it decreases, and an acid of 50 per cent. has about the same specific gravity as the anhydrous acid. When the acid contains water, it does not crystallize even at 0°.

Potassium acetate, $C^2H^3O^2K$. A white, very deliquescent salt, also soluble in alcohol. From a solution of this salt in concentrated acetic acid is deposited, on evaporating, a salt, $C^2H^3O^2K + C^2H^4O^2$, in laminæ, possessing a mother-of-pearl lustre. This salt fuses at 148°, and at 200° is resolved into acetic acid and potassium acetate.—*Sodium acetate*, $C^2H^3O^2Na + 2H^2O$. Clear, prismatic, easily soluble crystals.—*Ammonium acetate*, $C^2H^3O^2.NH^4$. White salt. Its solution loses ammonia when evaporated. Subjected to dry distillation, it yields acetamide.

Barium acetate, $(C^2H^3O^2)^2Ba$, crystalline, easily soluble salt.

Iron acetate. The salt of the suboxide, $(C^2H^3O^2)^2Fe$, forms green, easily soluble prisms. The salt of the oxide does not crystallize; it forms a deep red solution, from which all the iron is precipitated as a basic salt by boiling.

Lead acetate, $(C^2H^3O^2)^2Pb + 3H^2O$. *Sugar of lead.* Is prepared on the large scale by dissolving ground litharge in distilled acetic acid.—Colorless, lustrous prisms of a disagreeable, sweet taste; poisonous. Easily soluble in water and also in alcohol. Fuses at 75° in its water of crystallization, loses this at 100° and congeals. At a high temperature it fuses again and loses one-third of its acetic acid, which escapes as carbonic anhydride and acetone. The solidified residue is a basic salt, which at a still higher temperature decomposes, yielding lead oxide, carbonic anhydride and acetone. Basic salts can also be obtained by digesting a solution of sugar of lead with lead oxide. It combines with lead chloride, iodide, and bromide, forming easily soluble compounds.

Copper acetate, $(C^2H^3O^2)^2Cu+H^2O$. Dark green, untransparent, rhombohedral crystals. Difficultly soluble in water. Crystallizes at a low temperature with 5 molecules of water in transparent, blue crystals, which are converted, at 30°, into crystalline aggregates of a green salt. *Verdigris*, a mixture of several basic salts, is obtained by the action of vinegar or acid grape skins on sheet-copper. Blue or bluish-green, fine, crystalline mass, only partially soluble in water. Copper acetate combines directly with other acetates and also with salts of other acids. *Schweinfurt green* is such a compound with copper arsenite.

Silver acetate, $C^2H^3O^2Ag$. Lustrous, pliant needles or laminæ, difficultly soluble in water.

Methyl acetate, $C^2H^3O.O.CH^3$. Is present in crude wood-spirit; and is obtained by distilling acetates with methyl alcohol and sulphuric acid.—A liquid of pleasant odor, soluble in water and alcohol; boiling at 55°. Treated with chlorine, there is formed a series of liquid substitution-products, which crystallize with water.* Bromine does not act upon it at ordinary temperatures; at 150°, however, are formed methyl bromide, acetic acid, mono- and dibromacetic acids. Towards sodium it conducts itself the same as the ethyl ether.

Ethyl acetate (Acetic ether), $C^2H^3O.O.C^2H^5$. By distilling 10 parts of sodium acetate with a mixture of 15 parts of sulphuric acid, and 6 parts of alcohol.—Thin liquid, very pleasant, refreshing odor; specific gravity, 0.905 at 17°; boils at 72.1°; very inflammable. Soluble in 11 parts of water; is converted, however, by it into acetic acid and alcohol. It conducts itself towards bromine and chlorine the same as the methyl ether.—

* The same compounds are formed by the action of chlorine on citric acid and several other organic compounds. They were formerly considered as chlorinated acetones.

Sodium is dissolved by it, giving rise to the formation of sodium ethylate and of a compound, $C^6H^9NaO^3$ (sodium ethyldiacetate or sodium acetonecarbonic ether), which is decomposed into sodium carbonate, carbonic anhydride, alcohol, and acetone by the boiling of its watery solution, and yields a colorless compound, boiling at 181°, $C^6H^{10}O^3$ (acetyl-acetic ether, ethyl-acetone carbonate, ethyl-diacetic acid), on being heated in a current of dry carbonic anhydride or hydrochloric acid gas. By the successive action of an excess of sodium and ethyl iodide on acetic ether, there result, in addition to the ethyl compound, $C^6H^9O^3.C^2H^5$ (boiling point, 198°), corresponding to the above sodium compound, the ethyl ethers of *diethyldiacetic acid* (diethacetone carbonic acid), $C^8H^{13}O^3.C^2H^5$ (boiling point, 210–212°), of *butyric acid*, $C^4H^7O^2.C^2H^5$ (boiling point, 119°), and of *diethylacetic acid*, $C^6H^{11}O^2.C^2H^5$ (boiling point, 151°; isomeric with ethyl caproate).—Analogous products result by the successive action of sodium and the iodides of other alcohol radicles.

Amyl acetate, $C^2H^3O^2.C^5H^{11}$. A liquid, boiling at 140°, with a fruity odor.

Monochloracetic acid, $C^2H^3ClO^2 = CH^2Cl.CO.OH$, results from the action of chlorine on concentrated acetic acid, particularly in the presence of iodine, and by the decomposition of chlorinated acetyl chloride (which see) with water.—Rhombic plates or prisms, fusing at 62°; boiling at 185–187°. Yields glycolic acid by boiling with the alkalies in aqueous solutions or with silver oxide and water; by heating with ammonia, glycocol.

The *potassium salt*, $C^2H^2ClO^2.K + 1\frac{1}{2}H^2O$, crystallizes in laminæ; the *silver salt*, $C^2H^2ClO^2Ag$, in small scales of a mother-of-pearl lustre.

Ethyl monochloracetate, $C^2H^2ClO.O.C^2H^5$. A solution of chloracetic acid in absolute alcohol is saturated with hydrochloric acid gas, then heated gently for some time on a water-bath, the ether precipitated with water and purified by means of distillation.—Colorless liquid, boiling at 143.5°, but slightly soluble in water.

Dichloracetic acid, $C^2H^2Cl^2O^2 = CHCl^2.CO.OH$, is formed by the further action of chlorine on monochloracetic acid in the presence of iodine.—A liquid, boiling at 195°, forming, when perfectly pure, rhombohedral crystals.

Ethyl dichloracetate, $C^2HCl^2O.O.C^2H^5$, is obtained by conducting dried hydrochloric acid gas into a solution of the acid in absolute alcohol; is also formed by heating carbon chloride, C^2Cl^4, with sodium ethylate.—Heavy liquid, boiling at 153–158°. Is decomposed when kept for any length of time, or when agitated with caustic soda, forming oxalic acid and hydrochloric acid.

Trichloracetic acid, $C^2HCl^3O^2$, is formed by the action of an excess of chlorine on acetic acid in direct sunlight; by the decomposition of trichloracetyl chloride by water, and by the action of chlorine in direct sunlight on carbon chloride, C^2Cl^4, in presence of water; and is prepared most readily by the oxidation of chloral with fuming nitric acid.—Colorless, rhombohedral crystals, deliquescent; fuses at 46°; boils at 195–200°. Combines with bases forming crystalline salts. When boiled with ammonia, it is resolved into chloroform and potassium carbonate, potassium formate, and potassium chloride.

Mono- and Dibromacetic acids, $C^2H^3BrO^2$ and $C^2H^2Br^2O^2$, are formed when acetic acid or acetic ether is heated with bromine in sealed tubes at 180°. Monobromacetic acid forms deliquescent rhombohedral crystals, and boils at 208°; dibromacetic acid, a crystalline mass, fusing at 45–50° and boiling at 232–234°. The salts of both acids are somewhat unstable. *Ethyl monobromacetate* is a colorless liquid, boiling at 159° with partial decomposition. Its vapor attacks the eyes violently.—*Tribromacetic acid*, $C^2HBr^3O^2$, results by the action of water on tribromacetyl bromide (which see).—Crystals, which fuse at 130°, and boil at 245°.

Iodoacetic acid, $C^2H^3IO^2$. Is produced when a mixture of acetic anhydride, iodine, and iodic acid is heated to boiling (140°), a violent reaction taking place.—Ethyl bromacetate is decomposed by potassium iodide, forming potassium bromide and ethyl iodoacetate, and this, when heated with baryta water, gives barium iodoacetate, which, treated with sulphuric acid, yields the acid.—Colorless plates, which fuse at 82° with partial decomposition. When heated with hydriodic acid, it is reconverted into acetic acid. Most of its salts are decomposed, when merely boiled with water.—*Diiodoacetic acid*, $C^2H^2I^2O^2$, is obtained in a similar manner.

Cyanacetic acid, $C^3H^3NO^2 = CH^2(CN),CO.OH$. Monochloracetic acid (5 parts) is boiled with potassium cyanide (6 parts) and water (24 parts) until the smell of prussic acid can no longer be detected; the liquid is then neutralized exactly with sulphuric acid, evaporated down to a small volume, filtered, supersaturated with sulphuric acid, and by agitating with ether the cyanacetic acid extracted. The crude acid, that remains behind after the evaporation of the ether, can be purified by conversion into its lead salt and decomposition of this with sulphuretted hydrogen.—Colorless, crystalline mass. Its salts, with the exception of the silver and mercury salts, are easily soluble in water.

Amidoacetic acid (*Glycin*, *Glycocol*), $C^2H^5NO^2 = CH^2(NH^2)CO.OH$. Is produced from chlor- and bromacetic acids by heating with ammonia. Hippuric acid (which see), when boiled with acids or alkalies, is resolved into glycocol and benzoic acid. Glycocholic acid (which see), treated in the same manner, yields glycocol and cholic acid. It is produced further by boiling glue with sulphuric acid or potassa.—It is prepared most practicably by boiling hippuric acid for an hour with four times its weight of concentrated hydrochloric acid, allowing to cool, filtering the benzoic acid off, and evaporating the filtrate. Glycocol hydrochlorate remains behind. To an aqueous solution of this, lead or silver oxide is added, the lead or silver chloride filtered off, and, after the removal of any lead which may remain dissolved, by means of sulphuretted hydrogen, the solution is evaporated to crystallization.

Large crystals, stable in the air, soluble in 4 parts of water, but little in alcohol. Fuses at 170°; not volatile without decomposition. The watery solution possesses an acid reaction. It combines with bases, acids, and salts.

The *copper salt* $(C^2H^4NO^2)^2Cu + H^2O$, prepared by dissolving copper oxide in a hot solution of glycocol, separates on cooling in needles of a deep-blue color.—The *silver salt* $C^2H^4NO^2.Ag$ is obtained by allowing a solution of glycocol, which is saturated with silver oxide, to evaporate slowly over sulphuric acid.

Ethyl ether of glycocol, $CH^2(NH^2).CO.O.C^2H^5$. The hydriodate of this ether is obtained by heating an alcoholic solution of glycocol with ethyl iodide at 115–120°.—Clear, rhombic crystals, soluble in water, alcohol, and ether. Silver oxide removes the hydriodic acid from this compound, but the free ether decomposes, when its solution is evaporated, yielding glycocol and alcohol.

Glycocol combines with hydrochloric acid, forming two crystallizing salts, $C^2H^5NO^2.HCl$ and $2(C^2H^5NO^2).HCl$.—*Glycocol nitrate* $C^2H^5NO^2.HNO^3$ crystallizes in prisms.

In addition to these there are a number of crystallizing compounds with chlorides, sulphates, and nitrates known.

Heated with dry caustic baryta, glycocol yields carbonic anhydride and methylamine. When its aqueous solution is treated with nitrous acid, glycolic acid is produced.

Methylglycocol (*Sarcosine*), $C^3H^7NO^2 = CH^2(NH.CH^3).CO.OH$. Is produced by the action of methylamine on chloracetic acid; by the evaporation of a solution of creatine (which see) with barium hydroxide; and by heating caffeine for several hours with barium hydroxide.—Colorless, rhombic prisms, easily soluble in water, less in alcohol, fuses somewhat above 100°, and sublimes undecomposed. Yields salts with acids and with bases.

Ethylglycocol, $CH^2(NH.C^2H^5).CO.OH$ (isomeric with the ethyl ether of glycocol), is formed from ethylamine and monochloracetic acid.—Small, laminated crystals, which deliquesce in the air, become brown at 150–160°, and fuse at a higher temperature, undergoing decomposition. Like glycocol, it combines with acids, bases, and salts.—*Diethylglycocol* $C^2H^3[N(C^2H^5)^2]O^2$ is obtained from monochloracetic acid by the action of diethylamine.—Deliquescent crystals, which sublime under 100°.

Acetylglycocol (Aceturic acid), $CH^2(NH.C^2H^3O).CO.OH$, results by the action of acetyl chloride on glycocol silver.—Small, white needles, soluble in water and alcohol, which turn brown at 130°. Monobasic acid; forms easily soluble salts.

In the preparation of glycocol from monochloracetic acid and ammonia, there are formed as secondary products: *Diglycolamidic acid* $C^4H^7NO^4$ and *triglycolamidic acid* $C^6H^9NO^6$. Both compounds crystallize well and unite with bases and acids.

Sulphoacetic acid, $CH^2\begin{cases} SO^2.OH \\ CO.OH, \end{cases}$ is formed by the action of sulphuric anhydride on acetic acid with the aid of heat. Its salts with the alkalies are produced by heating monochloracetic acid with concentrated solutions of alkaline sulphites.—Colorless, deliquescent prisms; fusing point, 62°. Bibasic acid.—The *barium salt* $CH^2\begin{cases} SO^2.O \\ CO.O \end{cases}Ba + H^2O$ crystallizes in laminæ.—When heated with sulphuric acid, it is converted into disulphometholic acid and carbonic anhydride.

Thiacetic acid, $C^2H^4OS = CH^3.CO.SH$, is produced by distilling concentrated acetic acid with phosphorus tersulphide or pentasulphide.—Colorless liquid, which turns yellow when left for any length of time; smells of acetid acid and sulphuretted hydrogen; boils at 93°; and is soluble in water and in alcohol. Its salts are soluble in water.

The *lead salt*, $(C^2H^3O.S)^2Pb$, forms colorless needles, which are decomposed easily, sulphur being thrown down.

Acetic anhydride, $(C^2H^3O)^2O$, is obtained by distilling 3 parts anhydrous sodium acetate and 1 part phosphorus oxichloride; or, better, by distilling equal parts by weight of acetyl chloride and anhydrous sodium acetate.—Colorless liquid, boiling at 138°, heavier than water, decomposed rapidly by it, forming acetic acid. With hydrochloric acid it yields acetic acid and acetyl chloride; with chlorine, monochloracetic acid and acetyl chloride. Bromine acts the same as chlorine. Phosphorus sulphide converts it into *thiacetic anhydride* $(C^2H^3O)^2S$, a yellowish liquid, boiling at 121°.

Acetyl hyperoxide, $(C^2H^3O)^2O^2$, is obtained by adding barium peroxide to an ethereal solution of acetic anhydride. After distilling off the ether at a very low temperature, washing with water and potassium carbonate, it remains as a thick, consistent liquid. Is rapidly decomposed in sunlight; explodes when gently heated, like nitrogen chloride. Powerful oxidizing agent, decolorizes indigo, separates iodine from potassium iodide, and converts potassium ferrocyanide into the ferricyanide.

Acetyl chloride, $C^2H^3O.Cl = CH^3.COCl$, is formed when dry hydrochloric acid gas is allowed to act upon acetic acid in the presence of phosphoric anhydride; by the action of phosphorus terchloride, pentachloride, or oxichloride on acetic acid or dry acetates. Is most readily prepared by carefully distilling a mixture of 9 parts acetic acid and 6 parts phosphorus terchloride on a water-bath.—Colorless liquid, boiling at 55°; is decomposed by water, forming acetic and hydrochloric acids. Dry chlorine gas converts it, in sunlight or in the presence of iodine, into substitution-products: $C^2H^2ClO.Cl$ (boiling point, 106°), $C^2HCl^2O.Cl$ and $C^2Cl^3O.Cl$ (boiling point, 118°). The same substances are produced by heating mono-, di-, or trichloracetic acids with phosphorus terchloride.

Acetyl bromide, $C^2H^3O.Br$, is produced by the action of phosphorus bromide on acetic acid.—Colorless liquid, boiling at 81°. Yields with bromine liquid substitution-products: $C^2H^2BrO.Br$ (boiling point, 149–151°), $C^2HBr^2O.Br$ (boiling point, 194°), $C^2Br^3O.Br$ (boiling point, 220–225°).

Acetyl iodide, $C^2H^3O.I$, is obtained by the action of iodine and phosphorus on acetic anhydride.—Liquid, boiling at 108°.

Acetyl cyanide, $C^2H^3O.Cy$, is formed by heating the chloride with silver cyanide.—Liquid, boiling at 93°. Conducts itself towards water like the chloride. By being preserved in imperfectly closed vessels and by treating with solid potassium hydroxide or sodium hydroxide it is converted into a polymeric crystalline compound $(C^2H^3O)^2Cy^2$, which fuses at 69° and boils at 208–209°.

Acetamide, $C^2H^3O.NH^2 = CH^3.CO.NH^2$, is formed by distilling ammonium acetate and by decomposing acetic ether by means of ammonia. The latter formation takes place slowly without the aid of heat, rapidly when the substances are heated to 120–130°.—Colorless crystals, easily soluble in water and alcohol; fuses at 78°; boils at 222°. Combines with metals $(C^2H^4NO)^2Hg$ and with acids $(C^2H^5NO.HCl.)$, forming unstable compounds.

Chloracetamide, $CH^2Cl.CO.NH^2$. Is produced from ethyl chloracetate and ammonia at the ordinary temperature.—Colorless, thick prisms; fusing point, 119.5°.

Amidoacetamide, $CH^2(NH^2)CO.NH^2$. The hydrochlorate is formed by heating ethyl chloracetate with an excess of an alcoholic solution of ammonia to 60–70°, the free compound by heating glycocol with alcoholic ammonia to 155–156°.—White mass, very easily soluble in ammonia; strongly alkaline; undergoes a partial spontaneous decomposition into glycocol and ammonia, when its aqueous solution is allowed to evaporate in contact with the air. Takes up carbonic anhydride from the air. It is hence difficult to obtain it in a free condition.—The *hydrochlorate* $C^2H^6N^2O.HCl$ consists of easily soluble prisms.

Diacetamide, $(CH^3.CO)^2NH$. Is formed, together with other bodies, by heating acetamide in a current of dry hydrochloric acid, and by heating acetonitrile with concentrated acetic acid up to 200°.—Colorless crystals, easily soluble in water; fusing point, 59°; boiling point, 210–215°.

Triacetamide, $(CH^3.CO)^3N$. Is formed when acetonitrile is heated for a long time with acetic anhydride to 200°.—Small, colorless crystals; fusing point, 78–79°.

3. *Propionic Acid.*

$C^3H^6O^2 = CH^3.CH^2.CO.OH.$

Formation and preparation. In small quantity, together with acetic acid, by the dry distillation of wood. From metacetone (see cane-sugar) and other acetones by oxidation. From sugar by the action of concentrated potassa. Sodium ethylate combines with carbonic anhydride, forming sodium propionate. Carbonic oxide and sodium alcoholate unite, forming sodium propionate.—Is prepared most practicably by boiling propionitrile (see p. 47) for a long time with an alcoholic solution of potassa, evaporating, and distilling the residue with sulphuric acid.

Properties. Colorless, clear liquid, with an odor resembling that of acetic acid; specific gravity, 0.992 at 18°; boiling point, 139°. Mixes with water in all proportions, can be separated from this solution by means of calcium chloride.

Its salts are all soluble in water.

The *silver salt*, $C^3H^5O^2Ag$, crystallizes in small needles, which are difficultly soluble in cold water.

Ethyl propionate, $C^3H^5O.O.C^2H^5$. Is prepared like acetic ether.—Colorless liquid, boiling at 100°.

Substitution-products of propionic acid. Of each of the simple substitution-products, formed by the displacement of one hydrogen atom by a monovalent element or a monovalent group, two varieties can exist. Their difference results from the difference in position of the substituted hydrogen atoms; it being in the one case in the group CH^3, in the other in the centre group CH^2. The direct action of chlorine, etc., appears only to cause the substitution of hydrogen, that is in combination with the central carbon atom.

α-Chlorpropionic acid, $C^3H^5ClO^2 = CH^3.CHCl.CO.OH$. Is prepared by the decomposition of lactyl chloride (see Lactic Acid) with water.—Colorless liquid, boiling at 186°; specific gravity, 1.28.—The *ethyl ether* of this acid $C^3H^5ClO.O.C^2H^5$ is obtained by bringing lactyl chloride together with alcohol, and by the action of phosphorus terchloride on lactic ether.—Liquid, boiling at 144°.

β-Chlorpropionic acid, $C^3H^5ClO^2 = CH^2Cl.CH^2.CO.OH$. The crystalline chloride of this acid ($C^3H^5O.ClO.Cl$) is formed by the action of 3 molecules phosphorus pentachloride on lead glycerate or glyceric acid. Yields with alcohol *ethyl β-chlorpropionate* $C^3H^5ClO.O.C^2H^5$, a liquid that boils at 150–160°. From this is obtained the free acid by treating with baryta water, and decomposing the salt formed by means of sulphuric acid. Can also be prepared by boiling iodopropionic acid with chlorine water.—Fibrous, fascicular crystals, which fuse at 65°. The ethers of the two acids when boiled with potassium cyanide yield two different cyanpropionic acids.

α-**Brompropionic acid,** $CH^3.CHBr.CO.OH$, is produced together with dibrompropionic acid by heating propionic acid with bromine in sealed tubes; also by heating lactic acid with concentrated hydrobromic acid. It is a liquid, boiling at 202°, congealing at —17°.—*Dibrompropionic acid*, $C^3H^4Br^2O^2$, is also formed by the oxidation of allylalcohol bromide.—Colorless crystals that fuse at 65° and boil at 227°.

β-**Brompropionic acid,** $CH^2Br.CH^2.CO.OH$. Is obtained by heating β-iodopropionic acid with bromine and water.—Colorless crystals, fusing at 61.5°.

α-**Iodopropionic acid,** $CH^3.CHI.CO.OH$. Obtained by the action of phosphorus iodide upon lactic acid. —Thick oil, scarcely soluble in water.

β-**Iodopropionic acid,** $CH^2I.CH^2.CO.OH$. Is formed by treating glyceric acid with hydriodic acid (phosphorus iodide and water).—Colorless crystal-plates, easily soluble in hot water; fusing point, 82°. Yields propionic acid, when heated to 180° with hydriodic acid.

α-**Amidopropionic acid** (Alanin), $C^3H^7NO^2 = CH^3.CH(NH^2).CO.OH$. Is produced by heating α-chlorpropionic acid or ethyl α-chlorpropionate with an aqueous solution of ammonia. Can be most readily prepared by boiling an aqueous solution of aldehyde-ammonia (2 parts) for a long time with hydrocyanic acid (1 part anhydrous) and an excess of hydrochloric acid. Sal-ammoniac is separated from the concentrated solution by means of alcohol, and, from the alanin hydrochlorate in solution, the alanin is obtained in the same manner as glycocol (p. 85) is obtained from its hydrochlorate.

—Hard, fascicular needles, soluble in 5 parts cold water, more easily in hot water and alcohol. When carefully heated it sublimes; when rapidly heated it decomposes, yielding ethylamine and carbonic anhydride. It combines, like glycocol, with bases, acids, and salts.

β-**Amidopropionic acid,** $CH^2(NH^2)CH^2.CO.OH$. Is obtained, like alanin, from β-iodopropionic acid.—Colorless, transparent, oblique rhombic prisms. Easily soluble in water, but slightly in absolute alcohol. When heated it fuses and decomposes, carbon being deposited. When very carefully heated to 170° it sublimes partially in needles.

The remaining derivatives of propionic acid are prepared in the same manner as the corresponding derivatives of acetic acid.

Propionyl chloride, $C^3H^5O.Cl$. Liquid. Boiling point, 80°.

Propionyl bromide, $C^3H^5O.Br$. Liquid. Boiling point, 96–98°.

Propionyl iodide, $C^3H^5O.I$. Liquid. Boiling point, 127–128°.

Propionylamide, $C^3H^5O.NH^2$. Colorless prisms, fusing at 75–76°.

4. *Butyric Acids.*

$$C^4H^8O^2 = C^3H^7.CO.OH.$$

Theoretically there are two acids of this composition possible (p. 75). Both are known.

1. Normal butyric acid (butyric acid of fermentation), $CH^3.CH^2.CH^2.CO.OH$. Is contained in a great many animal juices, and in the form of the glycerin ether in butter.—Is produced by the oxidation of normal butyl alcohol and, in small quantity together with acetic acid and other acids, by the dry distillation of wood. Its ethyl ether is produced together with other substances by the successive action of sodium and ethyl iodide on acetic ether (p. 82).—Most readily obtained by the fermentation of sugar. 3 kilogrammes cane-sugar and 15 grm. tartaric acid are dissolved in 13 kilogrammes boiling water and allowed to stand for a few days; then about 120 grm. rotten cheese, suspended in 4 kilogrammes sour milk, and $1\frac{1}{2}$ kilogrammes chalk, are added, and the whole allowed to remain unmolested in some place, where the temperature is kept at 30–35°. In ten days the mass becomes pulpy from the presence of calcium lactate, which has separated; at a later period hydrogen is evolved together with carbonic anhydride, the mass again becomes a thin liquid, and in the course of five or six weeks the fermentation is completed. Now the same volume of water and 4 kilogrammes crystallized sodium carbonate are added, the calcium carbonate filtered off, the filtrate evaporated to about 5 kilogrammes, and then mixed with $2\frac{3}{4}$ kilogrammes sulphuric acid previously diluted with water. The principal amount of butyric acid separates as an oily layer. It is removed, desiccated by means of calcium, chloride and then rectified. By distilling the residual solution of the salt, the dissolved acid can be obtained from this.

Colorless liquid, boiling at 157°; specific gravity, 0.988 at 0°; mixes with water in every proportion; is, however, thrown down from its watery solution by easily soluble salts. It is not acted upon by potassium bichromate and sulphuric acid; by continued oxidizing with nitric acid, a small portion is converted into succinic acid.

Its salts are soluble in water.

Calcium butyrate, $(C^4H^7O^2)^2Ca$, is less soluble in hot water than in cold. A solution, saturated at the ordinary temperature, on being heated, throws down nearly all the dissolved salt, in the form of lustrous laminæ.

Silver butyrate, $C^4H^7O^2Ag$, crystallizes from hot water in microscopic prisms.

Ethyl butyrate, $C^4H^7O.O.C^2H^5$. Colorless liquid of a pleasant odor, boiling at 119°.

Butyroacetic acid, $C^6H^{12}O^4$, a remarkable compound of butyric with acetic acid, is produced by the fermentation of crude calcium tartrate. It forms salts, but the free acid, when subjected to partial distillation, is decomposed into equal molecules of butyric and acetic acids.

Substitution-products. Of each substitution-product, in which one hydrogen atom is replaced by a monovalent group, there are three modifications possible. Up to the present, but few of them have been prepared, and their constitution is not well known.

Monochlorbutyric acid, $C^4H^7ClO^2$. By the action of chlorine on butyric acid in the presence of iodine.—Fine, pliant needles. Easily soluble in hot water. Fuses at 98–99°, and sublimes at 80°.

A non-crystalline, viscid acid, isomeric with this, is produced by the decomposition of chlorbutyryl chloride with water.

Monobrombutyric acid, $C^4H^7BrO^2$ (a liquid, which does not congeal at —15°; boiling at about 217°, not, however, without undergoing decomposition), and *dibrombutyric acid* (colorless, long, thin prisms, fusing at 45–48°) are produced by heating butyric acid with bromine.

Amidobutyric acid, $C^4H^7(NH^2)O^2$. From monobrombutyric acid and ammonia.—Small laminæ or needles, easily soluble in water, difficultly in alcohol, insoluble in ether.

2. Isobutyric acid, $\left.\begin{matrix}CH^3\\CH^3\end{matrix}\right\}CH.CO.OH$. Is contained in the Carob bean (the fruit of *Ceratonia siliqua*). Is obtained from pseudopropyl cyanide (p. 67) by heating with alkalies and by the oxidation of isobutyl alcohol

(p. 68).—A liquid very similar to butyric acid; is, however, more difficultly soluble in water (in 3 parts at the ordinary temperature); boils at 153–154°.

Calcium isobutyrate, $(C^4H^7O^2)^2Ca + 5H^2O$, crystallizes in long prisms, and is much more easily soluble in hot than in cold water.

Silver isobutyrate, $C^4H^7O^2Ag$, crystallizes from hot water in lustrous laminæ.

Monobromisobutyric acid, $C^4H^7BrO^2 = \left.\begin{matrix}CH^3\\CH^3\end{matrix}\right\}$ CBr.CO.OH. By heating isobutyric acid with bromine to 140°.—Colorless crystals, fusing at 45°, not volatile without decomposition. Becomes oily on being mixed with water; over sulphuric acid in a vacuum, it congeals again. But slightly soluble in cold water, soluble in every proportion in hot water.

5. *Valeric Acids.*

$$C^5H^{10}O^2 = C^4H^9.CO.OH.$$

Of the four acids of this composition, which are theoretically possible, only two are well known.

1. Normal valeric acid, $CH^3.CH^2.CH^2.CH^2.CO.OH$. Is prepared from butyl cyanide like propionic acid. Is also obtained by the oxidation of the mixture of alcohols from the amyl hydrides of petroleum.—Colorless liquid, with an odor like that of butyric acid. Boiling point, 184–185°; specific gravity at 0°, 0.9577.

The *barium salt* $(C^5H^9O^2)^2Ba$ crystallizes in small anhydrous laminæ.

2. Ordinary valeric acid (Isopropylacetic acid), $\left.\begin{matrix}CH^3\\CH^3\end{matrix}\right\}$ CH.CH².CO.OH. Is contained in the root of *Valeriana* and *Angelica officinalis* and of *Athamanta oreoselinum;* in the berries and bark of *Viburnum* *opulus;* in the oil of *Delphinum globiceps.*—Is produced by the oxidation of amyl alcohol; by warming isobutyl cyanide with potassa; the ethyl ether is produced by the successive action of sodium and isopropyl iodide on acetic ether. It is produced further by the oxidation of fats and of leucine; by the putrefaction of albuminoid substances (hence contained in old cheese).—To prepare it, valerian roots are distilled with water.—More practicably from ferment amyl alcohol. To 5 parts potassium bichromate and 4 parts water in a retort, which is united with a condensing apparatus in such a manner that the condensed vapors are returned to it, is gradually added a mixture of 1 part amyl alcohol and 4 parts concentrated sulphuric acid. At first the liquid becomes heated spontaneously, afterward it is kept at the boiling temperature, until oily streaks (of valeric aldehyde) are no longer observable in the neck of the retort, then distilled off. The distillate is neutralized with sodium carbonate, the amyl valerate, which separates, drawn or distilled off, and the dried salt decomposed with $\frac{4}{5}$ its weight of sulphuric acid, previously diluted with $\frac{1}{2}$ its weight of water. The valeric acid, which separates, is drawn off, desiccated and rectified.—Colorless liquid, with a peculiar, pungent, acid odor; specific gravity, 0.9468; boiling point, 171–172°. Soluble in 30 parts of water. Can be separated from this solution by means of easily soluble salts.

The *valerates of the alkalies* are deliquescent salts.

Barium valerate, $(C^5H^9O^2)^2Ba$, easily soluble, lustrous prisms of the triclinic system, or laminæ.—*Zinc valerate,* $(C^5H^9O^2)^2Zn$, separates, on the evaporation of its solution, in the form of lustrous scales.—*Silver valerate,* $C^5H^9O^2Ag$, white precipitate, crystallizing from boiling water.

Methyl valerate, $C^5H^9O.O.CH^3$. Liquid, boiling at 115°; insoluble in water. The *ethyl ether,* $C^5H^9O.O.C^2H^5$, boils at 133°; the *amyl ether,* $C^5H^9O.O.C^5H^{11}$, boils at 188°.

The derivatives of valeric acid are perfectly analogous to those of acetic and propionic acid.

Amidovaleric acid (Butalanin), $C^5H^{11}NO^2$, occurs in the spleen and in the pancreas of the ox. Is formed by heating bromvaleric acid with ammonia.—Colorless laminæ; easily soluble in water. When carefully heated, sublimable without decomposition, without previous fusion. Combines, like glycocol, with bases and acids.

The valeric acid from valerian root, as well as that obtained by oxidation of amyl alcohol and of leucine from different albuminoid substances, appears almost always to consist of two acids in varying proportions, one of which is the optically inactive isobutylformic (isopropylacetic) acid, while the other is optically active. The barium salt of the latter is distinguished by being more easily soluble and by crystallizing less readily.

The *optically active valeric acid* (probably methylethylacetic acid) boils, at the most, **1** to **1.5°** lower than the inactive acid; its specific gravity is higher, 0.9505 at 0°. Its power to act upon polarized light is completely removed by heating it with a few drops of concentrated sulphuric acid.

6. *Caproic Acids.*

$$C^6H^{12}O^2 = C^5H^{11}.CO.OH.$$

Of the eight acids of this composition, whose existence is indicated by the theory, four are known.

1. Normal caproic acid, $CH^3.CH^2.CH^2.CH^2.CH^2.CO.OH$. Is prepared by treating normal amyl cyanide with alcoholic potassa.—Clear liquid, of a sharp, acid taste; does not mix with water; boiling point, 204.5–205°; specific gravity, 0.9499 at 0°.

The caproic acid, obtained by oxidation of the mixture of alcohols from the hexyl hydrides of petroleum and mannite, is probably also normal caproic acid.

2. Ordinary caproic acid (Isobutylacetic acid), $\left.\begin{matrix}CH^3\\CH^3\end{matrix}\right\}$ CH.CH²CH².CO.OH. Occurs, sometimes in a free state, sometimes in the form of the glycerin ether, in a number of plants (for instance, in the blossoms of *Satyrium hircinum*, in the fruit of *Gingko biloba*, in cocoa-nut oil), further in butter and many other fats; and results from the oxidation of fats and a number of albuminoid bodies. It is most readily obtained by boiling amyl cyanide with an alcoholic solution of potassa.

Colorless liquid, but slightly soluble in water, with a sudorific odor; congealing at +5°; boiling at 195–198°.

Leucine (Amidocaproic acid), $C^6H^{13}NO^2$, is extensively distributed throughout the animal organism, is formed by the putrefaction of urine, glue, and protein substances, and by boiling them with dilute sulphuric acid. Results from valeric aldehyde, the same as alanine (p. 91) from acetic aldehyde.—Lustrous, colorless crystalline laminæ; fuses at 170°; sublimes when very carefully heated; when rapidly heated, it is decomposed, yielding carbonic anhydride and amylamine. Soluble in 27 parts cold water; but slightly in cold alcohol, more easily in hot alcohol.

3. Isocaproic acid, $\left.\begin{matrix}CH^3\\CH^3\end{matrix}\right\}CH.CH\left\{\begin{matrix}CH^3\\CO.OH,\end{matrix}\right.$ is obtained by heating the cyanide corresponding to amylenehydrate (p. 71) with potassa.—An oil but slightly soluble in water.

4. Pseudocaproic acid (Diethylacetic acid), $\left.\begin{matrix}C^2H^5\\C^2H^5\end{matrix}\right\}$ CH.CO.OH. The ether of this acid (boiling point, 151°) is produced from acetic ether by the action of sodium and ethyl iodide (p. 82). The acid separated from this is liquid.

The remaining acids of this series have been but very slightly investigated in regard to their constitution.

Of most of them but one modification is as yet known; but whether the acids of the same composition of different origins are really identical or not, is a question still to be answered, as the researches on the subject have not the necessary exactness.

7. *Œnanthylic acid*, $C^7H^{14}O^2$. Is produced by the oxidation of a number of fats, especially castor oil, and by oxidation of the mixture of alcohols prepared from the heptyl hydrides of petroleum. It is obtained most conveniently by the oxidation of its aldehyde (p. 108).—Liquid, of an agreeable, aromatic odor, but slightly soluble in water, boiling at 219–222°.

8. *Caprylic acid*, $C^8H^{16}O^2$. In the fusel-oil of wine. As the glycerin ether in butter and other fats. Is produced by the oxidation of primary octyl alcohol.—Crystallizes in fine needles or laminæ; fuses at 16–17°; boils at 232–234°.

9. *Pelargonic acid*, $C^9H^{18}O^2$. In the volatile oil of *Pelargonium roseum*. Results from the oxidation of oleic acid and oil of rue.—Crystalline mass; fusing at 7°; boiling at 248–250°.

An acid called *nonylic acid*, probably identical with the preceding compound, is obtained from the cyanide of the alcohol derived from the volatile oil of *Heracleum spondylium* and other species of *Heracleum*. Fusing point, 253–254°; specific gravity, 0.9065 at 17°.

10. *Capric acid*, $C^{10}H^{20}O^2$. In the fusel-oil of wine. As the glycerin ether in a number of fats (butter, cocoa-nut oil).—Crystalline mass, of a sudorific odor; fusing at 30°; boiling at 268–270°.

11. *Lauric acid*, $C^{12}H^{24}O^2$. In the form of the glycerin ether in the fruit of *Lauris nobilis*, in pichurim beans, in cocoa-nut oil.—Needles of a silky lustre; fusing point, 43.6°.

12. *Myristic acid*, $C^{14}H^{28}O^2$. In nutmeg-butter and in spermaceti.—Crystalline scales; fusing at 53.8°.

13. *Palmitic acid*, $C^{16}H^{32}O^{2}$. Palmitic and stearic acids, in the form of glycerin compounds, constitute the principal ingredients of most solid fats. It is present in large quantity, and partially in a free condition, in palm oil. In order to prepare it from fats, these are heated with caustic potassa (saponified), the soap (potassium palmitate and stearate) precipitated from the solution and decomposed by hydrochloric acid. The acids are now dissolved in alcohol, and separated from each other by means of partial precipitation with magnesium acetate. If only $\frac{1}{4}$ of the amount of the magnesium salt necessary for complete precipitation is added, magnesium stearate falls down almost free of the palmitate; the succeeding precipitations contain the stearate mixed with palmitate; the last precipitations are almost pure magnesium palmitate. The precipitates are now decomposed separately by hydrochloric acid, and the free acids treated a few times more in the same manner.

Fine white needles, which congeal after fusion in the form of a scaly, crystalline mass. Fusing point, 62°.

14. *Margaric acid*, $C^{17}H^{34}O^{2}$. Probably does not occur in nature. That which was formerly designated as such has proven to be a mixture of palmitic and stearic acids. It is prepared artificially by boiling cetyl cyanide with caustic potassa. It resembles palmitic acid.

15. *Stearic acid*, $C^{18}H^{36}O^{2}$. On the occurrence and preparation see Palmitic Acid. Crystallizes from alcohol in laminæ; fuses at 69.2°, and congeals in crystalline scales.

16. *Arachidic acid*, $C^{20}H^{40}O^{2}$, is contained in oil of earth-nut and in the fruit kernels of *Nephelium lappaceum*.

17. *Benic acid*, $C^{22}H^{44}O^{2}$. In the oil expressed from the nuts of *Moringa nux Behen*.

18. *Hyänic acid*, $C^{25}H^{50}O^{2}$. In the anal glands of *Hiæna striata*.

19. *Cerotic acid*, $C^{27}H^{54}O^2$. In beeswax as a free acid. As ceryl ether in Chinese wax.

20. *Melissic acid*, $C^{30}H^{60}O^2$. Results from heating myricyl alcohol with soda-lime. Has not been detected in nature.

D. Aldehydes, $C^nH^{2n}O$.

Aldehydes are compounds which occupy an intermediate position between the primary alcohols and the acids. They contain two atoms of hydrogen less than the alcohols, and one atom of oxygen less than the acids. They are produced by careful oxidation of the primary alcohols, the group $CH^2.OH$ being hereby transformed into the group CHO, which is common to all aldehydes. Hence, the aldehydes can also be considered as derivatives of the hydrocarbons, formed by the displacement of a hydrogen-atom by the monovalent group CHO. By the action of hydrogen in *statu nascendi*, they are reconverted into the primary alcohols; under the influence of oxidizing agents, they are readily changed to acids. They possess strong reducing properties.

1. *Formic Aldehyde* (*Methyl Aldehyde*).
$CH^2O = H.CHO$.

Formation. Is produced when the vapors of methyl alcohol, together with air, are conducted over a platinum spiral, which at first is heated. The spiral becomes red and continues so during the operation. Is further formed by subjecting glycolic acid, calcium formate, or glycolate to dry distillation; by treating methylene iodide (see p. 36) with silver oxide or silver oxalate.

Properties. It appears that it can only exist at a high temperature in the form of gas. At the ordinary temperature, several (probably three) molecules combine, forming a white, indistinctly crystalline mass, *oxymethylene* $C^3H^6O^3$, which sublimes below 100°, fuses at 152°, and at a somewhat higher temperature is converted into gas. The specific gravity of the vapor is 1.06, corresponding to the simpler formula CH^2O.

Formylsulphaldehyde, $C^3H^6S^3$. When the liquid obtained by oxidizing methyl alcohol, or when pure oxymethylene, is saturated with sulphuretted hydrogen, a substance, having an alliaceous odor, separates, which dissolves, when the liquid is heated with half its volume of concentrated hydrochloric acid, and crystallizes out on cooling. Is also produced by the action of hydrogen in *statu nascendi* (zinc and hydrochloric acid) on carbon bisulphide, sulphocyanic acid, ethyl and allyl mustard-oils; and by treating methylene iodide with an alcoholic solution of sodium sulphide.—Fine, needly crystals, which fuse at 218°, and are volatile without decomposition. Difficultly soluble in boiling water, more readily in alcohol and ether.

Combines with silver nitrate, mercury chloride, and platinum chloride, forming crystalline compounds.

Hexamethyleneamine, $(CH^2)^6N^4$. Is formed when ammonia is conducted over oxymethylene, at first at the ordinary temperature and finally with the aid of gentle heat.—Clear, colorless, lustrous, rhombohedric crystals. Sublimable when very carefully heated. Easily soluble in water and boiling alcohol, but slightly soluble in cold alcohol and in ether. Has an alkaline reaction, and yields salts with acids.

The *hydrochlorate*, $C^6H^{12}N^4.HCl$, crystallizes in long, colorless needles, which are easily soluble in water.

2. *Acetic Aldehyde.*

$$C^2H^4O = CH^3.CHO.$$

Preparation. By imperfect oxidation of alcohol. 2 parts of alcohol are distilled with 3 parts manganese peroxide, 3 parts sulphuric acid, and 2 parts water until that which passes over begins to have an acid reaction. The distillate is rectified over calcium chloride, then mixed with an equal volume of ether, and saturated with dry ammonia. Or a mixture of 4 parts sulphuric acid, 12 parts water, and 3 parts alcohol is poured upon 3 parts potassium bichromate, care being taken to cool the vessel in which the reaction takes place; the vapors, freed of water as well as possible, are taken up by ether and saturated with ammonia. The crystalline ammonia compound, when distilled with sulphuric acid, yields pure aldehyde, which can be obtained free of water by rectifying again over calcium chloride.

Properties. Colorless liquid, of a suffocating odor; specific gravity, 0.807 at 0°; boiling point, 21°; mixes with water, alcohol, and ether in every proportion. Acts as a powerful reducing agent; from a silver solution it separates the metal, which forms a beautiful specular coating on the sides of the vessel. Combines with the bisulphites of the alkalies, forming crystallizing compounds. Phosphorus pentachloride converts it into ethylidene chloride (p. 64). It unites with hydrogen in *statu nascendi*, forming ethyl alcohol and butylene glycol (which see). All oxidizing agents convert it into acetic acid. Alkalies decompose it, forming resinous bodies.

Polymeric aldehydes. Small quantities of various substances (chlorcarbonic oxide, hydrochloric acid, sulphurous acid, zinc chloride, a drop of concentrated sulphuric acid) cause aldehyde to become transformed into polymeric compounds of entirely different properties. At the ordinary temperature *paraldehyde* $C^6H^{12}O^3$ is produced. This is a colorless liquid, boiling at 124°, but slightly soluble in water; congeals at a low temperature and fuses again at 10.5°.—At a temperature below 0°, *metaldehyde* is principally formed. This is a white, finely crystallizing body, which, without previously fusing, sublimes at 112–115°, at the same time being partially decomposed into aldehyde. When heated in fused tubes to 112–115°, it is completely reconverted into aldehyde. Both of these compounds, when distilled with dilute sulphuric acid, hydrochloric acid, etc., are reconverted into aldehyde;

and, with phosphorus chloride, hydrochloric acid, etc., they yield the same products as alhehyde.

Aldehyde-ammonia, $CH^3.CH\left\{\begin{matrix} OH \\ NH^2. \end{matrix}\right.$ Is formed when aldehyde, either alone or in ethereal solution, is brought together with dry ammonia.—Colorless, lustrous rhombohedrons. Fusing point, 70–80°; easily soluble in water; more difficultly soluble in alcohol; insoluble in ether.

Hydracetamide, $C^6H^{12}N^2 = (CH^3.CH)^3N^2$. Is formed when a solution of aldehyde in alcoholic ammonia is allowed to stand for some time.—Amorphous, easily soluble powder. Diatomic base. When boiled with water or dilute acids, it is resolved into ammonia and *oxytrialdine* $C^6H^{11}NO$, an amorphous, brown substance, possessing basic properties.

Aldehyde-hydrocyanate, $CH^3.CH\left\{\begin{matrix} OH \\ CN. \end{matrix}\right.$ Is produced by the direct combination of aldehyde with anhydrous hydrocyanic acid.—Colorless liquid; soluble in water and alcohol in all proportions; boiling point, 183°; is, however, partially resolved at this temperature into hydrocyanic acid and aldehyde. Concentrated hydrochloric acid decomposes it at the ordinary temperature, forming ammonium chloride and lactic acid.

Aldehyde-acetate, $CH^3.CH\left\{\begin{matrix} O.C^2H^3O \\ O.C^2H^3O. \end{matrix}\right.$ Is formed by direct union of aldehyde with acetic anhydride at 180°.—Colorless liquid, boiling at 169°. Does not mix with water.

Acetal, $C^6H^{14}O^2 = CH^3.CH\left\{\begin{matrix} O.C^2H^5 \\ O.C^2H^5. \end{matrix}\right.$ Is produced by the slow oxidation of alcohol (hence contained in crude spirits of wine), and is a secondary product in the preparation of aldehyde; and can be prepared by heating alcohol with aldehyde to 100°; or by double decomposition of ethylidene bromide (p. 46) and sodium ethylate.—Clear liquid, boiling at 104°. Is converted by oxidation into aldehyde and acetic acid; by concentrated hydrochloric acid into ethyl chloride.

Dichloracetal, $CHCl^2.CH(O.C^2H^5)^2$ (colorless liquid, boiling at 180°) and *trichloracetal* $CCl^3.CH(O.C^2H^5)^2$ (lustrous needles; fusing point, 72°; boiling point, 230°) are formed by the action of chlorine on ethyl alcohol. The former is also produced by the action of chlorine on acetal.

Monobromacetal, $CH^2Br.CH(O.C^2H^5)^2$. Is formed by the action of bromine on acetal.—Colorless liquid that boils at 170° without undergoing decomposition.

Dimethyl-acetal, $C^4H^{10}O^2 = CH^3.CH\left\{\begin{matrix} O.CH^3 \\ O.CH^3. \end{matrix}\right.$ Is contained in crude wood-spirit. Is produced by the action of black oxide of manganese and sulphuric acid on a mixture of ethyl and methyl alcohols; and by heating aldehyde with methyl alcohol to 100°.—Colorless liquid, boiling at 64°.

Ethylidene oxichloride, $C^4H^8Cl^2O = (CH^3.CHCl)^2O$ (isomeric with the second substitution-product resulting from ethylether, p. 49), is obtained by the action of hydrochloric acid on aldehyde.—Colorless liquid, boiling at 116–117°, which, when heated with water, yields aldehyde and hydrochloric acid.

Dichloraldehyde, $CHCl^2.CHO$. Is obtained by distilling dichloracetal with concentrated sulphuric acid.—Colorless liquid, boiling at 88–90°. Insoluble in water. When preserved it is gradually transformed into a solid, polymeric substance, which, at 120°, is reconverted into the original body. Yields dichloracetic acid by oxidation.

Trichloraldehyde (Chloral), $CCl^3.CHO$. Difficult to prepare directly from aldehyde. Is obtained by thoroughly saturating ethyl alcohol with dry chlorine, and distilling the crystalline product after the addition of concentrated sulphuric acid.—Colorless liquid, of a penetrating odor; specific gravity, 1.502, boils at 94.4°. When kept for a time it is changed to a solid polymeric body, from which it can be regenerated by heating. Like aldehyde it combines with ammonia and the bisulphites of the alkalies. Yields trichloracetic acid by oxidation, dichloracetic acid by treatment with silver oxide and water, and is resolved into chloroform and formic acid by treatment with alkalies in aqueous solutions. When taken internally in small quantities it causes sleep. Combines with water, forming *chloral-hydrate* $C^2HCl^3O + H^2O$, a substance that crystallizes well (fusing point, 46°; boiling point, 96–98°; insoluble in water); with alcohol forming *chloral-alcoholate* $C^2HCl^3O + C^2H^6O$. Colorless crystals; fusing point, 56°; boiling point, 114–115°. This compound is the final product of the action of chlorine on alcohol.

Dibromaldehyde, $CHBr^2.CHO$. By bringing bromine and aldehyde together carefully.—Colorless, long needles, of a penetrating odor, exciting to tears.

Tribromaldehyde (Bromal), $CBr^3.CHO$. Is prepared in the same manner as chloral, and resembles it in every respect.

Sulphaldehyde, $C^6H^{12}S^3 = (CH^3.CHS)^3$. When sulphuretted hydrogen is conducted into an aqueous solution of aldehyde, an oil $C^2H^4S + C^2H^4O$, of a disagreeable odor, congealing at —8°, is precipitated, which, when distilled, or, better, when treated with hydrochloric acid, yields sulphaldehyde.—White needles, insoluble in water, easily soluble in alcohol and ether. Begins to sublime at 45°.

Trialdine, $C^6H^{13}NS^2$, crystallizes from a watery solution of aldehyde-ammonia when sulphuretted hydrogen is conducted into it.—Large, colorless crystals, fusing at 43°, which are decomposed by keeping. It is a base, and yields crystallizing salts with acids.

The remaining aldehydes of this series are prepared either by carefully oxidizing the corresponding alcohols with potassium bichromate and dilute sulphuric acid, the apparatus being so arranged that the aldehyde formed may distil over immediately; or by subjecting an intimate mixture of the calcium salt of the corresponding acid with calcium formate to dry distillation. To purify them and separate them from foreign substances, they are shaken with a concentrated solution of potassium or sodium bisulphite. The aldehydes combine with these substances, forming crystalline compounds, which are difficultly soluble in cold water. These are then pressed, washed with alcohol, or recrystallized from a little warm water, and, finally, decomposed by distillation with a solution of an excess of sodium carbonate, the pure aldehyde now passing over.

The most important aldehydes are the following:—

3. *Propionic aldehyde*, $C^3H^6O = CH^3.CH^2.CHO$. A liquid, possessing a suffocating odor, very similar to acetic aldehyde. Specific gravity, 0.8327; boiling point, 49.5°.

4. *Butyric aldehyde*, $C^4H^8O = CH^3.CH^2.CH^2.CHO$. Colorless liquid; specific gravity, 0.834 at 0°; boiling point, 75°; soluble in 27 parts water.

Chlorbutyric aldehyde, C^4H^7ClO. Is produced by direct combination of crotonic aldehyde (which see) with hydrochloric acid gas.—Colorless needles; fusing point, 96–97°; insoluble in water, scarcely soluble in alcohol.

Isobutyric aldehyde, $C^4H^8O = \left.\begin{matrix} CH^3 \\ CH^3 \end{matrix}\right\} CH.CHO$. Colorless liquid; specific gravity, 0.8226 at 0°; boiling point, 62°.

5. *Normal valeric aldehyde*, $C^5H^{10}O = CH^3.CH^2.CH^2.CH^2.CHO$. A liquid, boiling at 102°.

Ordinary valeric aldehyde, $C^5H^{10}O = \left.\begin{matrix} CH^3 \\ CH^3 \end{matrix}\right\} CH.CH^2.$ CHO. Colorless liquid, of a pleasant, fruity, slightly suffocating odor; specific gravity, 0.822 at 0°; boiling point, 92.5°.

6. *Caproic aldehyde*, $C^6H^{12}O$. Liquid of a disagreeable odor; boiling point, 121°.

7. *Œnanthylic aldehyde* (œnanthol), $C^7H^{14}O$. Is most readily obtained by dry distillation of castor oil. The aldehyde is separated from the distillate by the process above described.—Liquid, of an unpleasant odor; specific gravity, 0.827; boiling point, 152°.

8. *Palmitic aldehyde*, $C^{16}H^{32}O$. From cetylic alcohol. —White, indistinctly crystalline mass; fusing point, 46–47°.

E. KETONES (ACETONES).

Ketones are compounds, which consist of two monovalent hydrocarbon groups, held together by the bivalent group (CO). They stand in close relation to the aldehydes, and can be considered as aldehydes, in which the hydrogen-atom of the group COH is replaced by a monovalent residue of hydrocarbon. They are produced by careful oxidation of the secondary alcohols, the group CH.OH, common to these alcohols, being hereby converted into CO, by a loss of two hydrogen atoms; further, by subjecting the salts of the fatty acids to dry distillation, by the action of zinc methyl, zinc ethyl, etc., on acetyl chloride and the homologous chlorides, etc.—Most of them form crystallizing compounds with the bisulphides of the alkalies. Nascent hydrogen converts them into secondary alcohols (the group CO being changed to CH.OH). —When oxidized with potassium bichromate and sulphuric acid, they are resolved into simpler compounds, in the following manner. One of the hydrocarbon residues (where the residues are different, that which has the largest number of carbon-atoms) is oxidized, yielding the fatty acid with the same number of carbon-atoms, while the other residue remains in combination with the CO and, together with this, is converted into the corresponding acid. Thus dimethylketone $CH^3.CO.CH^3$ yields formic and acetic acids; ethylmethylketone $CH^3.CO.C^2H^5$, only acetic acid; ethylbutylketone $C^2H^5.CO.C^4H^9$, propionic and butyric acids, etc.

1. *Acetone* (*Dimethylketone*).

$$C^3H^6O = CH^3.CO.CH^3.$$

Formation and preparation. By the action of zinc-methyl on acetyl chloride; by the destructive distillation of acetates; by the oxidation of isopropyl alcohol and propylene; by boiling the substance (p. 82) obtained from acetic ether by means of sodium with water; by heating monobrompropylene with mercury acetate and glacial acetic acid for several days at 100°; by heating citric acid; by distillation of wood (hence contained in crude wood spirit), and sugar or gum, mixed with lime.—Is prepared most expediently by distilling calcium acetate or a mixture of lead acetate with lime.—It is obtained on the large scale, as a secondary product in the manufacture of anilin with acetic acid and iron.

Properties. Clear liquid, of a pleasant odor, miscible with water, alcohol, and ether; specific gravity, 0.814; boils at 58°.

Acetone combines, like aldehyde, with the bisulphites, forming crystallizing compounds. Nascent hydrogen converts it into isopropyl alcohol with an accompanying formation of pinacone (s. hexylene alcohol). Concentrated sulphuric acid, alkalies, and caustic lime eliminate water from acetone, and convert it into *mesityl oxide* $C^6H^{10}O$ (colorless liquid, boiling at 130°), *phoron* $C^9H^{14}O$ (slightly colored crystals; fusing point, 28°; boiling point, 196°), and *mesitylene* C^9H^{12} (s. Aromatic Compounds).

Methylchloracetol (Acetone-chloride), $C^3H^6Cl^2 = CH^3.CCl^2.CH^3$. Is produced by the action of phosphorus chloride on acetone.—Colorless liquid, boiling at 69°. Treated with alcoholic potassa or ammonia, it is converted into monochlorpropylene C^3H^5Cl.

Methylbromacetol (Acetone-bromide), $C^3H^6Br^2 = CH^3.CBr^2.CH^3$. By the action of phosphorus bromide or phosphorus chlorobromide (PCl^3Br^2) on acetone.—Colorless liquid, boiling at 113–116°; specific gravity, 1.815 at 0°.

Monochloracetone, C^3H^5ClO. Is produced when an electric current is conducted through a mixture of acetone and hydrochloric acid; and by the action of hypochlorous acid on monochlor- or monobrompropylene.—Colorless liquid, exciting to tears; boiling point, 119°.

Dichloracetone, $C^3H^4Cl^2O$. Is produced by saturating acetone with chlorine.—A liquid, boiling at 120°. With phosphorus chloride it yields

Dichloracetone chloride, $C^3H^4Cl^4$. Heavy liquid, boiling at 153°.

By the action of bromine and chlorine-iodide on acetone, there are produced bromine and iodine substitution-products.

Sulphacetone, $C^6H^{12}S^2$. Is produced by the action of phosphorus trisulphide on acetone.—Yellowish liquid, of a very unpleasant odor; boiling point, 183–185°. Does not mix with water.

2. *Propione (Diethylketone).*

$$C^5H^{10}O = C^2H^5.CO.C^2H^5.$$

Formation and preparation. By destructive distillation of propionates; or by the action of zincethyl on propionyl chloride; further, by bringing together sodium ethylate and carbonic oxide; by oxidizing diethoxalic acid with potassium bichromate and dilute sulphuric acid; and by heating the same acid with concentrated hydrochloric acid to 130–150°.

Properties. Colorless, pleasant smelling liquid, of specific gravity, 0.815; boiling point, 101°. Oxidized with potassium, bichromate, and dilute sulphuric acid, it yields acetic and propionic acids.

---

3. *Methylethylketone*, $C^4H^8O = CH^3.CO.C^2H^5$. By the oxidation of secondary butyl alcohol (p. 68). By the action of zincethyl on acetyl chloride. In small quantity in the preparation of acetone on a large scale. From ethyl-methyl acetone carbonate (obtained by the successive action of sodium and methyl iodide on acetic ether) by heating with potassa ley.—Colorless liquid, boiling at 81°; of 0.8125 specific gravity. Combines with alkaline bisulphites.

4. *Methylpropylketone*, $C^5H^{10}O = CH^3.CO.C^3H^7$. Is produced by the distillation of a mixture of calcium butyrate and acetate; as a secondary product by the distillation of calcium butyrate and by the oxidation of isoamyl alcohol (p. 71). By the oxidation of the mixture of alcohols from the heptyl hydrides of petroleum.—Colorless liquid; boils at 102–105°; specific gravity, 0.807.

5. *Ethylpropylketone*, $C^6H^{12}O = C^2H^5.CO.C^3H^7$. By the distillation of calcium butyrate; by the action of butyl chloride on zincethyl.—Boiling point, 126°; specific gravity, 0.818 at 17.5°.

6. *Dipropylketone* (Butyrone), $C^7H^{14}O = C^3H^7.CO.C^3H^7$. Is formed as the principal product by the distillation of calcium butyrate.—Boiling point, 144°; specific gravity at 20°, 0.82.

7. *Methylbutylketone*, $C^6H^{12}O = CH^3.CO.C^4H^9$. By the oxidation of secondary hexyl alcohol; by the oxidation of the mixture of alcohols obtained from the hexyl hydrides of petroleum.—Boiling point, 127°; specific gravity, 0.8298.

8. *Methylamylketone*, $C^7H^{14}O = CH^3.CO.C^5H^{11}$. By the oxidation of the mixture of alcohols obtained from the heptyl hydrides of petroleum.—Boiling point, 150–152°.

9. *Methylhexylketone*, $C^8H^{16}O = CH^3.CO.C^6H^{13}$. By the oxidation of secondary octyl alcohol; and by the distillation of a mixture of calcium œnanthylate and acetate.—Boiling point, 171°; specific gravity, 0.818.

10. *Methylnonylketone*, $C^{11}H^{22}O = CH^3.CO.C^9H^{19}$. Forms the principal constituent of oil of rue (from *Ruta graveolens*); and is produced by the distillation of a mixture of calcium caprate and acetate.—Colorless liquid, with a peculiar, bluish fluorescence. Boiling point, 225–226°; specific gravity, 0.8268; congeals at +6°, forming a laminated crystalline mass, which fuses again at 15°.

SECOND GROUP.

A. Hydrocarbons, C^nH^{2n} (Ethylene Series).

The hydrocarbons of this series differ from those of the marsh gas series in containing two hydrogen atoms less. They may be considered as non-saturated compounds; it is, however, more probable that two of the carbon atoms contained in them are united by means of so-called double-union. A characteristic property of these hydrocarbons is that of combining directly with two monovalent atoms (Cl^2, Br^2, I^2, HI, etc.), and thus yielding compounds which may be looked upon as substitution-products of the hydrocarbons of the marsh gas series, and are either identical or isomeric with the products obtained from the latter.

The first member of this series CH^2 is not known, and is apparently not capable of existence.

1. *Ethylene (Elayl, Olefiant Gas).*
$C^2H^4 = CH^2.CH^2$.

Formation and preparation. By the destructive distillation of the salts of a great many fatty acids; by distillation of fats, resins, wood, of anthracite coal, and a large number of other organic bodies. Prepared most easily by heating a mixture of 1 part alcohol and 4 parts concentrated sulphuric acid, to which has been added sand enough to form a thick pulp, in order to prevent foaming. The gas, which is cooled, is conducted through soda ley and sulphuric acid, in order to free it of carbonic and sulphurous anhydrides, alcohol and ether vapors.

Properties. Colorless gas; specific gravity, 0.978; not congealing above —110°; burns with a luminous flame, and is absorbed but little by water. When conducted through an ignited tube, it is decomposed into carbon, marsh gas, hydrogen, and acetylene. It combines with sulphuric anhydride to form carbyl sulphate; English sulphuric acid absorbs it very slowly, forming ethylsulphuric acid. Hydrochloric, hydrobromic, and hydriodic acids combine slowly with it, forming ethyl chloride, bromide, and iodide. A solution of platinum chloride in hydrochloric acid absorbs it slowly and, on the addition of potassium chloride, lemon-colored crystals of $C^2H^4.PtCl^2.KCl + H^2O$ are deposited.

Ethylene chloride (Elayl chloride), $C^2H^4Cl^2 = CH^2Cl.CH^2Cl$, is formed from ethylene and chlorine by direct combination. In order to prepare it, ethylene gas is conducted into a gently heated chlorine mixture and the chloride finally distilled off.—Colorless liquid, of an ethereal odor; of specific gravity 1.271 at 0°; boiling point, 85°. When boiled with alcoholic potassa, it is converted into *chlorethylene*, C^2H^3Cl, water and potassium chloride being formed at the same time. A gas condensable at —18°. Chlorine acts upon ethylene chloride, yielding substitution-products, and, according to the length of time occupied in the action, one, two, three, or all the hydrogen atoms can be replaced by chlorine. These chlorinated products conduct themselves towards alcoholic potassa like ethylene chloride; a molecule of hydrochloric is given off, and in this way are formed the two series:—

$C^2H^4Cl^2$;	boiling	point,	85°
$C^2H^3Cl^3$	"	"	115°
$C^2H^2Cl^4$	"	"	137°
C^2HCl^5	"	"	158°
C^2Cl^6	"	"	182°

And

C^2H^3Cl	"	"	—18°
$C^2H^2Cl^2$	"	"	37°
C^2HCl^3	"	"	87–90°
C^2Cl^4	"	"	117°

The first three members are different from the substitution-products (p. 46) obtained from ethyl hydride or ethyl chloride by the action of chlorine; for the fourth and fifth members there is but one kind of constitution possible.

Ethylene bromide, $C^2H^4Br^2$, a colorless liquid, boiling at 129°, congealing at a temperature below +9°, is formed by conducting ethylene into bromine under water. It conducts itself towards an excess of bromine and towards an alcoholic solution of potassa like the chloride. In this way are obtained the two series:—

$C^2H^4Br^2$; boiling point, 129°
$C^2H^3Br^3$ " " 186.5°
$C^2H^2Br^4$, liquid } not distillable without decomposition.
C^2HBr^5 } solid }
C^2Br^6 }

And

C^2H^3Br;	boiling	point,	23–24°
$C^2H^2Br^2$	"	"	88°
C^2HBr^3	"	"	130°
C^2Br^4;	fusing	point,	50°

Ethylene iodide, $C^2H^4I^2$, is produced by conducting ethylene gas over iodine in sunlight at an elevated temperature. Can be most easily prepared by conducting the gas into an alcoholic solution of iodine, to which is added an excess of iodine.—Colorless crystals, fusing at 73°, which, when kept in a dark place, slowly turn yellow. Exposed to the light, this discoloration takes place rapidly. Sublimable; above 80°, however, it is resolved into iodine and ethylene.

Ethylene nitrite, $C^2H^4(NO^2)^2$. Is produced by direct combination of ethylene with hyponitric acid. —Four-sided prisms, which fuse at 37.5°.

Chlornitrocarbon, $C^2Cl^4(NO^2)^2$. By heating carbon chloride C^2Cl^4 with hyponitric acid in sealed tubes to 110–120°.—Colorless, crystalline mass. At 140° it is resolved into C^2Cl^4 and hyponitric acid.

Ethylene cyanide, $C^4H^4N^2 = C^2H^4(CN)^2$. Is obtained by heating ethylene chloride, or better, bromide, in an alcoholic solution with potassium cyanide.—Crystalline mass, fusing at about 37°. Not volatile, without undergoing decomposition. Heated with an alcoholic solution of potassa, it yields ammonia and potassium succinate; it combines with nascent hydrogen, forming butylene diamine $C^4H^8(NH^2)^2$.

2. *Propylene.*

$$C^3H^6 = CH^3.CH:CH^2.$$

Is produced together with the homologous hydrocarbons from a large number of organic bodies, when these are distilled either alone or mixed with lime. It can be obtained pure by heating glycerin with phosphorus iodide, or by heating the allyl iodide C^3H^5I, which is formed simultaneously with it, with mercury or zinc and fuming hydrochloric acid; or, most readily, by heating isopropyl iodide with an alcoholic solution of potassa on a water-bath. It is also produced together with ethyl bromide by the action of zincethyl on bromoform.—Colorless gas, with an odor similar to that of ethylene; condensable by pressure, but is still gaseous at —40°. It is but slightly absorbed by water, more easily by alcohol (12 volumes). It conducts itself like ethylene, and combines, like this gas, directly with chlorine, bromine, and iodine. It combines with hydriodic acid, forming isopropyl iodide. By agitating it with concentrated sulphuric acid, a sulpho-acid is formed, which, when distilled with water, yields isopropyl alcohol.

Propylene chloride, $C^3H^6Cl^2 = CH^3.CHCl.CH^2Cl$ (isomeric with methylchloracetol, p. 110), is also produced by the action of chlorine on propyl hydride.—Colorless liquid, boiling at 93–98°.

Monochlorpropylene, $C^3H^5Cl = CH^3.CCl:CH^2$. Is produced from propylene chloride, and also from the isomeric methylchloracetol by the action of alcoholic potassa.—Liquid, boiling at 23°.

Propylene bromide, $C^3H^6Br^2$. Is also obtained by allowing bromine to act upon isopropyl bromide; and, together with the following compound, by the action of hydrobromic acid on allyl alcohol.—Colorless liquid, boiling at 142°.

Trimethylene bromide, $C^3H^6Br^2 = CH^2Br.CH^2.CH^2Br$. A liquid, boiling at 160–163°; specific gravity, 2.0177 at 0°.

Monobrompropylene, C^3H^5Br. Liquid; boils at 56.5°.

3. *Butylenes.*
C^4H^8.

There are three butylenes of different constitution known.

1. Butylene (Methylallyl), $CH^3.CH^2.CH:CH^2$. Is obtained by decomposing a mixture of allyl iodide and methyl iodide with sodium.—Colorless liquid, boiling between —4 and +8°. Combines with hydriodic acid, forming secondary butyl iodide (p. 69).

The *bromide*, $CH^3.CH^2.CHBr.CH^2Br$, boils at 156–159°.

2. Isobutylene, $\left.\begin{matrix}CH^3\\CH^3\end{matrix}\right\}C:CH^2$. Is produced by the action of alcoholic potassa on the iodides of isobutyl alcohol and tertiary butyl alcohol; by the decomposition of amyl alcohol at red heat; and by the electrolysis of potassium valerate.—Boiling point —6°. Unites with hydriodic acid, forming tertiary butyl iodide. Is absorbed by sulphuric acid, a sulpho-acid being formed, which yields tertiary butyl alcohol (p. 69), when subjected to distillation with water.

The *bromide*, $\left.\begin{matrix}CH^3\\CH^3\end{matrix}\right\}CBr.CH^2Br$, boils at 160°.

3. Pseudobutylene, $CH^3.CH:CH.CH^3$. Is obtained by the action of alcoholic potassa, silver oxide and water, or silver acetate on secondary butyl iodide; and by heating secondary butyl alcohol to 250°.—Boiling point, +3°; congeals as a crystalline mass when cooled down to a very low temperature. Combines with hydriodic acid, regenerating secondary butyl iodide.

The *bromide*, $CH^3.CHBr.CHBr.CH^3$, boils at 159°.

A fourth butylene, the constitution of which is not well known, results from the action of zincethyl on monobromethylene. It boils at —5°; yields a bromide that boils at 166°; and, as it seems, combines with hydriodic acid, forming isobutyl iodide.

4. *Amylenes.*
C^5H^{10}.

1. Ethylallyl, $C^5H^{10} = CH^3.CH^2.CH^2.CH:CH^2$. Is obtained by the action of zincethyl on allyl iodide.—Colorless liquid, boiling at 37°. Combines with hydriodic acid, forming isoamyl iodide (p. 71); with bromine yielding a bromide, $C^5H^{10}Br^2$, that boils at about 175°.

2. Amylene, $C^5H^{10} = \left.\begin{matrix}CH^3\\CH^3\end{matrix}\right\}CH.CH{:}CH^2$. Is produced, together with diamylene, triamylene, and small quantities of other hydrocarbons, by the distillation of amyl alcohol of fermentation over zinc chloride.—Colorless liquid; boiling point, 35°; specific gravity, 0.663 at 0°.—When shaken with concentrated sulphuric acid, it is converted into polymeric compounds (diamylene, triamylene); combines with hydriodic acid, forming the iodide of amylenehydrate (p. 71.)

Amylene bromide, $C^5H^{10}Br^2$. Colorless liquid, boiling at 170–180°, not, however, without a slight decomposition.

3. Isoamylene, $C^5H^{10} = \left.\begin{matrix}CH^3\\CH^3\end{matrix}\right\}C{:}CH.CH^3$ or $\left.\begin{matrix}CH^3.CH^2\\CH^3\end{matrix}\right\}C{:}CH^2$. Is formed from tertiary amyl iodide by the action of very concentrated alcoholic potassa.—Colorless liquid, that boils at 35°. Unites with hydriodic acid, forming tertiary amyl iodide.

Other isomeric hydrocarbons are produced by the action of sodium on chlorinated amyl chloride (alpha-amylene, boiling point, 28–30°) and by the action of zincethyl on chloroform.

---

5. *Hexylene,* C^6H^{12}. Is produced from the iodide of secondary hexyl alcohol (p. 72) by boiling with alcoholic potassa.—Colorless liquid, boiling at 68–70°. Combines with hydriodic acid regenerating secondary hexyl iodide.

It is not known exactly, whether the hexylene, obtained in the same manner from the iodide of the primary alcohol, and also boiling at 68–70°, is identical with that mentioned, or not. Another hexylene (alpha-hexylene, boiling point, 68–71°) is obtained by the action of sodium on bichlorinated hexyl hydride (from petroleum).

The remaining hydrocarbons of this series—*Heptylene* C^7H^{14}, boiling point, 94–96°; *octylene* C^8H^{16}, boiling point, 118–120°; *nonylene* C^9H^{18}, boiling point, 140°; *decatylene* $C^{10}H^{20}$, boiling point, 158–160°—are obtained in the same manner by treating the alcohol chlorides or iodides with alcoholic potassa, or like *diamylene* $C^{10}H^{20}$, boiling point, 150–153°; *dihexylene* $C^{12}H^{24}$, and *triamylene* $C^{15}H^{30}$, etc., by polymerisation of the simpler hydrocarbons by means of sulphuric acid.

B. Monatomic Alcohols, $C^nH^{2n}O$.

The alcohols of this series bear the same relation to the hydrocarbons of the ethylene series, as the alcohols $C^nH^{2n+2}O$ bear to the hydrocarbons of the marsh gas series. There is at present but one alcohol belonging to this group well known.

Allyl Alcohol.

$C^3H^6O = CH^2{:}CH.CH^2.OH.$

Formation and preparation. Four parts glycerin are heated slowly with one part crystallized oxalic acid (an addition of quarter to half per cent. ammonium chloride is advantageous) to 220–230°, finally to 260°. At first an aqueous solution of formic acid passes over, afterward allyl alcohol. The receiver is changed when the temperature of the mixture has reached 195°; that which passes over from 195–260° is redistilled, the operation being continued until potassium carbonate precipitates no oil drops from a specimen of the distillate. The allyl alcohol is then precipitated from the whole distillate with potassium carbonate, purified by treatment with powdered potassium hydroxide, freed of water, by means of barium hydroxide, and rectified.—Or allyl iodide is transformed into allyl oxalate by digestion with silver oxalate; this is then decomposed by means of dry ammonia gas and the alcohol distilled off.—It is also produced by the action of sodium on dichlorhydrine (see Glycerin).

Properties. Colorless liquid of a pungent odor; specific gravity, 0.858 at 0°; boiling point, 96–97°; congeals at —50°. Mixes with water in all proportions. Combines with two atoms of chlorine or bromine without elimination of hydrogen; does not combine with hydrogen. Heated with potassium hydroxide to 100–150°, it yields propyl alcohol, ethyl alcohol, formic acid, and other products.

Allyl chloride, $C^3H^5Cl = CH^2{:}CH.CH^2Cl$ (isomeric with monochlorpropylene). By allowing hydrochloric acid or phosphorus terchloride to act upon allyl alcohol; by bringing an alcoholic solution of allyl iodide together with mercury chloride; and by heating allyl oxalate with an alcoholic solution of calcium chloride to 100°. —Colorless liquid, boiling at 46°; specific gravity, 0.954 at 0°.

Allyl bromide, C^3H^5Br. Colorless liquid; boiling point, 70–71°; specific gravity, 1.461 at 0°.

Allyl iodide, C^3H^5I. Prepared from allyl alcohol, like ethyl iodide from ethyl alcohol (p. 46). Most expediently by adding 6 parts phosphorus gradually to a mixture of 15 parts glycerin and 10 parts iodine. After the reaction, which is frequently very violent, is finished, the substance is distilled off, the distillate washed with water and caustic soda, dehydrated and rectified; that portion which passes over between 98–103° is pure allyl iodide.—Colorless liquid, of an unpleasant, leeky odor; boiling at 101°; specific gravity, 1.789.—When its alcoholic solution is shaken with mercury, *mercurallyl iodide* C^3H^5IHg is formed. Colorless laminæ, difficultly soluble in alcohol; when distilled with iodine, yields allyl iodide; when treated with hydrochloric or hydriodic acids, yields propylene. Hydriodic acid converts allyl iodide into isopropyl iodide.

Allyl cyanide, $C^4H^5N = C^3H^5.CN$, is contained in the mustard-oil of commerce, and is prepared from this by distilling repeatedly with water. (On its preparation from mustard-seed, see Glucosides, Myronic Acid.)—Colorless liquid, of an agreeable, leeky odor, boiling at 117–118°.

By the action of silver cyanide on allyl iodide, a compound is formed, which is isomeric with allyl cyanide.

Allylether, $(C^3H^5)^2O$, results from the action of allyl iodide on silver or mercury oxide, and from the decomposition of potassium allylate by means of allyl iodide.—Colorless liquid, insoluble in water, boiling at 82°. A body of the same composition (allyl oxide), which is perhaps identical with allylether, occurs in crude oil of garlic. The compound ethers of allyl alcohol are formed by the action of allyl iodide on the silver salts of the respective acids.

Allyl formate, $CHO.O.C^3H^5$, is formed as a secondary product in the preparation of formic acid from oxalic acid and glycerin (p. 76).—Liquid, of a sharp odor; specific gravity, 0.932; boiling point, 81–83°.

Allyl acetate, $C^2H^3O.O.C^3H^5$. A liquid, with a penetrating odor, boiling at 98–100°.

Allyl valerate, $C^5H^9O.O.C^3H^5$, boils at 162°.

Allyl sulphide (oil of garlic), $(C^3H^5)^2S$. By the distillation of garlic (bulbs of *Allium sativum*) with water, a heavy, yellow oil is obtained, which contains allyl sulphide as its principal ingredient. By repeated rectification, finally over potassium, it is obtained in a free condition. It is further contained in the leaves of *Alliaria officinalis*, and in the seeds and green portions of a great many other plants of the cruciferous order. —It can be prepared artificially by the action of an alcoholic solution of potassium sulphide on allyl iodide.—Colorless oil, of a repulsive odor, which boils at 140°. Gives a crystalline precipitate of $(C^3H^5)^2S + 2AgNO^3$ with an alcoholic solution of silver acetate. This precipitate crystallizes from alcohol in needles.

Allyl-mercaptan, $C^3H^5.SH$, is formed from allyl iodide and an alcoholic solution of potassium sulphydrate.—A liquid, boiling at 90°, very similar to ethylmercaptan.

Allylamine, $C^3H^5.NH^2$. Is produced when allyl mustard-oil is treated with zinc and hydrochloric acid, or, better, with concentrated sulphuric acid.—Liquid, boiling at 58°.—By the action of ammonia on allyl iodide, the principal product formed is *tetrallylammonium iodide* $(C^3H^5)^4NI$, a crystalline body, which, when heated with silver oxide, yields *tetrallylammonium hydroxide* $(C^3H^5)^4N.OH$, a strongly alkaline liquid.

C. Monobasic Monatomic Acids, $C^nH^{2n-2}O^2$.

1. *Acrylic Acid.*

$$C^3H^4O^2 = CH^2{:}CH.CO.OH.$$

Formation and preparation. Is produced from its aldehyde, acrolein, when the latter, mixed with 3 parts H^2O, is left for a few days in contact with silver oxide. The liquid is then heated to boiling, sodium carbonate added until it shows an alkaline reaction, evaporated to dryness, the residue decomposed with dilute sulphuric acid, and filtered. By distilling the filtrate the pure acid is obtained still containing water. Can be prepared in any quantity from allyl alcohol. The alcohol is combined with bromine, the resulting alcohol oxidized, and the dibrompropionic acid thus obtained, freed of bromine by the action of zinc-dust and water. Can only be obtained free of water by the decomposition of its silver or lead salt by means of sulphuretted hydrogen. It is also produced by heating β-iodopropionic acid with alcoholic potassa or with milk of lime.

Properties. Clear liquid, boiling above 100°; has an odor similar to that of acetic acid; miscible with water in all proportions. Oxidizing agents resolve it into acetic and formic acids. Treated with sodium amalgam and water, it is converted into propionic acid. It combines with two atoms of bromine, without elimination of hydrogen, forming an exceedingly unstable acid.

Its salts are all easily soluble in water, with the exception of the *silver salt* $C^3H^3O^2Ag$.—*Lead acrylate*, $(C^3H^3O^2)^2Pb$, crystallizes in thin needles of a silky lustre.

2. *Crotonic Acid.*

$C^4H^6O^2 = C^3H^5.CO.OH.$

There are three isomeric acids of this composition known.

1. Crotonic acid, $CH^2{:}CH.CH^2.CO.OH$. Is produced by the oxidation of its aldehyde (p. 129); from allyl cyanide by boiling with caustic potassa; and by the destructive distillation of β-oxybutyric acid.—Fine, fleecy needles or large plates, which fuse at 72° and boil at 180–182°. Nascent hydrogen converts it into butyric acid, fusing potassium hydroxide into acetic acid.

Monochlorcrotonic acid, $C^4H^5ClO^2$. Is produced by the action of zinc and hydrochloric acid on trichlorcrotonic acid, and, together with another acid, when phosphorus chloride is allowed to act on acetylacetic ether (p. 82) and the product then treated with water.—Colorless crystals, easily soluble in water; fusing point, 94–95°.

Trichorcrotonic acid, $C^4H^3Cl^3O^2$. Is formed by the action of cold concentrated nitric acid on trichlorcrotonic aldehyde (p. 129).—Colorless, radiating needles; fusing point, 44°; soluble in 25 parts water.

Monobromcrotonic acid, $C^4H^5BrO^2$, is formed by boiling citradibrompyroracemic acid (see Pyroracemic Acid) with watery solutions of the alkalies.—Long, flat needles, but slightly soluble in cold water. Fusing point, 65°; boiling point, 228–230°; combines directly with 1 molecule bromine; and is converted into butyric acid by the action of sodium amalgam and water.

---

2. Isocrotonic acid, $CH^3.CH:CH.CO.OH$. Is obtained by the action of sodium amalgam on chlorisocrotonic acid.—Colorless liquid; boils at 172°; does not congeal at —15°.

Chlorisocrotonic acid, $C^4H^5ClO^2$. Is formed from acetylacetic ether together with chlorcrotonic acid. —Colorless crystals, difficultly soluble in water. Sublimes at the ordinary temperature. Fusing point, 59.5°; boiling point, 195°.

---

3. Methacrylic acid, $CH^2:C\left\{\begin{matrix}CH^3\\CO.OH.\end{matrix}\right.$ The ether of this acid, $C^4H^5O.O.C^2H^5$, is produced by the action of phosphorus terchloride on ethyl isoxybutyrate.—The free acid is liquid, does not congeal at 0°, and is resolved into formic and propionic acids by fusing with caustic potassa.

3. *Angelic Acid.*

$$C^5H^8O^2 = C^4H^7.CO.OH.$$

Occurrence and preparation. In the roots of *Angelica archangelica.* In order to prepare the acid from these, they are boiled with lime, filtered, the filtrate decomposed by sulphuric acid, and distilled; the distillate saturated with sodium carbonate, evaporated to dryness, and the residue again distilled with sulphuric acid. Acetic, valeric, and angelic acids pass over; by means of cooling, the latter separates from the mixture in crystalline form. It is also produced by the action of caustic potassa on the essential oil of chamomile (volatile oil of *Anthemis nobilis*), which appears to contain an ether of angelic acid; and by heating laserpitium and peucedanin with an alcoholic solution of potassa.

Properties. Colorless needles, fusing at 45°, but

slightly soluble in cold water, more easily in hot water and in alcohol; boiling point, 191°. Sodium amalgam produces no change in the acid in aqueous solution, but when it is heated for a long time with hydriodic acid and a little amorphous phosphorus at 180–200°, it is completely converted into valeric acid. Fusing caustic potassa resolves it into acetic and propionic acids.—Combines directly with bromine, forming the dibromide $C^5H^8Br^2O^2$. Crystals fusing at 76°. This dibromide is converted into angelic acid by the action of sodium amalgam and water.

Methylcrotonic acid, $C^5H^8O^2$ (isomeric with angelic acid), is produced from isoxyvaleric acid in the same manner as methacrylic acid.—Colorless needles, fusing at 62°; conducts itself towards fusing potassa the same as angelic acid.

4. *Hydrosorbic Acid.*

$$C^6H^{10}O^2 = C^5H^9.CO.OH.$$

Is produced by the action of sodium amalgam and water on sorbic acid.—Colorless liquid, of a sweaty odor, but slightly soluble in water; specific gravity, 0.969; boiling point, 201°. Does not congeal at —18°; melting potassa resolves it into butyric and carbonic acids.

The following acids are isomeric:—

Pyroterebic acid, $C^6H^{10}O^2$. Is produced by the destructive distillation of terebic acid (see Oil of Turpentine).—Oily liquid, boiling at 210°. Is broken up by means of fusing potassa, yielding acetic and butyric acids.

Ethylcrotonic acid, $C^6H^{10}O^2$, is produced from ethyl diethoxalate the same as methacrylic acid; also by heating ethyl diethoxalate for several hours with concentrated hydrochloric acid at 130–150°.—Quadratic prisms, fusing at 41.5°. Conducts itself towards potassa the same as pyroterebic acid.

5. *Cimicic Acid.*

$$C^{16}H^{28}O^{2} = C^{14}H^{27}.CO.OH.$$

Occurs in a leaf bug, *Rhapigaster punctipennis.*—Yellowish, crystalline mass. Fusing point, 44°.

6. *Hypogæic Acid.*

$$C^{16}H^{30}O^{2} = C^{15}H^{29}.CO.OH.$$

In the form of the glycerin ether in ground-nut oil (the oil of the fruit of *Arachis hypogæa*).—Colorless, needly crystals, fusing at 33°, which become yellow in the air. Combines with bromine to form the dibromide $C^{16}H^{30}Br^{2}O^{2}$ (solid, uncrystalline mass, fusing at 29°), which, when treated with alcoholic potassa at 100° yields *monobromhypogæic acid* $C^{16}H^{29}BrO^{2}$.—When carefully heated with nitric acid, hypogæic acid is converted into an acid of the same composition, *gaïdic acid.* Crystals, fusing at 39°.

7. *Oleic Acid* (*Elaïc Acid*).

$$C^{18}H^{34}O^{2} = C^{17}H^{33}.CO.OH.$$

Contained in nearly all fats as glycerin ethers (elain); in the largest proportion in the liquid fats, for instance, olive oil, oil of almonds, whale and seal oils. —In order to prepare it from these, the liquid fat is digested with lead oxide, the lead salts washed, and from these the lead oleate extracted with ether, and the ethereal solution decomposed with hydrochloric acid. The solution, poured off from lead chloride, leaves impure oleic acid behind when evaporated. It is dissolved in ammonia, barium oleate precipitated by adding barium chloride, and the salt, after having been repeatedly recrystallized from alcohol, is decomposed by means of tartaric acid.

Colorless oil, congeals at 4° and fuses at 14°; inodorous and tasteless. Alone, it cannot be distilled, the distillation can, however, be effected by means of overheated vapor of water at 250°. In a pure condition pretty stable; in an impure condition it takes up oxygen rapidly from the air, turns yellow, and then emits a rancid odor. Fusing potassa decomposes it into palmitic and acetic acids.

Of its salts only those of the alkalies (soaps) are soluble in water; these are, however, separated from their solutions by easily soluble salts. The *lead salt*, $(C^{18}H^{33}O^2)^2Pb$, forms the principal ingredient of ordinary lead-plaster.

Oleic acid combines directly with bromine, forming a liquid *dibromide*, $C^{18}H^{34}Br^2O^2$, which, when treated with alcoholic potassa, at the ordinary temperature, is converted into crystalline monobromoleic acid $C^{18}H^{33}BrO^2$, difficult to prepare in a pure condition.

By treatment with nitrous acid, oleic acid is converted into *elaidic acid*, which is isomeric with it. This crystallizes in laminæ, which fuse at 44–45°, and yield a crystalline dibromide with bromine; fusing point, 27°.

8. *Erucic Acid.*

$$C^{22}H^{42}O^2 = C^{21}H^{41}.CO.OH.$$

Is contained in mustard-oil and in rape-seed oil in the form of the glycerin ether.—Rape-seed oil is decomposed with litharge; the resulting lead-plaster, after being repeatedly extracted with ether, leaves behind pure lead erucate, which, when decomposed with hydrochloric acid, yields pure erucic acid.—Long, thin needles, insoluble in water, easily soluble in alcohol and ether; fusing point, 33–34°.

Unites with bromine to form the *dibromide*, $C^{22}H^{42}Br^2O^2$, which crystallizes in verrucose crystals, fuses at 42–43°, and yields *monobromerucic acid* $C^{22}H^{41}BrO^2$ (fusing point, 33–34°), when treated with alcoholic potassa at the ordinary temperature.

When carefully heated with dilute nitric acid to 60–70°, erucic acid is converted into *brassidic acid*, which is isomeric with it. This acid crystallizes in white, lustrous laminæ, fusing at 60°, and yields a dibromide (fusing point, 54°) with bromine.

Linoleic acid ($C^{16}H^{28}O^{2}$?), in linseed oil, and *ricinic acid*, $C^{18}H^{34}O^{3}$, in castor oil, are similar to, but not homologous with, oleic acid. Both are contained in the oils as glycerin compounds, and are prepared by saponifying the oils and decomposing the alkali salts with hydrochloric acid. Ricinic acid can be purified by dissolving its lead salt in ether. It is an almost colorless liquid, congealing at 0°; not volatile without decomposition. Like oleic acid, it is transformed into an isomeric, crystalline acid, *ricinelaïdic acid*, fusing at 50°. Conducts itself towards bromine like oleic acid.

D. Aldehydes, $C^{n}H^{2n-2}O$.

1. *Acrolein.*

$$C^{3}H^{4}O = CH^{2}{:}CH.CHO.$$

Formation and preparation. From allyl alcohol by careful oxidation; by the distillation of glycerin and the fats. Can be most readily prepared by distilling 1 part glycerin with 2 parts potassium bisulphate.

Properties. Colorless liquid, the vapor of which attacks the eyes and nose violently. Boiling point, 52°. Lighter than water and but slightly soluble in it. When kept it becomes changed, sometimes in a very short time, into a white, amorphous substance. It forms no crystallizing compounds with alkaline bisulphites; with nascent hydrogen it yields allyl alcohol. Alkalies convert it into a resinous mass. It combines directly with hydrochloric acid, forming $C^{3}H^{5}ClO$ (colorless needles, fusing at 32°, insoluble in water), which, when subjected to distillation, are converted into acrolein and hydrochloric acid.

Metacrolein (polymeric acrolein, probably $C^{9}H^{12}O^{3}$) is formed when the compound of acrolein with hydrochloric acid is distilled with caustic potassa.—Colorless crystals; fusing point, 50°; boiling point, 170°; insoluble in water, easily soluble in alcohol and ether. When distilled it is partially reconverted into acrolein.

Acrolein chloride, $C^3H^4Cl^2 = CH^2{:}CH.CHCl^2$. Is produced, together with the isomeric dichlorglycide, (see Glycerin), by the action of phosphorus pentachloride on acrolein or metacrolein.—Colorless liquid, boiling at 84°.

Acrolein-ammonia, C^6H^9NO. Is produced when the vapor of acrolein is conducted into ammonia, and when an alcoholic solution of acrolein is mixed with alcoholic ammonia.—Yellowish-white mass, which, when dried, becomes brownish-red. Combines with acids, forming amorphous salts. When subjected to dry distillation it is resolved into water and picoline.

2. *Crotonic Aldehyde.*

$C^4H^6O = CH^2{:}CH.CH^2.CHO$.

Formation. Is produced from acetic aldehyde (p. 103), when this is heated for some time at 100° with watery solutions of potassium formate or acetate, or with a little zinc chloride.

Properties. Colorless liquid, of an exceedingly pungent odor. Boiling point, 103–105°. In contact with the air, and under the influence of oxidizing agents, it is converted into crotonic acid.—Phosphorus chloride converts it into a fluid chloride $C^4H^6Cl^2$, boiling at 125–126°. It combines with hydrochloric acid directly, forming chlorbutyric aldehyde (p. 107).

Trichlorcrotonic aldehyde (Crotonchloral), $C^4H^3Cl^3O$. Is formed when acetic aldehyde, either alone or dissolved in carbon tetrachloride, is saturated with chlorine.—Colorless, oily liquid; boiling point, 163–165°. Combines with water, forming a hydrate $C^4H^3Cl^3O + H^2O$, which crystallizes in colorless, very thin laminæ, fusing at 78°. Caustic potassa decomposes it without the aid of heat, forming potassium formate and chloride, and dichlorallylene $C^3H^2Cl^2$. Nitric acid oxidizes it, forming trichlorcrotonic acid.

Crotonal-ammonia (Oxytetraldin), $C^8H^{13}NO$. Is produced when an alcoholic solution of acetic aldehyde-ammonia is heated to 90–100°.—Amorphous brown mass, very much like acrolein-ammonia. Combines, like the latter, with acids, yielding amorphous salts; and is resolved by heat into water and collidine.

Pyridine bases, $C^nH^{2n-5}N$. When acrolein-ammonia and crotonal-ammonia are heated, there result liquid bases, picoline and collidine, which belong to an homologous series, the single members of which are formed by the dry distillation of anthracite coal, peat, and particularly of bones. They are extracted from the distillation-products (coal-tar, bone-oil) by treating with dilute sulphuric acid; set free again by means of alkalies; and separated from each other by means of fractional distillation.

1. Pyridine, C^5H^5N. Colorless liquid, of a penetrating odor. Boiling point, 116.7°; specific gravity, 0.986 at 0°. Soluble in water. Strong base.—The *hydrochlorate* $C^5H^5N.HCl$ is deliquescent, and gives with platinum chloride a yellow double salt $(C^5H^5N.HCl)^2PtCl^4$, which is difficultly soluble in water.

In the presence of metallic sodium, pyridine is changed, gradually at the ordinary temperature, more rapidly when heated, into a polymeric base, *dipyridine* $C^{10}H^{10}N^2$, which crystallizes in colorless needles, fuses at 108°, and sublimes without decomposition.

2. Picoline, C^6H^7N. Is formed by the distillation of acrolein-ammonia and also when tribromhydrine is heated for several days with alcoholic ammonia to 250°.—Colorless liquid, mixes with water; specific gravity, 0.96; boiling point, 135°. Strong base. Is converted into a polymeric base by sodium, the same as pyridine.

3. Lutidine, C^7H^9N. Colorless liquid; specific gravity, 0.946; boiling point, 155.5°. More easily soluble in cold water than in hot.

4. Collidine (Aldehydine), $C^8H^{11}N$. Is obtained by heating an alcoholic solution of acetic aldehyde-ammonia to 120–130°, or of ethylidene chloride (p. 46) with alcoholic or aqueous ammonia to 160°.—Colorless, liquid, but slightly soluble in water; specific gravity, 0.944; boiling point, 176°.

In addition to these the following bases have been separated from coal-tar, but not carefully investigated: *Parvoline* $C^9H^{13}N$, boiling point, 188°; *corindine* $C^{10}H^{15}N$, boiling point, 211°; *rubidine* $C^{11}H^{17}N$, boiling point, 230°, and *viridine* $C^{12}H^{19}N$, boiling point, 251°.

THIRD GROUP.

A. Hydrocarbons, C^nH^{2n-2} (Acetylene Series).

The hydrocarbons of this series differ from those of the ethylene series, in that they contain two hydrogen atoms less; and are produced from these when their bromides are heated in sealed tubes with alcoholic potassa. They contain either two carbon atoms united by triple union (acetylene CH:CH) or twice two carbon atoms united by double union (diallyl CH^2:CH. CH^2CH^2.CH:CH^2).

1. *Acetylene.*
C^2H^2.

Formation and preparation. Is formed directly from its elements under the influence of an electric flame, which is produced in a current of pure hydrogen between points of purified carbon; is also formed by the decomposition of carbon-calcium with water; by the action of heat on ethylene and marsh gas (hence contained in coal-gas); by the decomposition of the latter by electrical sparks; by imperfect combustion of a great many organic bodies; by heating ethylene bromide or monobromethylene with alcoholic potassa; and in many other ways.

Properties. Colorless gas; somewhat soluble in water; of a characteristic unpleasant odor; burns with a very luminous flame.—It is absorbed in large quantity by an ammoniacal solution of copper subchloride; the resulting red precipitate, which is exceedingly explosive and evolves pure acetylene gas when hydrochloric acid is poured upon it, is *cuprosoacetyl oxide* $(C^2CuH)^2O$. In an ammoniacal solution of silver, it produces a white precipitate with similar properties. By the aid of this property acetylene can be separated from other gases and prepared in a pure condition.—By the action of nascent hydrogen (when the copper compound is brought in contact with zinc and ammonia) it is transformed into ethylene.

Acetylene dichloride, $C^2H^2Cl^2$. Cannot be prepared by direct action. Acetylene detonates when brought in contact with chlorine gas. Acetylene is entirely absorbed by antimony chloride ($SbCl^5$), large crystalline laminæ $C^2H^2.SbCl^5$ being formed, which, when heated, are resolved into antimony terchloride ($SbCl^3$) and acetylene dichloride. Colorless liquid; boiling point, 55°. Is decomposed when heated to 360°, yielding carbon and hydrochloric acid; when heated with alcoholic potassa to 100°, it yields potassium chloride and acetate.

Acetylene tetrachloride, $C^2H^2Cl^4$. Is formed when the compound $C^2H^2.SbCl^5$ is distilled with an excess of antimony chloride.—Colorless liquid, boiling at 147°.

Acetylene unites directly with bromine, forming $C^2H^2Br^2$ and $C^2H^2Br^4$. Both compounds are liquids.—When heated with iodine to 100°, it yields a crystalline iodide $C^2H^2I^2$, fusing at about 70°. It combines with hydriodic acid, forming liquid compounds: C^2H^3I, boiling point, 62°, and $C^2H^4I^2$, boiling point, 182°. The latter compound is isomeric with ethylene iodide.

---

2. *Allylene*, C^3H^4, is produced by the action of sodium ethylate on monochlor- or monobrompropylene, and by

the action of sodium on dichloracetone chloride (p. 110).—Gaseous; produces a yellow precipitate in an ammoniacal solution of copper subchloride; a white precipitate (C^3H^3Ag) in an ammoniacal solution of silver. Conducts itself towards bromine, iodine, and hydriodic acid like acetylene.

By the action of alcoholic potassa on monobrompropylene bromide, tribrom- or trichlorhydrine (see Glycerin), dichlorglycid, allylenbromide, and some other similar compounds, is produced propagylic ether $C^3H^3.O.C^2H^5$, a liquid boiling at 72°, which causes a yellow precipitate in a solution of copper subchloride; in solutions of silver, a white crystalline precipitate $C^3H^2Ag.O.C^2H^5$ or $C^6H^4Ag^2.O^2.(C^2H^5)^2$.

3. *Crotonylene*, C^4H^6. From monobrombutylene with alcoholic potassa at 100°.—Liquid, boiling at 18°.

4. *Valerylene*, C^5H^8. From monobromamylene, like crotonylene.—Liquid; boiling point, 45°. Gives no precipitates in solutions of copper subchloride or of silver.

Propylacetylene (isomeric with valerylene), C^5H^8. From methylpropylketone chloride ($CH^3.CCl^2.C^3H^7$) with alcoholic potassa.—A liquid boiling at 50°, which gives a yellow precipitate in an ammoniacal solution of copper subchloride, and a white precipitate in a silver solution.

5. *Hexoylene*, C^6H^{10}. From monobromhexylene. Boiling point, 76–80°.

Diallyl, C^6H^{10} (isomeric with the preceding compound). Is formed by the action of sodium on allyl iodide, and by the distillation of mercurallyl iodide (p. 120). Liquid, boiling at 59°.

The hydrocarbons, with a larger number of carbon atoms, are produced in a similar manner.

Alcoholic derivatives of these hydrocarbons are not known.

B. Monobasic, Monatomic Acids, $C^nH^{2n-4}O^2$.

The acids of this series are formed, like the hydrocarbons, by heating the dibromides of the acids $C^nH^{2n-2}O^2$ with alcoholic potassa.

1. *Sorbic Acid.*

$$C^6H^8O^2 = C^5H^7.CO.OH.$$

Occurrence and preparation. Together with malic acid in the juice of the unripe berries of the mountain-ash. If this is subjected to distillation after being partially neutralized with milk of lime, impure sorbic acid passes over with the vapors of water in the form of a yellow oil. The pure acid is obtained from this by heating gently with potassa or with concentrated sulphuric acid, or by boiling with concentrated hydrochloric acid.

Properties. Long, colorless needles, inodorous, almost insoluble in cold water, more easily soluble in hot water and alcohol; fuses at 134.5°; cannot be distilled alone without decomposition, readily with water vapor.

Barium sorbate, $(C^6H^7O^2)^2Ba$. Laminæ of a silvery lustre, easily soluble in water, scarcely more in boiling than in cold water.—*Silver sorbate* $C^6H^7O^2Ag$. White, insoluble, scarcely crystalline precipitate.

Ethyl sorbate, $C^6H^7O.O.C^2H^5$. Liquid, of a pleasant, aromatic odor, boiling at 195.5°; lighter than water.

Sorbic acid combines with nascent hydrogen, forming hydrosorbic acid (p. 125); with bromine forming a *tetrabromide* $C^6H^8Br^4O^2$, which crystallizes well, fuses at 178–179°, and is but slightly soluble in water.

2. *Palmitolic Acid.*

$$C^{16}H^{28}O^2 = C^{15}H^{27}.CO.OH.$$

Results from heating the dibromide of hypogœic acid or gaïdic acid with alcoholic potassa to 170°.—

Fine needles, of a silvery lustre, insoluble in water, easily soluble in alcohol and ether. Fusing point, 42°. Combines directly with 1 and with 2 molecules of bromine, but not with hydrogen.

Palmitoxylic acid, $C^{16}H^{28}O^4$, is formed, together with suberic acid and suberic aldehyde, by the action of fuming nitric acid on palmitolic acid.—Crystalline laminæ, insoluble in water, easily soluble in alcohol and ether. Fusing point, 67°; monobasic acid.

3. *Stearolic Acid.*

$$C^{18}H^{32}O^2 = C^{17}H^{31}.CO.OH.$$

Is produced, like the preceding acid, from the dibromide of oleic acid or elaïdic acid.—Long, colorless prisms. Fusing point, 48°; can be distilled, almost entirely without decomposition; insoluble in water, but slightly in cold alcohol, easily soluble in ether and hot alcohol.—Yields salts that crystallize well. Is not changed by the action of nascent hydrogen; combines, however, with bromine, forming a liquid *dibromide* $C^{18}H^{32}Br^2O^2$, and a crystalline *tetrabromide* $C^{18}H^{32}Br^4O^2$, fusing at about 70°.

Stearoxylic acid, $C^{18}H^{32}O^4$, produced like palmitoxylic acid.—Lustrous laminæ; fusing point, 86°. Very similar to palmitoxylic acid.

4. *Behenolic Acid.*

$$C^{22}H^{40}O^2 = C^{21}H^{39}.CO.OH.$$

Is produced from the dibromide of erucic acid by heating it with alcoholic potassa to 140–150°, and from the dibromide of brassinic acid by heating with alcoholic potassa to 210–220°.—White, lustrous, fascicular needles. Fusing point, 57.5°. Conducts itself towards hydrogen and bromine, the same as stearolic acid.

Behenoxylic acid, $C^{22}H^{40}O^4$. Lustrous scales; fusing point, 90–91°.

FOURTH GROUP.

A. Diatomic Alcohols, $C^nH^{2n+2}O^2$ (Glycols).

The diatomic alcohols are derived from the hydrocarbons of the marsh gas series by the replacement of two hydrogen atoms by means of two hydroxyl-groups. They are formed from the chlorides, bromides, and iodides of the hydrocarbons C^nH^{2n} by the exchange of the chlorine, bromine, or iodine atoms for hydroxyl.

The first member of this series, *methylene alcohol* $CH^2(OH)^2$, is not known and can probably not exist. Methylene iodide (p. 36), when treated with silver acetate, yields, besides silver iodide, methylene acetate $CH^2\left\{\begin{array}{l}O.C^2H^3O\\O.C^2H^3O,\end{array}\right.$ a liquid, that boils at 170°. If, however, the attempt is made to isolate the alcohol from this ether by means of heating with water or alkalies, formic aldehyde (oxymethylene) is obtained instead. It appears to be a general fact, that such diatomic alcohols as contain both hydroxyl groups in combination with the same carbon atom, cannot exist. Two diatomic alcohols can theoretically be derived from ethyl hydride $CH^3.CH^3$., viz., $\begin{array}{l}CH^2.OH\\CH^2.OH\end{array}$ and $CH^3.CH\left\{\begin{array}{l}OH\\OH.\end{array}\right.$ Only the first of these can, however, be isolated; the second, the acetic ether $CH^3.CH\left\{\begin{array}{l}O.C^2H^3O\\O.C^2H^3O\end{array}\right.$ of which can readily be prepared (p. 104), is resolved into aldehyde and water when the attempt is made to isolate it.

1. *Ethylene Alcohol* (*Ethylglycol*).

$$C^2H^6O^2 = \begin{array}{l}CH^2.OH\\\dot{C}H^2.OH.\end{array}$$

Preparation. Ethylene bromide is boiled for a few hours with potassium acetate and alcohol, then distilled; that portion of the distillate boiling between 140–200° (which consists mainly of monacetic glycol ether), is separated from the rest and decomposed with potassium or barium hydroxide.

Properties. Colorless, inodorous, somewhat viscid liquid, of specific gravity 1.125; boiling point, 197.5°; mixes with water and alcohol.—Sodium dissolves in it, hydrogen being evolved and *sodium-glycol* $C^2H^4 \left\{ \begin{matrix} ONa \\ OH, \end{matrix} \right.$ a crystalline mass, resulting, which, heated up to 190° with sodium, yields *disodium-glycol* $C^2H^4 \left\{ \begin{matrix} ONa \\ ONa. \end{matrix} \right.$ From these compounds *ethylglycol ether* $C^2H^4 \left\{ \begin{matrix} O.C^2H^5 \\ OH \end{matrix} \right.$ (a liquid of a pleasant odor) and *diethylglycol ether* $C^2H^4 \left\{ \begin{matrix} O.C^2H^5 \\ O.C^2H^5 \end{matrix} \right.$ (a liquid boiling at 123.5°, insoluble in water, isomeric with acetal) are formed by heating with ethyl iodide.

Oxidizing agents convert ethylene alcohol into glycolic acid and oxalic acid.

Ethylene chlorhydrine (Glycol hydroclorate), $C^2H^5ClO = CH^2Cl.CH^2.OH$, is formed by the direct union of ethylene with an aqueous solution of hypochlorous acid, or when ethyl alcohol is saturated with hydrochloric acid and then heated.—A liquid, boiling at 128°. When heated with potassium iodide, yields *glycol iodohydrine* $C^2H^4I.OH$, a heavy undistillable liquid; when heated with potassium cyanide, *glycol cyanhydrine* $C^2H^4Cy.OH$, a yellow syrup, which, treated with potassa, yields paralactic acid, together with some ordinary lactic acid.

Ethylene oxide (Glycol ether), $C^2H^4O = \begin{matrix} CH^2 \\ | \\ CH^2 \end{matrix}\!\!>O$ (isomeric with acetic aldehyde). Is produced by the action of potassium hydroxide on glycol chlorhydrine. —Colorless liquid, boiling at 13.5°; specific gravity at 0°, 0.898; mixes with water in all proportions; does not enter into combination with the bisulphites of the alkalies. It possesses basic properties; combines directly with acids to form ethylene ethers; and precipitates the hydroxides from solutions of metallic salts. It unites with water, when heated with it to 100° in sealed tubes, forming ethylene alcohol; with the latter, forming *diethylene alcohol* $C^4H^{10}O^3 = CH^2.OH.CH^2.O.CH^2.CH^2.OH$ (a liquid, boiling at 250°) and *triethylene alcohol* $C^6H^{14}O^4 = CH^2.OH.CH^2.O.CH^2.CH^2.O.CH^2.CH^2.OH$ (boiling point, 285–289°).—Is converted into ethyl alcohol by nascent hydrogen (from sodium-amalgam and water).

Ethylene sulphydrate (Glycolmercaptan), $C^2H^4(SH)^2$, is formed by the action of ethylene chloride or bromide on an alcoholic solution of potassium sulphydrate.—Colorless oil, of a penetrating odor. It forms salts with metallic oxides, like ethylmercaptan.—Ethylene chlorhydrine gives a similar compound with potassium sulphydrate, *ethylene monosulphydrate* $C^2H^4\begin{cases} OH \\ SH. \end{cases}$

Ethylene sulphide. By the action of ethylene chloride or bromide on an alcoholic solution of potassium sulphide, a crystalline substance, *diethylene sulphide* $(C^2H^4)^2S^2$ is formed, together with an amorphous yellow powder C^2H^4S, which is prepared most readily by double decomposition of mercurio-glycolmercaptan $C^2H^4.S^2Hg$ with ethylene bromide at 150°.—Fuses at 111°, and boils undecomposed at 200°. It unites directly with chlorine, bromine, iodine, with oxygen and several salts. Amorphous ethylene sulphide is converted into diethylene sulphide by being heated alone or with carbon bisulphide.

Ethylene monacetate, $C^2H^4\begin{cases} O.C^2H^3O \\ OH. \end{cases}$ Ethylene bromide (1 part) is heated on a water-bath for a length of time with potassium acetate (1 part) and alcohol (2 parts) in a flask connected with an inverted condensing apparatus. It is separated and purified by means of distillation.—A liquid, boiling at 182°, mixes with water and alcohol. Hydrochloric acid gas decomposes it at 100°, yielding water and *glycolchloracetin* $C^2H^4Cl.O.C^2H^3O$, a liquid, boiling at 145°.

Ethylene diacetate, $C^2H^4\left\{\begin{array}{l}O.C^2H^3O\\O.C^2H^3O,\end{array}\right.$ is formed by mixing dry silver acetate with ethylene iodide.—A liquid, boiling at 186°, soluble in 7 parts water.

Ethyleneamine bases. By the action of ethylene bromide on an alcoholic solution of ammonia, the crystalline hydrobromates of the three bases: *Ethylenediamine* $C^2H^4(NH^2)^2$, *diethylenediamine* $(C^2H^4)^2(NH)^2$, and *triethylenediamine* $(C^2H^4)^3N^2$, are formed. These can be separated from each other by means of crystallization. From these salts the volatile bases can be set free by means of silver oxide or by distillation with potassa. They are liquid. Ethylenediamine, which can also be produced by conducting cyanogen into a mixture of tin and hydrochloric acid, boils at 123°. Its formula is $C^2H^4(NH^2)^2 + H^2O$, and it does not give off the water even by repeated distillation over caustic baryta. Diethylenediamine boils at 170°; triethylenediamine boils at 210°.

Oxethylamine bases. When ethylene oxide is heated with aqueous ammonia, heat is evolved, and a mixture of three bases is formed: *Oxethylamine* $(C^2H^4.OH)NH^2 = \begin{array}{l}CH^2.OH\\ \dot{C}H^2.NH^2\end{array}$ (isomeric with aldehyde-ammonia), *dioxethylamine* $(C^2H^4.OH)^2NH$, and *trioxethylamine* $(C^2H^4.OH)^3N$. Their hydrochlorates are also produced when ethylene chlorhydrine is heated with aqueous ammonia to 100°. The difference in the solubility of the hydrochlorates and the platinum double salts of the three bases in alcohol affords a means of separation. They are of a syrupy consistence, easily soluble in water, strongly alkaline, and yield crystallizing salts.

Similar bases of more complicated constitution are formed by the union of ethylene oxide or ethylene chlorhydrine with organic bases. The most important of these is—

Trimethyloxethylammonium hydroxide (Bilineurine, choline, sinkaline), $C^5H^{15}NO^2 = C^2H^4.OH.(CH^3)^3N.OH$. Is contained in bile; is produced from sinapine (see Alkaloids) by gently heating with barium or potassium hydroxide; and can be most readily prepared by mixing a concentrated solution of trimethylamine with ethylene oxide. The chloride $C^2H^4.OH.(CH^3)^3N.Cl$ is produced by the direct union of ethylene chlorhydrine and trimethylamine.—The free base is colorless, crystalline, very easily soluble in water, and possesses very strong basic properties. Its solution in hydrochloric acid, when treated with platinum chloride and absolute alcohol, gives a yellow precipitate $(C^5H^{14}NO.Cl)^2PtCl^4$, which crystallizes from water in hexagonal plates; with gold chloride a yellow crystalline precipitate $C^5H^{14}NO.Cl + AuCl^3$.—Hydriodic acid converts it into *trimethyliodethylammonium iodide* $C^2H^4I(CH^3)^3NI$, a substance that crystallizes well and is difficultly soluble in cold water. When treated in aqueous solution with silver oxide, the latter compound is converted into *trimethylvinylammonium hydroxide* (*neurine*) $C^2H^3(CH^3)^3N.OH$. This is a very easily soluble base, which is also formed by boiling the substance of brain (lecithine, protagon) with baryta water.

Sulphoglycolic acid (*Glycolsulphuric acid*), $C^2H^6SO^5 = \begin{matrix} CH^2.OH \\ \dot{C}H^2.O.\dot{S}O^2.OH \end{matrix}$, is formed by heating equal molecules of ethylene alcohol and concentrated sulphuric acid to 150°. The *barium salt* $(C^2H^5SO^5)^2Ba$ is very easily soluble in water and crystallizes with difficulty.

Isethionic acid, $C^2H^6SO^4 = \begin{matrix} CH^2.OH \\ \dot{C}H^2.SO^2.OH \end{matrix}$ (isomeric with ethylsulphuric acid), is formed when sulphuric anhydride is conducted into well-cooled alcohol or ether, and at the end of the reaction the mass diluted with four times its volume of water and then boiled for a few hours. By neutralizing the liquid with barium carbonate, the soluble barium salt is prepared. Is also formed by mixing barium ethylsulphate with sulphuric anhydride, evaporating the excess of the anhydride and boiling for a long time with water. The sodium salt is formed by direct union of ethylene oxide with sodium bisulphite, and by treating ethylene chlorhydrine with a concentrated solution of sodium sulphite.—The free acid can be evaporated to a syrup; decomposes, however, when further evaporated. Monobasic acid. Its salts crystallize well and are very stable.—The *potassium salt*, when distilled with phosphorus chloride, yields *isethion chloride* (chlorethyl sulpho-chloride) $CH^2Cl.CH^2.SO^2Cl$, a liquid, boiling at 200°, which when heated with water is decomposed into *chlorethylsulphurous acid* $CH^2Cl.CH^2.SO^2.OH$ and water.

Taurin (Amidoisethionic acid), $C^2H^7NSO^3 = \begin{matrix} CH^2.NH^2 \\ CH^2.SO^2.OH. \end{matrix}$ Occurs free and in combination with cholic acid, as taurocholic acid, in the animal organism, in bile, in the contents of the alimentary canal, in the lung tissue, in the kidneys. Can be best prepared by evaporating bile to which has been added hydrochloric acid, removing the resinous substance which is thrown down, and mechanically separating the crystals of taurin and sodium chloride, which make their appearance on cooling. Is produced artificially by heating *ammonium isethionate* to 210°, and by heating silver chlorethylsulphite with aqueous ammonia to 100°.—Large, clear crystals, easily soluble in hot water, but slightly in cold water, insoluble in alcohol. It fuses, and decomposes at a high temperature. It does not yield well characterized compounds with bases nor with acids.

Disulphetholic acid, $C^2H^6S^2O^6 = \begin{matrix} CH^2.SO^2.OH \\ CH^2.SO^2.OH, \end{matrix}$ is produced by the action of fuming sulphuric acid on ethyl cyanide or propionamide; by the oxidation of ethylene sulphydrate with nitric acid; the sodium salt is formed by heating ethylene bromide with a concentrated solution of sodium sulphite.—Easily soluble crystals; fusing point, 94°. Bibasic, very stable acid.

Carbyl sulphate (*ethionic anhydride*), $C^2H^4S^2O^6 = \begin{matrix} CH^2.SO^2 \\ CH^2.O.SO^2 \end{matrix}>O$, is formed by the direct union of ethylene with sulphuric anhydride.—Colorless crystals, fusing at 80°; deliquesces in the air, and combines with water, forming

Ethionic acid, $C^2H^6S^2O^7 = \begin{matrix} CH^2.O.SO^2.OH \\ CH^2.SO^2.OH. \end{matrix}$ This acid is formed particularly when sulphuric anhydride is conducted into alcohol, which is cooled by means of ice.—Bibasic acid, which is resolved into isethionic acid and sulphuric acid when its aqueous solution is evaporated. Its salts are also decomposed by boiling their aqueous solutions.

2. *Propylene Alcohol (Propylglycol).*
$C^3H^8O^2 = C^3H^6(OH)^2$.

Taking for granted that alcohols, which contain two hydroxyl groups in combination with the same carbon atom, cannot exist, there are only two diatomic alcohols $C^3H^8O^2$ possible, viz.: $CH^2(OH).CH^2.CH^2(OH)$ and $CH^2(OH).CH(OH).CH^3$. The first is a primary alcohol, the second half primary, half secondary. Only the second alcohol is known as yet.

Preparation. From propylene bromide in the same manner as ethylene alcohol from ethylene bromide.

Properties. Colorless, viscid liquid; specific gravity, 1.051 at 0°; boiling point, 188–189°; mixes with alcohol and water in all proportions. When heated with concentrated hydriodic acid, it is converted into isopropyl alcohol and isopropyl iodide.

Propylene chlorhydrine, $C^3H^6Cl.OH$, is prepared like the analogous ethylene compound.—Colorless liquid, boiling at 127°. When carefully ozidized it is converted into monochloracetone (p. 110).

Propylene bromhydrine, $C^3H^6Br.OH$. Colorless liquid, boiling at 145–148°.

Propylene oxide, $C^3H^6.O$. Liquid, boiling at 35°. Combines with nascent hydrogen, forming isopropyl alcohol.

---

3. *Butylene alcohol* (Tetrylene alcohol), $C^4H^{10}O^2 = C^4H^8(OH)^2$, prepared from butylene (obtained from amyl alcohol), is a colorless, inodorous, thick liquid, of specific gravity 1.048 at 0°; boiling point, 183–184°; mixes with water and alcohol in all proportions.

A substance isomeric with this is—

Butyleneglycol, $C^4H^{10}O^2 = CH^3.CH.OH.OH^2.CH^2.CH$. Formed in small quantity, together with ethyl alcohol, by treating aldehyde, very much diluted with water, with sodium-amalgam, in a weakly acid solution.—Clear, viscid liquid, of a sweet, slightly pungent taste; boiling point, 203.5–204°. When oxidized, it yields carbonic, acetic, and oxalic acids, and crotonic aldehyde, the latter in very small quantity.

4. *Amylene alcohol,* $C^5H^{12}O^2 = C^5H^{10}(OH)^2$. From amylene bromide.—Colorless liquid; does not mix with water; specific gravity, 0.987; boiling point, 177°.

5. *Hexylene alcohol,* $C^6H^{14}O^2 = C^6H^{12}(OH)^2$. From hexylene bromide.—Colorless liquid; mixes with water; specific gravity, 0.967; boiling point, 207°.

The two following compounds are isomeric with this—

Diallylhydrate, $C^6H^{14}O^2 = C^6H^{12}(OH)^2$. The iodide of this alcohol $C^6H^{12}I^2$, a thick liquid that does not boil without undergoing decomposition, is produced by direct union of diallyl (p. 133) with hydriodic acid. The alcohol is obtained from this in the same manner as ethylene alcohol. It boils at 212–215°.

Pinacone, $C^6H^{14}O^2 = \left.\begin{matrix}CH^3\\CH^3\end{matrix}\right\} C(OH).C(OH) \left\{\begin{matrix}CH^3\\CH^3.\end{matrix}\right.$ Is formed, together with isopropyl alcohol, by the action of sodium-amalgam on acetone containing water.—Colorless, fine crystalline mass. Fusing point, 35–38°. Boiling point, 171–172°. Combines with water, forming a hydrate $C^6H^{14}O^2 + 6H^2O$, which crystallizes from water in large quadratic plates, fusing at 42°, and, when heated with dilute sulphuric acid or hydrochloric acid, or with concentrated acetic acid, is converted into *pinacoline* $C^6H^{12}O$, a colorless liquid, boiling at 105°; insoluble in water.

6. *Octylene alcohol*, $C^8H^{18}O^2 = C^8H^{16}(OH)^2$. From octylene bromide.—Colorless liquid, does not mix with water; specific gravity, 0.932; boiling point, 235–240°.

B. Monobasic, Diatomic Acids, $C^nH^{2n}O^3$.

The primary diatomic alcohols, when subjected to oxidation, conduct themselves like the monatomic alcohols. The groups $CH^2.OH$, contained in them, are oxidized, forming carboxyl:—

$CH^2.OH$	$CH^2.OH$	$CO.OH.$
$\dot{C}H^2.OH$	$\dot{C}O.OH$	$\dot{C}O.OH.$
Ethylene alcohol.	Glycolic acid.	Oxalic acid.

In this way are produced two series of acids. The acids of the first series are still half alcoholic in their character, and must hence play the part of monatomic alcohols and at the same time of monobasic acids. They stand in close relation to the fatty acids, and can be easily prepared from them by replacing hydrogen in the latter by hydroxyl. Of each acid of this series there can exist just as many isomeric modifications, as of the monochlorine- or monobromine-substitution-products of the corresponding fatty acid; of the first member only one; of the second two, $CH^2(OH).CH^2.CO.OH$ and $CH^3.CH\begin{cases} OH \\ CO.OH, \end{cases}$ etc.

1. *Glycolic Acid* (*Oxyacetic Acid*).

$$C^2H^4O^3 = CH^2\begin{cases} OH \\ CO.OH. \end{cases}$$

Occurrence. In unripe grapes.

Formation and preparation. By heating potassium chlor- or bromacetate with water, or by the addition of silver oxide to a hot aqueous solution of chlor- or bromacetic acid; by the action of nascent hydrogen (from zinc and sulphuric acid) on oxalic acid or oxalic ether; by treating glycocol with nitrous acid; and by careful oxidation of ethylene alcohol.—Can be most readily obtained by slow oxidation of ethyl alcohol. A mixture of 500 grms. alcohol and 440 grms. nitric acid is allowed to stand in cylinders, which are imperfectly closed, until small gas bubbles begin to appear in the liquid: the cylinders are then placed in water of 20°. In a few days the action is completed. The solution is now evaporated in small portions to a syrupy consistence, dissolved in water, neutralized with chalk and allowed to crystallize. The calcium glycolate, thus obtained, must be again dissolved and boiled for some time with milk of lime, for the purpose of decomposing any secondary products which may be present (glyoxal, glyoxylic acid). The solution is treated with oxalic acid in order to set the acid free, the filtrate from calcium oxalate almost neutralized with lead carbonate, and the solution of the lead salt evaporated to crystallization. From the solution of this salt, the lead is removed by means of sulphuretted hydrogen or, still better, sulphuric acid, which is added in not quite sufficient quantity to complete the decomposition, the filtrate evaporated and the glycolic acid extracted by means of anhydrous ether.

Properties. Deliquescent crystals; easily soluble in water, alcohol, and ether; fuses at 78–79°. When subjected to distillation, it undergoes decomposition, yielding formic aldehyde (oxymethylene, p. 101).

The *calcium salt* $(C^2H^3O^3)^2Ca$ forms fine, needly crystals, difficultly soluble in cold water; the *silver salt* $C^2H^3O^3Ag + \frac{1}{2}H^2O$, lustrous crystals, also difficultly soluble.

Glycolic acid is acetic acid in which one atom of hydrogen is replaced by hydroxyl. The hydrogen of this OH cannot be replaced by metals by treatment with bases, but easily by alcohol and acid radicals. A number of such compounds, for instance, *methylglycolic acid* $CH^2(O.CH^3)CO.OH$ (from sodium chloracetate and sodium methylate; colorless, thick liquid, boiling at 198°), *ethylglycolic acid* $CH^2.(O.C^2H^5).CO.OH$ (liquid, boiling at 206–207°), are known; and all these compounds, like glycolic acid, are monobasic acids.

Diglycolic acid, $C^4H^6O^5 + H^2O = \begin{matrix} CH^2.O.\,CH^2 \\ \dot{C}O.OH\dot{C}O.OH. \end{matrix}$

Is produced as a secondary product in the preparation of glycolic acid from monochloracetic acid and by the oxidation of diethylene alcohol (p. 138).—Large, colorless, monoclinic crystals. Easily soluble in water and alcohol. Fuses below 150°; bibasic acid; isomeric with malic acid.

Glycolid (Glycolic anhydride), $C^2H^2O^2 = CH^2\left\{\begin{matrix} O \\ CO \end{matrix}\right>$ is formed by heating glycolic acid, or potassium chloracetate; or by heating tartronic acid to 180°.—White, amorphous powder; is converted into glycolic acid by boiling with water or alkalies; by heating with ammonia, into glycolamide $C^2H^5NO^2 = CH^2.OH.CO.NH^2$ (isomeric with glycocol).—Colorless crystals, fusing at 120°.

2. *Oxypropionic Acids.*

$$C^3H^6O^3 = C^2H^4 \begin{cases} OH \\ CO.OH. \end{cases}$$

Both of the acids, possible according to the theory, are known.

1. Lactic acid (Ethylidenelactic acid) = $CH^3.CH \begin{cases} OH \\ CO.OH. \end{cases}$ Is produced by the souring of milk by fermentation of the sugar of milk contained in it. In the same way it is formed from cane-sugar, grape-sugar, gum, starch, when these are left for some time in contact with water and old cheese or similar protein substances at a temperature of 20–50° (lactic fermentation). It is hence contained in large quantity in acidified vegetable juices (for instance, in beet juice, in saurkraut), and its presence has also been proven in animal liquids, particularly in the gastric juice.—It is produced from α-chlor- or α-brompropionic acid, and from alanin in the same manner as the homologous glycolic acid is prepared from chloracetic acid and glycocol; further, by the action of hydrochloric acid on aldehyde hydrocyanate (see p. 104), and of nascent hydrogen on pyroracemic acid.

Most practically prepared in the following manner: 3 kilogrammes cane-sugar and 15 gr. tartaric acid are dissolved in 17 litres boiling water and allowed to stand several days; 100 grms. old cheese, suspended in 4000 grms. sour milk, and 1200 grms. zinc white are then added, and the temperature retained as nearly as possible at 40–45° during the period of fermentation. In eight to ten days the fermentation is ended. The whole mass is now heated to boiling, filtered, evaporated, and allowed to crystallize. The separated* zinc lactate is crystallized again from hot water, then

* The fermentation is prevented by any large amount of free acid, and hence ceases as soon as this is formed, long before all the sugar is decomposed. This can, however, be avoided by neutralizing the acid, from time to time, by means of a base, or by adding a base at the commencement.

dissolved in boiling waler and decomposed with sulphuretted hydrogen. The liquid filtered from zinc sulphide is evaporated on a water-bath. The acid thus obtained still contains mannite, as an impurity. It is separated from this by dissolving the residue in a little water and agitating with ether in which mannite is insoluble, and, after the separation of the two liquids, evaporating the ethereal solution.

Colorless, syrupy liquid, of 1.215 specific gravity; mixes with water, alcohol, and ether in all proportions. Not volatile without decomposition. Is decomposed by distillation into water, aldehyde, carbonic oxide, and lactide. When heated with dilute sulphuric acid to 130°, it is decomposed into aldehyde and formic acid. It is reduced by means of hydriodic acid, most readily by distillation with phosphorus iodide and a little water, to propionic acid. Heated with hydrobromic acid it is transformed into brompropionic acid. By oxidation with chromic acid, acetic and formic acids are formed.

The *lactates of the alkalies* do not crystallize.

Calcium lactate, $(C^3H^5O^3)^2Ca + 5H^2O$. White needles in verrucose combinations. Very easily soluble in hot water and alcohol, more difficultly in cold water (9½ parts).

Zinc lactate, $(C^3H^5O^3)^2Zn + 3H^2O$. Lustrous needles, or small crystals, in crusty formations, soluble in 6 parts hot and 58 parts cold water. Insoluble in alcohol.

Iron lactate, $(C^3H^5O^3)^2Fe + 3H^2O$. Can be prepared, like the zinc salt, directly from milk whey and iron filings; crystallizes in fine prisms, united together, forming an almost colorless crust; is difficultly soluble, and in solution undergoes a change in the air.

Ethyl lactate, $C^2H^4 \left\{ \begin{matrix} OH \\ CO.O.C^2H^5 \end{matrix} \right.$, results from heating lactic acid with alcohol to 170°.—Neutral liquid, boiling at 156°, which, in contact with water, is rapidly decomposed into lactic acid and alcohol.

Potassium and sodium are dissolved in it with evolution of hydrogen, and by the action of ethyl iodide on the resulting compounds are formed

Ethyl ethyllactate (lactic-diethylether), $C^2H^4\left\{\begin{array}{l}O.C^2H^5\\CO.O.C^2H^5.\end{array}\right.$ This is also produced by the decomposition of chlorpropionic ether with sodium ethylate.—Colorless liquid, insoluble in water, boiling at 126.5°. Treated with caustic potassa, only one atom of ethyl is replaced by potassium, and there is formed a potassium salt of

Ethyl lactic acid, $C^2H^4\left\{\begin{array}{l}O.C^2H^5\\CO.OH.\end{array}\right.$ This is a strong acid and isomeric with ethyl lactate.

When lactic acid is heated for a long time at 140–145°, it is converted into *dilactic acid* $C^6H^{10}O^5$, a yellow, amorphous substance, which, when boiled with alkalies and acids, is reconverted into lactic acid.

Lactide (lactic anhydride), $C^3H^4O^2 = CH^3.CH\left\{\begin{array}{l}O\\CO\end{array}\right>.$ The distillate from lactic acid is evaporated at 100°; the residue washed with cold absolute alcohol and crystallized from hot alcohol.—Rhombic plates, fusing at 107°, but slightly soluble in water, slowly uniting with it to form lactic acid.

Trichlorlactic acid, $CCl^3.CH\left\{\begin{array}{l}OH\\CO.OH.\end{array}\right.$ When hydrocyanic acid is allowed to act upon chloral, the crystallizing compound $CCl^3.CH\left\{\begin{array}{l}OH\\CN\end{array}\right.$ is obtained, which yields the acid when digested with moderately concentrated hydrochloric acid.—Crystalline mass, consisting of small prisms; fusing point, 105–110°; yields crystallizing salts.

Lactyl chloride, $C^3H^4Cl^2O = CH^3.CHCl.CO.Cl$, is formed by the distillation of zinc lactate with double its weight of phosphorus chloride; phosphorus oxichloride is formed at the same time, and, for the separation of the two products, no means have been devised up to the present. Not distillable without partial decomposition. Is decomposed by water, yielding hydrochloric acid and α-chlorpropionic acid; by alcohol, yielding hydrochloric acid and ethyl α-chlorpropionate; when heated with alkalies, it yields lactic acid; and in contact with zinc and water, propionic acid.

Lactamide, $C^2H^4 \left\{ \begin{array}{l} OH \\ CO.NH^2, \end{array} \right.$ is formed by heating alanine in a current of hydrochloric acid gas at 180–200°.—Colorless, transparent needles or laminæ; fusing point, 275; easily soluble in water and alcohol.

2. Sarcolactic acid (Paralactic acid, ethylene-lactic acid), $\begin{array}{l} CH^2.OH \\ \dot{C}H^2.CO.OH, \end{array}$ is contained in the juice of flesh and in animal secretions, at times also in urine, probably together with ordinary lactic acid. It is produced by boiling ethylene cyanhydrine (p. 137) with alkalies, and, together with some acrylic acid, by boiling β-iodopropionic acid with milk of lime.—To prepare it, baryta water is added to an aqueous extract of chopped meat, the whole then boiled, filtered, and evaporated. Sulphuric acid is added to the syrupy residue, and the lactic acid extracted by means of ether.—The free acid is very similar to lactic acid of fermentation, but the corresponding salts of the two acids present differences in the degree of their solubility and in the amount of water of crystallization contained in them. The *calcium salt* is less soluble in water than that of ordinary lactic acid, and crystallizes with 4 molecules of water. The *zinc salt* contains only two molecules of water, is much more easily soluble in water (in five to six parts of cold water), and also easily soluble in alcohol.—Oxidized by means of chromic acid, sarcolactic acid is converted into malonic acid. Heated up to 130–140°, and the residue dissolved in water, it is converted into ordinary lactic acid.

3. *Oxybutyric Acids.*

$$C^4H^8O^3 = C^3H^6 \begin{cases} OH \\ CO.OH. \end{cases}$$

1. α-Oxybutyric acid. From monobrombutyric acid by boiling with barium hydroxide.—Colorless, stellate needles or flat prisms; fusing point, 43–44°. When carefully heated it can be sublimed. Deliquescent in the air.

2. β-Oxybutyric acid, $CH^3.CH(OH).CH^2.CO.OH$. Is produced by the action of hydrogen (sodium-amalgam and water) on ethyl acetylacetate (p. 82) and by boiling propylene cyanhydrine with caustic potassa.—Colorless, syrupy, very deliquescent liquid.

3. Oxyisobutyric acid, $\left.\begin{matrix} CH^3 \\ CH^3 \end{matrix}\right\} C(OH).CO.OH$. Is produced by boiling bromisobutyric acid with barium hydroxide; by the action of cyanhydric and hydrochloric acids on acetone (*acetonic acid*); by the action of dilute nitric acid on amylene alcohol (*butyllactinic acid*); by heating methyl oxalate with methyl iodide and zinc, and then treating the product with water (*dimethoxalic acid*).—Colorless prisms, easily soluble in water. Fusing point, 79°. Sublimes in long needles even at 50°, when carefully heated. When carefully oxidized with potassium bichromate and dilute sulphuric acid, it yields acetone, together with carbonic and acetic acids.

4. *Oxyvaleric Acids.*

$$C^5H^{10}O^3 = C^4H^8 \begin{cases} OH \\ CO.OH. \end{cases}$$

1. Oxyvaleric acid. From bromvaleric acid in a hot aqueous solution by treatment with silver oxide. Its ether is formed when ethyl oxalate is heated with isopropyl iodide and zinc, and the product treated with water.—Large, colorless, very easily soluble plates. Fusing point, 80°. Sublimes even below 100°.

2. Isoxyvaleric acid (Ethomethoxalic acid). The ether, boiling at 165°, is produced by heating ethyl oxalate with a mixture of methyl iodide and ethyl iodide and zinc, and afterwards treating the product with water. The free acid, separated from the ether, forms colorless, easily soluble crystals, fusing at 63°.

5. *Oxycaproic Acids.*

$$C^6H^{12}O^3 = C^5H^{10}\begin{cases} OH \\ CO.OH. \end{cases}$$

1. Leucic Acid. Is produced by the action of nitrous acid on leucine (p. 98).—Colorless, easily soluble needles; fusing point, 73°.

2. Isoleucic acid. (Diethoxalic acid). Is obtained, like isoxyvaleric acid, by heating ethyl oxalate with ethyl iodide and zinc.—Colorless, easily soluble crystals; fusing point, 74.5°; sublimes at 50°. When carefully oxidized it gives propione (p. 110); also yields propione, together with ethylcrotonic acid (p. 125), when heated with concentrated hydrochloric acid.

C. Bibasic, Diatomic Acids, $C^nH^{2n-2}O^4$.

The acids of this series are derived from the hydrocarbons of the marsh gas series by the replacement of two hydrogen atoms in the latter by two carboxyl groups; or from the fatty acids by the replacement of one hydrogen atom by the carboxyl group CO.OH. They are produced by the complete oxidation of the primary diatomic alcohols containing twice the group $CH^2.OH$; by heating the dicyan-substitution-products of the marsh gas hydrocarbons (cyanides of the hydrocarbons C^nH^{2n}) and the monocyan-substitution-products of the fatty acids with caustic potassa.

1. *Oxalic Acid.*

$$C^2H^2O^4 = \begin{matrix} CO.OH \\ | \\ CO.OH. \end{matrix}$$

Occurrence. Very widely distributed in nature; in the form of the acid potassium salt in the different varieties of *Oxalis;* in the form of the calcium salt in a number of plants; in urine (some of the urinary calculi consist entirely of this salt); in the form of the ammonium salt in guano.

Formation. By the action of finely divided sodium on dry carbonic anhydride at 350–360°; by heating sodium formate; by the decomposition of cyanogen with water; by the heating of cellulose (paper, linen) with potassium hydroxide; the most important method of formation is, however, by the oxidation of a great many organic substances with nitric acid, hypermanganic acid, etc.

Preparation. The expressed juice of oxalis plants is precipitated by means of a solution of sugar of lead, the precipitate decomposed with sulphuric acid or sulphuretted hydrogen, and the filtrate evaporated to crystallization.—Or 1 part sugar or starch is heated with 8 parts of nitric acid (specific gravity, 1.38) until action has ceased, and the solution then evaporated to dryness.—On the large scale it is also produced by heating sawdust with caustic potassa or soda.

Properties. Colorless prisms, soluble in 15 parts water, more easily soluble in alcohol. It contains 2 molecules water of crystallization, which are given off at 100°. When carefully heated up to 150°, the effloresced crystals can be completely sublimed; when rapidly heated, it is partially resolved into carbonic anhydride, carbonic oxide, formic acid, and water. Oxidizing agents transform it into carbonic anhydride and water. Sulphuric acid resolves it into water, carbonic anhydride, and carbonic oxide. Nascent hydrogen (zinc and hydrochloric acid) converts it into glycolic acid (p. 145) and acetic acid.

Strong, bibasic acid. Its salts, with the exception of

those of the alkalies, are very difficultly soluble in water, but soluble in mineral acids.

Potassium oxalate. The *neutral salt* $C^2O^4K^2 + H^2O$ forms easily soluble crystals, which effloresce at an elevated temperature. The *acid salt* C^2O^4HK is difficultly soluble in cold water. A still more acid salt $C^2O^4HK + C^2H^2O^4 + 2H^2O$ is the salt-of-sorrel of commerce.

Ammonium oxalate. The *neutral salt* $C^2O^4(NH^4)^2 + H^2O$, long prismatic crystals, easily soluble in cold water, is decomposed at a high temperature, forming oxamide, carbonic anhydride, carbonic oxide, ammonia, and hydrocyanic acid. The *acid salt* $C^2O^4H(NH^4) + H^2O$, prisms, more difficultly soluble than the neutral salt; when heated, yields oxamic acid.

Calcium oxalate, $C^2O^4Ca + H^2O$. A crystalline powder, insoluble in water. When allowed to crystallize slowly, it combines with three molecules of water of crystallization. It can only be obtained in an anhydrous state by heating it above 200°, and it then reabsorbs one molecule very rapidly, when exposed to the air.

Lead oxalate, C^2O^4Pb. White precipitate, insoluble in water.—*Silver oxalate* $C^2O^4Ag^2$. White powder, insoluble in water. Detonates when heated.

Methyl oxalate, $C^2O^4(CH^3)^2$, is produced by the distillation of acid potassium oxalate (2 parts) with a mixture of methyl alcohol (1 part) and concentrated sulphuric acid (1 part).—Colorless, rhomboidal plates, of a weak odor, fusing point, 51°; boiling point, 162°; soluble in water and alcohol; is decomposed, however, by water, particularly rapidly with the aid of heat, yielding oxalic acid and methyl alcohol. With aqueous ammonia, it yields oxamide and methyl alcohol; with dry ammonia, methyl oxamate. The ethers of oxyisobutyric acid (p. 151) and isoleucic acid (p. 152) are formed by the action of zinc on a mixture of this ether with methyl or ethyl iodide, and subsequent addition of water.—The acid methyl ether, *methyloxalic acid* $C^2O^4.H.CH^3$, is contained in the mother-liquor from the neutral ether. When in a free state, it decomposes easily.

Ethyl oxalate (Oxalic ether), $C^2O^4(C^2H^5)^2$, is formed like methyl oxalate; is prepared most readily in the following manner: A mixture of 3 parts oxalic acid, dehydrated at 100°, and 2 parts absolute alcohol, in a tubulated retort, is heated slowly in an oil-bath until the thermometer shows 125–130°; in the mean time the vapor of 2 parts absolute alcohol is conducted upon the bottom of the retort in an uninterrupted current. The product is then distilled, and that portion which boils at 182–186° collected separately.—Colorless liquid with a slight odor; specific gravity, 1.0824; boiling point, 186°; does not mix with water. Conducts itself towards water and ammonia, and zinc and the alcoholic iodides, like the methyl ether. Its solution in absolute alcohol gives a crystalline precipitate with an alcoholic solution of potassa. This is the potassium salt of ethyloxalic acid $C^2O^4.H.C^2H^5$, which, in a free state, is readily decomposed.

Ethyloxy-oxalylchloride, $C^4H^5O^3Cl = C^2O^2{<}^{O.C^2H^5}_{Cl.}$

Is formed by the action of phosphorus oxichloride on potassium ethyloxalate.—Colorless, clear, mobile liquid, of a suffocating odor; boils at 140°; specific gravity at 16°, 1.216; fumes in contact with air, and is converted into oxalic acid.

Oxamide, $C^2H^4N^2O^2 = \begin{matrix} CO.NH^2 \\ \dot{C}O.NH^2, \end{matrix}$ is formed by the decomposition of oxalic ether with ammonia; by conducting cyanogen into strong hydrochloric acid; further from cyanogen and water in the presence of a very small quantity of aldehyde; and, in small quantity, by mixing hydrocyanic acid with manganese superoxide and a little sulphuric acid.—White powder or long entangled needles; inodorous and tasteless; insoluble in cold water, slightly in hot water; partially sublimable without decomposition. Heated in closed vessels up to 200° with water, it is converted into neutral ammonium oxalate.

Diethyloxamide, $\begin{matrix} CO.NH.C^2H^5 \\ \dot{C}O.NH.C^2H^5. \end{matrix}$ Is produced by bringing together oxalic ether and ethylamine.—Colorless crystals, difficultly soluble in cold water, easily soluble in hot water. When distilled with caustic potassa, it yields potassium oxalate and pure ethylamine. Well adapted for the preparation of ethylamine in a pure condition and for separating it from di- and triethylamine.*

Oxamic acid, $C^2H^3NO^3 = \begin{matrix} CO.NH^2 \\ \dot{C}O.OH, \end{matrix}$ is produced by heating acid ammonium oxalate; and by continued boiling of oxamide with strong aqueous ammonia. It is separated from the ammonium salt, obtained in this way, by means of hydrochloric acid.—White, crystalline powder, difficultly soluble in cold water; is rapidly transformed into acid ammonium oxalate by boiling with water. Monobasic acid, gives crystallizing salts. Its ethers result from the action of dry or alcoholic ammonia on the ethers of oxalic acid. The *ethyl ether* (*oxamethan*) $\begin{matrix} CO.NH^2 \\ \dot{C}O.O.C^2H^5 \end{matrix}$ forms large, colorless, laminous crystals.

Glyoxal (Oxalic aldehyde), $C^2H^2O^2 = \begin{matrix} CHO \\ \dot{C}HO. \end{matrix}$ Is produced as a secondary product in the preparation of glycolic acid from alcohol (p. 145).—Solid, amorphous

* Diethylamine, when brought together with oxalic ether, yields *ethyl diethyloxamate* $\begin{matrix} CO.N(C^2H^5)^2 \\ \dot{C}O.O.C^2H^5, \end{matrix}$ a liquid, boiling at 250–254°. When this is distilled with caustic potassa, it yields pure diethylamine. Oxalic ether does not act upon triethylamine.

mass, easily soluble in water, alcohol, and ether. Combines with the bisulphites of the alkalies; with ammonia, water being eliminated, forming bases free of oxygen; *glycosine* $C^6H^6N^4$ and *glyoxaline* $C^3H^4N^2$. Very dilute nitric acid oxidizes it, forming glyoxylic acid; concentrated nitric acid converts it into oxalic acid.

Glycolacetal, $\begin{matrix} CH^2.OH \\ \dot{C}H.(O.C^2H^5)^2. \end{matrix}$ Is produced by the action of alcoholic potassa on bromacetal (p. 105).—Colorless liquid of a pleasant odor, boiling without decomposition at 167°.

Glyoxalacetal, $\begin{matrix} CH.(O.C^2H^5)^2 \\ \dot{C}H.(O.C^2H^5)^2. \end{matrix}$ Is obtained by allowing sodium alcoholate to act upon dichloracetal.—Colorless liquid, boiling at 180°.

Glyoxylic acid, $C^2H^2O^3 = \begin{matrix} CHO \\ \dot{C}O.OH. \end{matrix}$ Is produced by heating dibromacetic acid with water and silver oxide; by heating ethyl dichloracetate with water to 120°; and, together with glycolic acid, by slow oxidation of ethyl alcohol with nitric acid.—Tenacious, pale-yellow syrup; easily soluble in water; volatile with water vapor without undergoing decomposition. Unites with nascent hydrogen, forming glycolic acid. Its salts, with the alkalies, combine with the bisulphites of the alkalies, forming crystallizing compounds.

The *calcium salt* $(C^2HO^3)^2Ca + 2H^2O$ crystallizes in hard prisms, which are difficultly soluble. In its solution, lime-water gives a white precipitate, soluble in acetic acid. By boiling with lime-water, it is decomposed, yielding calcium glycolate and oxalate.

2. *Malonic Acid.*

$$C^3H^4O^4 = CH^2 \begin{cases} CO.OH \\ CO.OH. \end{cases}$$

Formation. By careful oxidation of malic acid with potassium bichromate; by heating cyanacetic acid (p.

14

84) with alkalies; by decomposing barbituric acid (see Uric Acid) with caustic potassa; and by oxidation of sarcolactic acid, of propylene and allylene.

Properties. Large, lamellar crystals, easily soluble in water and alcohol; fuses at 132° and breaks up at a higher temperature into carbonic anhydride and acetic acid. The *barium salt* $C^3H^2O^4Ba+H^2O$ forms silky tufts; the *calcium salt* $(C^3H^2O^4Ca)^2+3\frac{1}{2}H^2O$, small transparent needles. Both salts are difficultly soluble in cold water.

Nitroso-malonic acid, $C^3H^3(NO)O^4$, is formed by heating potassium violurate (see Uric Acid) with caustic potassa.—Shining, prismatic needles, very easily soluble in water. Fuses, when heated, and then decomposes with a sharp report.

Amido-malonic acid, $C^3H^3(NH^2)O^4$, is formed by the action of nascent hydrogen (sodium-amalgam and water) on nitroso-malonic acid.—Large, shining prisms, easily soluble in water. Is decomposed, by heating alone or by warming its aqueous solution, into glycocine and carbonic anhydride.

Mesoxalic acid, $C^3H^2O^5=CO\left\{\begin{matrix}CO.OH\\CO.OH.\end{matrix}\right.$ The barium salt is produced by boiling barium alloxanate (see Uric Acid) for five or ten minutes with a great deal of water (to 15 grms. of the salt 1 litre water). From this is obtained the free acid by heating with the required amount of sulphuric acid, at 40–50°, and evaporating the filtrate. It is further formed by adding iodine to a solution of amido-malonic acid, which contains potassium iodide.—Prismatic crystals, very deliquescent, also easily soluble in alcohol. The acid dried at 100° still contains a molecule of water of crystallization. It fuses at 150° without giving off this water, and decomposes at a somewhat higher temperature.

Barium mesoxalate, $C^3O^5Ba + 1\frac{1}{2}H^2O$. Colorless, microscopic crystals, almost insoluble in water, soluble with difficulty in hot water.—*Lead mesoxalate* C^3O^5Pb+

PbO,H^2O. White insoluble precipitate. *Silver mesoxalate* $C^3O^5Ag^2+H^2O$. Colorless, amorphous precipitate, which is rapidly transformed into yellowish colored crystals. Becomes discolored quickly in the air, and is decomposed by boiling with water into carbonic anhydride, silver oxalate, metallic silver, and free mesoxalic acid.

3. *Succinic Acids.*

$$C^4H^6O^4=C^2H^4\begin{cases}CO.OH\\CO.OH.\end{cases}$$

There are two differently constituted acids of this composition possible. Both are known.

1. Succinic acid (Ethylene succinic acid), $\begin{matrix}CH^2.CO.OH\\ \vert \\ CH^2.CO.OH.\end{matrix}$ Is contained in amber, in some bituminous coals, in a number of plants, in the animal organism, and also in urine. Is produced by continued action of nitric acid on fats, and a great many other bodies; by fermentation of crude calcium malate; and, in small quantity, by the fermentation of sugar; from malic and tartaric acids by means of hydriodic acid; from ethylene cyanide and β-cyanpropionic acid by heating with alkalies.

It is prepared by the dry distillation of amber. The acid that passes over is purified by pressing and recrystallization.—Most readily by fermentation of crude calcium malate with water and old cheese at 30–40°. The calcium succinate, formed in the course of several days, is decomposed by means of sulphuric acid, filtered, evaporated, and the acid purified by means of crystallization.—Clear crystals, fusing at 180° and boiling at 235°, being decomposed at this temperature into water and succinic anhydride. Soluble in 23 parts of cold water, more easily in hot, less in alcohol. In the presence of a salt of uranium in sunlight it breaks up, in aqueous solution, into carbonic anhydride and propionic acid.

The *salts of the alkalies* are easily soluble in water; the *calcium salt* $C^4H^4O^4Ca$ is difficultly soluble, separates from a cold solution with 3 molecules, from a hot solution with 1 molecule, of water of crystallization. The *ferric salt* is a brown precipitate, perfectly insoluble in water. The *silver salt* is an amorphous white precipitate.

Ethyl succinate, $C^4H^4O^4(C^2H^5)^2$, is obtained by conducting hydrochloric acid gas in a hot alcoholic solution of succinic acid.—Colorless oil, insoluble in water; of specific gravity, 1.037; boiling point, 217°.

Monobromsuccinic acid, $C^4H^5BrO^4$, and **Dibromsuccinic acid,** $C^4H^4Br^2O^4$, result, when succinic acid is heated with bromine and water in sealed tubes up to 150–180°. The former is produced in the presence of a great deal of water, the latter when but little water (equal parts water and succinic acid) is employed. Both are crystalline acids; monobromsuccinic acid is much more easily soluble in water than the dibrominated acid.—A dibromsuccinic acid with somewhat different properties (*isodibromsuccinic acid*) is formed from maleïc acid by combination with bromine.

Amidosuccinic acid (Aspartic acid), $C^4H^7NO^4 = C^2H^3.NH^2.(CO.OH)^2$. Is contained in beet-molasses; is produced by boiling albuminous bodies with dilute sulphuric acid; and by boiling asparagine with water, acids, or alkalies.—Small, rhombic crystals, very difficultly soluble in cold water, more easily soluble in hot water. Combines, like glycocol, with bases and with acids, forming crystallizing salts.

Asparagine (Amidosuccinamide), $C^4H^8N^2O^3 + H^2O = C^2H^3(NH^2)\begin{cases} CO.NH^2 \\ CO.OH. \end{cases}$ Occurs in a great many plants, in asparagus, in liquorice root, in marshmallow root, in the root of *scorzonera hispanica*, in beets, in grain sprouts, in the plants of the pea, vetch, and bean before the time of blossoming.—It crystallizes from the expressed juice of these plants by evaporation.—Colorless, large crystals, difficultly soluble in alcohol. Not volatile without decomposition. It unites with acids and bases, and is decomposed when heated with them; also decomposed slowly when boiled with water, yielding amidosuccinic acid and ammonia.

Sulphosuccinic acid, $C^2H^3\left\{\begin{matrix}(CO.OH)^2\\SO^2.OH.\end{matrix}\right.$ Is formed by the action of sulphuric anhydride on succinic acid and treatment of the product with water. Its salts with the alkalies are formed by direct union of fumaric and maleïc acids with the bisulphites of the alkalies, when concentrated aqueous solutions of both are mixed, and then boiled for several hours.—Indistinct, deliquescent crystals. Tribasic acid. When fused with potassa it yields potassium fumarate.

Succinic anhydride, $C^4H^4O^3$, is produced by the distillation of succinic acid either alone or with phosphorus chloride (equal molecules).—White crystalline mass. When it is heated with another molecule of phosphorus chloride, there passes over

Succinyl chloride, $C^4H^4O^2Cl^2$. An oil which solidifies at 0°, forming tabular crystals.

Succinamide, $C^4H^4O^2(NH^2)^2$, is formed, like oxamide, from the ethers.—Fine, white needles, difficultly soluble in cold water, easily soluble in hot water, insoluble in alcohol. Carefully heated to 200° it is resolved into ammonia and

Succinimide, $C^4H^4O^2.NH$ (isomeric with cyanpropionic acid). This is also formed by heating the anhydride in ammonia gas and by distilling ammonium succinate.—Crystallizes with one molecule of water in rhombic plates; is easily soluble in water and alcohol; sublimable; with barium hydroxide, it gives the barium salt of *succinamic acid* $C^4H^7NO^3$.

2. Isosuccinic acid, $CH^3.CH\left\{\begin{array}{l}CO.OH\\CO.OH,\end{array}\right.$ is formed by heating cyanpropionic acid with alkalies. The ethyl ether is formed by boiling formic and lactic ethers with phosphoric anhydride.—Crystals, fusing at 130°; sublimes below 100°; but, when heated to 150°, is resolved into propionic acid and carbonic anhydride. It is more easily soluble in water (in 5 parts), than succinic acid. The *sodium salt* gives no precipitate with iron chloride.

4. *Pyrotartaric Acid.*

$$C^5H^8O^4 = C^3H^6\left\{\begin{array}{l}CO.OH\\CO.OH\end{array}\right. = \begin{array}{l}CH^3.CH.CO.OH\\\quad\;\; \dot{C}H^2.CO.OH.\end{array}$$

Formation. By dry distillation of tartaric acid (best when the acid is previously mixed with an equal weight of pumice-stone) and evaporation of the distillate to crystallization; or by heating with concentrated hydrochloric acid at 180°; from propylene cyanide by heating with acids or potassa; from itaconic, citraconic, and mesaconic acids by the action of nascent hydrogen; from gamboge by fusing with potassa.

Properties. Small, transparent crystals, which fuse at 112° and decompose partially at a higher temperature into water and the anhydride. Easily soluble in water, alcohol, and ether. Decomposes in an aqueous solution in the presence of a salt of uranium in direct sunlight into carbonic anhydride and butyric acid.

Its salts are crystallizable and almost all soluble in water. The *acid potassium salt* $C^5H^6O^4.HK$ and the *neutral calcium salt* $C^5H^6O^4Ca + 2H^2O$ are difficultly soluble; the silver salt $C^5H^6O^4Ag^2$ is an insoluble, white, curdy precipitate.

The substitution-products of pyrotartaric acid are formed by direct addition of chlorine, bromine, or hydrogen acids to itaconic acid and the isomeric citraconic, and mesaconic acids, The substitution-products obtained in this way from these acids are somewhat different from each other; they have been designated, according to the acid from which they are produced, by means of the prefixes ita, citra, and mesa.

Ita-, Citra-, and Mesamonochlorpyrotartaric acids, $C^5H^7ClO^4$, are produced by heating the three acids with strong hydrochloric acid. They all cyrstallize. The ita-acid melts at 140–145°. When heated with water or bases, it is transformed into itamalic acid. The citra-acid is very unstable, breaks up, in a dry condition, as well as in solution, by gentle heating, into hydrochloric acid and mesaconic acid; by heating with bases, into hydrochloric acid, carbonic anhydride, and crotonic acid (p. 123). The mesa-acid is more stable; fuses without decomposition at 129–130°; yields, however, the same products as the citra-acid, by boiling with water.

Ita-, Citra-, and Mesadibrompyrotartaric Acids, $C^5H^6Br^2O^4$, are crystalline, of different solubility in water; the citra-acid most easily soluble, the ita-acid most difficultly. When a solution of its sodium salt is heated, the ita-acid yields easily soluble, crystalline *aconic acid* $C^5H^4O^4$; the citra- and mesa-acids, on the other hand, yield bromcrotonic acid (p. 123) under like circumstances.

When treated with sodium-amalgam, all the substitution-products yield the same pyrotartaric acid.

Amidopyrotartaric acid (Glutamic acid), $C^5H^9NO^4 = C^3H^5.NH^2(CO.OH)^2$. Is contained, together with amidosuccinic acid, in beet-molasses; and is formed, together with the same acid, by boiling albuminous substances with dilute sulphuric acid.—Colorless, rhombic octahedral crystals; fusing point, 135–140°; difficultly soluble in cold water; almost insoluble in alcohol.

The succeeding acids of this series are formed simultaneously by the action of nitric acid on fats and fatty acids, and are best separated by means of partial crystallization from ethereal solutions; they are produced further by the action of fuming nitric acid on palmitolic acid and the acids homologous to it. In the latter case an acid is always formed, containing half as many carbon atoms as that from which it results; from palmitolic acid is obtained suberic acid; from stearolic acid, azelaic acid; and from behenolic, brassylic acid.

5. *Adipic acid*, $C^6H^{10}O^4$. Is produced from β-iodopropionic acid by heating with finely divided silver to 120-130°; and by the action of nascent hydrogen on muconic acid (see Mucic Acid); is prepared most readily by boiling sebacic acid with nitric acid, and separating it from succinic acid, which is formed at the same time by recrystallizing from alcohol.—Laminæ with a vitreous lustre, or flattened prisms; easily soluble in hot water, alcohol, and ether; fusing point, 148°.

A syrupy acid isomeric with this is obtained in the same way from α-brompropionic acid.

6. *Suberic acid*, $C^8H^{14}O^4$, is also formed, when cork is treated for a long time with nitric acid.—Needles or plates, often an inch in length. But slightly soluble in cold water and ether; easily soluble in alcohol.—Fusing point, 140°. When heated with barium hydroxide, hexyl hydride is formed.

Suberic aldehyde, $C^8H^{14}O^3$, is formed, together with suberic acid, by the action of fuming nitric acid on palmitolic acid.—Thick oil, boiling at 202°, at the same time undergoing decomposition.

7. *Azelaic acid* (Lepargylic acid), $C^9H^{16}O^4$. Large laminæ or flattened needles. Very difficultly soluble in cold water, easily soluble in hot water, ether, and alcohol. Fusing point, 106°; gives heptyl hydride, when heated with barium hydroxide.

8. *Sebacic acid*, $C^{10}H^{18}O^4$, is best obtained by boiling spermaceti or stearic acid with an equal weight of

nitric acid of specific gravity, 1.2.—Shining, lamellar crystals; more easily soluble in ether than suberic acid, more difficultly than azelaic acid; fusing point, 127–128°.

9. *Brassylic acid*, $C^{11}H^{20}O^4$. As yet it has only been obtained from behenolic acid together with the oily aldehyde $C^{11}H^{20}O^3$.—Colorless scales, difficultly soluble in water, even at the boiling temperature; fusing point, 108.5°.

10. *Roccellic acid*, $C^{17}H^{32}O^4$. As yet, it has not been prepared artificially. Occurs in *Roccella tinctoria*, and can be extracted from it by ammonia or ether.—Colorless prisms, insoluble in water, soluble in alcohol and ether; fusing point, 132°.

D. Bibasic, Diatomic Acids, $C^nH^{2n-4}O^4$.

The acids of this series are derived from the hydrocarbons of the ethylene series in the same way as the acids of the oxalic acid series are derived from the hydrocarbons of the marsh gas series.

1. *Fumaric Acid.*

$$C^4H^4O^4 = C^2H^2 \begin{cases} CO.OH \\ CO.OH. \end{cases}$$

Occurrence. In a great many plants; in *Fumaria officinalis*, *Corydalis bulbosa*, *Glaucium luteum*, in Iceland moss and several fungi.

Formation and preparation. When chlorous acid or potassium chlorate and dilute sulphuric acid are allowed to act on benzol, there is produced, together with other bodies, an acid, *trichlorphenomalic acid* $C^6H^7Cl^3O^5$, which crystallizes well and fuses at 131–132°. When this is boiled with baryta water, barium fumarate is formed, from which by exact precipitation with sulphuric acid the free acid is obtained. From dibromsuccinic acid and isodibromsuccinic acid by heating with potassium iodide; from sulphosuccinic acid by fusing with potassa; most readily from malic acid, which when heated to 150° is resolved almost completely into fumaric acid and water.

Properties. Colorless prisms, difficultly soluble in cold water, more easily in hot water. Fuses and volatilizes above 200°, partially without decomposition, for the greater part, however, being resolved into water and maleïc anhydride.—It is converted into succinic acid by nascent hydrogen, and when heated with hydriodic acid; combines directly with bromine, forming dibromsuccinic acid.

Barium fumarate, $C^4H^2O^4Ba$, and **Calcium fumarate,** $C^4H^2O^4Ca$, are difficultly soluble crystalline precipitates.—*Silver fumarate,* $C^4H^2O^4Ag^2$. White amorphous, insoluble precipitate. Detonates when heated.

Ethyl fumarate, $C^4H^2O^4.(C^2H^5)^2$, is produced by distilling ethyl malate either alone or with phosphorus chloride.—Colorless liquid, boiling at 225°.

Maleic acid, $C^4H^4O^4 = C^2H^2(CO.OH)^2$, (isomeric with fumaric acid). The anhydride of this acid ($C^4H^2O^3$, fusing point, 57°; boiling point, 196°), is produced by heating fumaric acid and by distilling malic acid rapidly. This is readily converted into the acid by assimilation of water.—Colorless prisms, very easily soluble in water and alcohol. Fusing point, 130°. At 160° it is resolved into its anhydride and water. When boiled with dilute sulphuric acid or treated with hydrobromic acid or hydriodic acid, it is converted into fumaric acid. It combines with hydrogen, forming succinic acid; with bromine forming isodibromsuccinic acid.

Chlormaleic acid, $C^2HCl(CO.OH)^2$. When 1 part tartaric acid is heated with 6 parts phosphorus chloride, there is formed chlormaleinchloride $C^2HCl(CO.Cl)^2$, an oily substance. This is decomposed by water, forming hydrochloric acid and chlormaleïc acid.—Colorless, small needles, soluble in water and alcohol. Fusing point, 171–172°. Its ether $C^2HCl(CO.O.C^2H^5)^2$, a thick liquid, boiling at 250–260°, is formed by the action of phosphorus chloride on ethyl tartrate.

Brommaleic and Isobrommaleic acids, $C^2HBr(CO.OH)^2$. The acid barium salts of these acids are formed by boiling barium dibrom- and isodibromsuccinates with water.—Both acids crystallize in easily soluble prisms, and are converted into succinic acid by nascent hydrogen. Brommaleïc acid fuses at 126°, isobrommaleïc acid at 160°.

2. *Itaconic Acid.*

$$C^5H^6O^4 = C^3H^4 \begin{cases} CO.OH \\ CO.OH. \end{cases}$$

Formation. Together with citraconic anhydride, by the distillation of citric and itamalic acids; by heating citric acid to 160°. Can be most readily obtained in a pure condition by heating a concentrated solution of citraconic acid (or the distillate from citric acid) to 120–130°.

Properties. Colorless rhombic octahedrons, soluble in 15 parts of water of ordinary temperature, more easily in hot water. Is resolved, by heating, into water and citraconic anhydride. Combines with nascent hydrogen, forming pyrotartaric acid; with chlorine, bromine, and hydrogen acids, forming substitution-products of citraconic acid.

The following acids are isomeric with itaconic acid.

Citraconic acid, $C^5H^6O^4 = C^3H^4(CO.OH)^2$. When citric and itaconic acids are distilled, an oily substance, citraconic anhydride, $C^5H^4O^3$, passes over, which is converted into this acid by water, and when left in contact with moist air.—Four-sided deliquescent prisms. Fuses at 80°. Is converted into itaconic acid slowly at 100°, completely when heated in aqueous solution to 120–130°. Conducts itself towards hydrogen, chlorine, etc., like itaconic acid.

Mesaconic acid, $C^5H^6O^4 = C^3H^4(CO.OH)^2$, is produced by boiling a dilute solution of citraconic acid with nitric acid for a long time; by heating citraconic acid with concentrated hydriodic acid to 100°, and by heating citramonochlorpyrotartaric acid (p. 163) alone or with water.—Thin, lustrous prisms, which fuse at 208° and sublime at a slightly higher temperature without decomposition. Difficultly soluble in cold water. Conducts itself towards hydrogen, chlorine, etc., like itaconic acid.

Paraconic acid, $C^5H^6O^4 = C^3H^4(CO.OH)^2$, is formed, together with itamalic acid (p. 178), by the decomposition of itamonochlorpyrotartaric acid with water, and can be separated from the itamalic acid by preparing the calcium salts and treating them with alcohol. From a concentrated aqueous solution calcium itamalate is precipitated by alcohol, whereas the paraconate remains in solution. Silver paraconate is also formed by the action of itachlorpyrotartaric acid on silver carbonate.—The free acid is crystalline, easily soluble; fuses at 70°; yields, when subjected to distillation, citraconic anhydride; combines with hydrobromic acid, forming itamonobrompyrotartaric acid; and is readily converted into itamalic acid by treatment with bases.

FIFTH GROUP.

A. Triatomic Alcohols, $C^nH^{2n+2}O^3$.

These alcohols are derived from the hydrocarbons of the marsh-gas series by the replacement of three atoms of hydrogen by three hydroxyl groups. Only one of these alcohols is known with any degree of exactness.

Glycerin.

$$C^3H^8O^3 = \begin{array}{l} CH^2.OH \\ \dot{C}H.OH \\ \dot{C}H^2.OH. \end{array}$$

Occurrence. Not in a free state. In combination with acids, as compound ethers, it forms most of the fats occurring in nature, in the vegetable as well as the animal kingdom. It is formed in small quantities as a product of the fermentation of sugar.

Preparation. When a fat is decomposed (saponified) by boiling with an excess of alkali or with lime, the salts of the acids (soaps) contained in the fat are thrown down, as the salts of the alkalies are insoluble in the alkaline liquid and the calcium salts are insoluble in water; the glycerin, however, remains dissolved, and, after saturating the alkali with sulphuric acid and evaporating, can be extracted from the mass. It is most easily obtained by boiling a fat with lead oxide and water.—The lead salts (lead plaster) formed are insoluble; in the water remains glycerin with a little lead oxide, which is removed by sulphuretted hydrogen.—Treatment with superheated steam also decomposes fats. In this case there is obtained an aqueous solution of glycerin, upon which the acids float.

Properties. Colorless, syrupy liquid, of a pure sweet taste, easily soluble in water and alcohol. Under certain circumstances, as it appears, by continued shaking at a low winter-temperature, it congeals, forming a solid crystalline mass. Can be distilled in a vacuum without decomposition; in the air only with partial decomposition.—When sodium is dissolved in alcohol and glycerin added, a crystallizing substance $C^3H^5(OH)^2(ONa) + C^2H^6O$ is formed.—In an aqueous solution, in contact with yeast, it is gradually converted at 20–30° into propionic acid, with rotten cheese and chalk at 40° into alcohol and butyric acid. Heated with dehydrating substances (concentrated sulphuric or phosphoric acid, potassium bisulphite) it is converted into acrolein (p. 128).

Chlorhydrine, $C^3H^7ClO^2 = C^3H^5Cl(OH)^2$, is produced when glycerin is saturated with hydrochloric acid and heated for some time at 100°; and by direct union of allyl alcohol with hypochlorous acid.—Oil of an ethereal odor, boils at 225–230°. Is converted into propylene alcohol by treatment with sodium-amalgam.

Dichlorhydrine, $C^3H^6Cl^2O = C^3H^5Cl^2.OH$, is formed by continued heating of glycerin with from twelve to fifteen times its volume of fuming hydrochloric acid. Is most easily prepared by saturating a mixture of equal volumes of glycerin and glacial acetic acid with hydrochloric acid at 100°. On subsequently distilling, that portion, which passes over between 140–200°, is collected separately and purified by washing with water and sodium carbonate, and partial distillation.—Oil of an ethereal odor. The crude product appears to consist of two isomeric modifications, one of which boils at 174–178°, the other at 182–184°. Is converted into isopropyl alcohol by treatment with sodium-amalgam. Treated with sodium it yields allyl alcohol. Heated with potassium cyanide it is converted into *dicyanhydrine* $C^3H^5(CN)^2OH$.—Is converted into *epichlorhydrine* (monochlorhydroglycide) C^3H^5ClO by the action of aqueous alkalies: a liquid boiling at 118–119°, which combines with hydrochloric acid with evolution of heat to reform dichlorhydrine, and yields *epicyanhydrine* $C^3H^5(CN)O$ (lustrous crystals, fusing at 162°), when heated with potassium cyanide.

Trichlorhydrine, $C^3H^5Cl^3$, is formed when the preceding compounds are treated with phosphorus chloride.—Colorless liquid, boiling at 158°.—Treated with sodium or potassium hydroxide, it yields *epidichlorhydrine* (dichlorhydroglycide) $C^3H^4Cl^2$, which is also produced, together with the isomeric substance acrolein chloride, by the action of phosphorus chloride on acrolein (p. 128).—Colorless liquid, boiling at 101–102°. Combines directly with 1 molecule chlorine or bromine and with hydrogen acids.

Hydrobromic acid and phosphorus bromide act on glycerin in the same way.

Bromhydrine, $C^3H^5Br(OH)^2$, and **Dibromhydrine,** $C^3H^5Br^2.OH$, are colorless liquids. The former boils at 180°, the latter at 219°.

Tribromhydrine (Tribromallyl), $C^3H^5Br^3$. Is also formed by bringing allyl iodide or bromide together with bromine.—Colorless, lustrous prisms; fusing point, 16°; boiling point, 219–220°. When heated with potassium cyanide and alcohol, it is converted into *tricyanhydrine* (tricyanallyl) $C^3H^5(CN)^3$.

Hydriodic acid and phosphorus iodide convert glycerin into allyl iodide and pseudopropyl iodide (p. 66).

Glycerindisulphuric acid, $C^3H^5\left\{\begin{matrix}OH\\(SO^2.OH)^2\end{matrix}\right.$. The potassium salt is obtained by heating dichlorhydrine with a concentrated solution of potassium sulphite. The free acid is a deliquescent syrup. Its salts crystallize well.

Glycerintrisulphuric acid, $C^3H^5(SO^2.OH)^3$. Is obtained from trichlorhydrine, like the previous acid. Tribasic.

Glycerinsulphuric acid, $C^3H^5\left\{\begin{matrix}(OH)^2\\O.SO^2.OH.\end{matrix}\right.$ Is formed by mixing glycerin and concentrated sulphuric acid.—Very easily decomposed. Monobasic acid.

The sulphur compounds, analogous to mercaptan, *monothioglycerin* $C^3H^8O^2S$, *dithioglycerin* $C^3H^8OS^2$, and *trithioglycerin* $C^3H^8S^3$, are produced by the action of potassium sulphydrate on mono-, di-, and trichlorhydrine. They are viscid bodies, very slightly soluble in water, more easily in alcohol. They combine with metals like mercaptan.

Compound ethers. Glycerin nitrate (Nitroglycerin). $C^3H^5(O.NO^2)^3$. Glycerin is added drop by drop to a mixture of concentrated nitric and sulphuric acids immersed in a freezing mixture, and the mixture afterward poured into water.—Colorless oil, heavier than water, and insoluble in it. Exceedingly explosive and poisonous.

Fats. The fats, which occur in nature, are partially solid (tallow, lard) and partially liquid bodies (fatty oils), and a number of varieties occur at the same time in plants and animals. In a pure condition they are all colorless, inodorous, and tasteless, but, in consequence of the presence of foreign substances, they are generally more or less yellowish colored and have a taste and smell. They float on water and are insoluble in it. Only a few are soluble in alcohol. Several of the liquid fats dry in the air, at the same time absorbing oxygen; others never become dry, but only more consistent and rancid from the formation of an acid. They are not volatile. Most fats are mixtures of varying proportions of neutral glycerin ethers of stearic acid (stearin), palmitic acid (palmitin), and oleic acid (olein, elain). Upon the relative quantity of these constituents depends the consistence of the fats; they are the more liquid the more olein they contain. Human fat, beef and mutton tallow, hog's lard, cocoa butter, palm-oil, and tree-oil consist essentially of these three compounds. In other fats, however, there are generally contained, in addition to these, glycerides of other acids; in butter, for instance, those of butyric, caproic, caprylic, and capric acids. Only a few of the natural fats are ethers of other alcohols; the principal ingredient of spermaceti, for instance, is cetyl palmitate (p. 74).

The simple fats can be prepared artificially by heating the fatty acid contained in them for a long time with glycerin in closed vessels to 200°. In the same way compounds analogous to fats can be prepared. As glycerin contains 3HO, it is capable of forming three series of ethers. Only a few of these compounds will be described here.

Monoformin, $C^3H^5 \left\{ \begin{matrix} (OH)^2 \\ O.CHO. \end{matrix} \right.$ Is formed by heating glycerin with oxalic acid to 190°, and can be extracted from the mass with ether.—Colorless liquid. Can only be distilled in a vacuum without decomposition. At 200° it is resolved into carbonic acid and allyl alcohol. (See Preparation of Allyl Alcohol, p. 119.)

Acetin, $C^3H^5 \left\{ \begin{matrix} (OH)^2 \\ O.C^2H^3O, \end{matrix} \right.$ is formed by continued heating of glacial acetic acid with glycerin to 100°.—A liquid with an ethereal odor; miscible with little water.

Diacetin, $C^3H^5 \left\{ \begin{matrix} OH \\ (O.C^2H^3O)^2, \end{matrix} \right.$ is formed when the heating is carried to 200°.—Neutral liquid, boiling at 280°.

Triacetin, $C^3H^5(O.C^2H^3O)^3$, by heating diacetin with an excess of glacial acetic acid to 250°. Is contained in the oil of *Evonymus Europæus*. Liquid, boiling at 268°.

Trilaurin, $C^3H^5(O.C^{12}H^{23}O)^3$, occurs in the fruit of laurel, in butter of cocoa, and in pichurim beans; and can be obtained by boiling these with alcohol.—Colorless, small needles, which fuse at 44–46°.

Tripalmitin, $C^3H^5(O.C^{16}H^{31}O)^3$. Contained in most fats, particularly abundantly in palm-oil. From this it can be obtained pure by strong pressure, repeated washing of the residue with alcohol, and recrystallization of the portion insoluble in alcohol from ether. Artificially, it is obtained by heating glycerin with an excess of palmitic acid to 270° for several hours.—Small colorless crystals, in alcohol, even at the boiling temperature, but slightly soluble; easily soluble in ether.

Tristearin, $C^3H^5(O.C^{18}H^{35}O)^3$. It can be obtained pure from solid fats by repeatedly extracting them with cold ether, pressing the residue, and crystallizing several times from ether. Artifically, it can be prepared like palmitin.—It crystallizes in laminæ, which fuse at 66.5°. The fusing point is, however, changed when the substance is heated only a few degrees above it and then allowed to solidify.

Triolein, $C^3H^5(O.C^{18}H^{33}O)^3$. It can be prepared from liquid oils, for instance olive oil, when these are only gently heated with concentrated caustic potassa.—Colorless oil, which becomes oxidized very easily in the air.

Succinin, $C^3H^5\left\{\begin{matrix}OH\\O^2.C^4H^4O^2\end{matrix}\right.$, is formed by heating equal parts of glycerin and succinic acid at 220°.—Brown, hard mass, insoluble in water, alcohol, and ether.

B. Monobasic, Triatomic Acids, $C^nH^{2n}O^4$.

Glyceric Acid.

$$C^3H^6O^4 = CH^2(OH).CH(OH).CO.OH.$$

Formation. From glycerin by oxidation, either with nitric acid or with bromine and water; from nitroglycerin by spontaneous decomposition; also probably by heating dibrompropionic acid with silver oxide and water.

Preparation. 1 part glycerin is mixed in a glass cylinder with somewhat more than an equal bulk of water, and 1 part nitric acid of specific gravity 1.5 then introduced below it by means of a long-necked funnel; it is allowed to stand for five or six days and then dissolved in a large amount of water, neutralized with lead oxide and filtered boiling hot. The crude lead salt, obtained by evaporating and allowing to cool, is purified and then decomposed in an aqueous solution by means of suphuretted hydrogen.

Properties. Thick, uncrystallizable syrup, of a faint yellowish color; soluble in water and alcohol in all proportions. Monobasic acid.

Calcium glycerate, $(C^3H^5O^4)^2Ca + 2H^2O$. Small, white, concentrically-grouped crystals, easily soluble in water, insoluble in alcohol.

Lead glycerate, $(C^3H^5O^4)^2Pb$. Hard, crystalline crusts, difficultly soluble in cold water, more easily in hot water.

Decompositions. The pure acid heated to 140° is converted into a brownish mass, resembling gum arabic; it absorbs water with avidity and breaks up at a higher temperature into water and pyroracemic acid. Pyrotartaric acid and pyrotartaric anhydride are formed as secondary products in this reaction.—By boiling glyceric acid with potassa, oxalic and lactic acids are produced; fusing caustic potassa resolves it into acetic and formic acids. Hydriodic acid (phosphorus iodide and water) converts it into iodopropionic acid (p. 91). By the action of phosphorus chloride on the free acid or the lead salt, β-chlorpropionyl chloride (p. 90) is formed.

Serine (Glyceramic acid, amidolactic acid), $C^3H^7NO^3 = C^2H^3\left\{\begin{matrix}OH\\NH^2\end{matrix}\right\}CO.OH$, is formed when glue is boiled with dilute sulphuric acid.—Hard, brittle crystals; easily soluble in hot water, more difficultly in cold (in 32 parts at 10°), insoluble in alcohol and ether. Combines with acids and bases; yields glyceric acid, when treated with nitrous acid.

Cystine, $C^3H^7NO^2S$ probably $= C^2H^3\left\{\begin{matrix}SH\\NH^2\end{matrix}\right\}CO.OH$. As yet, only found in urinary calculi of rare occurrence, and in urinary sediments.—As a stone, it is of a dirty yellowish color, translucent, crystalline. In a pure condition, crystallizing in colorless, transparent laminæ, insoluble in water, easily soluble in acids and alkalies. From a solution in hot aqueous potassa, it crystallizes slowly on the addition of acetic acid.

Pyroracemic acid (pyruvic acid), $C^3H^4O^3 = CH^3.CO.CO.OH$, is produced by the dry distillation of tartaric and glyceric acids.—Liquid, boiling at 165–170°, suffers partial decomposition, however, when distilled, carbonic anhydride and pyrotartaric acid being formed. Monobasic acid. Gives crystallizing salts, which are changed when their solutions are heated. Nascent hydrogen converts it into lactic acid. On boiling it with barium hydroxide for some time, it is broken up into oxalic acid, a crystalline substance, difficultly soluble in water, *uvitic acid* $C^9H^8O^4$ (see Aromatic Compounds), and a syrupy liquid *uvitonic acid* $C^9H^{12}O^7$.

Carbacetoxylic acid, $C^3H^4O^4 = CH^2(OH).CO.CO.OH$ (isomeric with malonic acid). Is produced, when an aqueous solution of β-chlorpropionic acid is boiled for a long time with a great excess of silver oxide.—Thick syrup, easily soluble in water. Monobasic. Nascent hydrogen converts it into glyceric acid; hydriodic acid at 200° into pyrotartaric acid.

C. Bibasic, Triatomic Acids, $C^nH^{2n-2}O^5$.

1. *Tartronic Acid* (*Oxymalonic Acid*).

$$C^3H^4O^5 = CH\left\{\begin{array}{l}OH\\(CO.OH)^2.\end{array}\right.$$

Formation. By spontaneous evaporation of an aqueous solution of nitrotartaric acid; by the action of nascent hydrogen on mesoxalic acid; and probably also by boiling grape-sugar with an alkaline solution of copper oxide.

Properties. Large, colorless prisms, which fuse at 175° and are resolved at this temperature into water, carbonic anhydride, and glycolid (p. 146).

2. *Malic Acid* (*Oxysuccinic Acid*).

$$C^4H^6O^5 = \left\{\begin{array}{l}CH.OH.CO.OH\\CH^2.CO.OH.\end{array}\right.$$

Occurrence. Is very widely distributed throughout the vegetable kingdom; partially united with metals, as, for instance, with potassium in sweet cherries, partially in a free state, as in the juice of unripe apples, unripe grapes, the berries of mountain-ash, etc.

Formation and preparation. By boiling monobromsuccinic acid with water and silver oxide; by the action of nitrous acid on asparagine.—Most practicably prepared from the berries of mountain-ash which are not quite ripe. The boiled and filtered juice of these berries is mixed with so much milk of lime, that it still remains slightly acid, and then boiled for some time. During the boiling the neutral calcium salt separates. This is dissolved in nitric acid (diluted with ten times its volume of water), a saturated boiling solution being formed. On cooling, acid calcium malate crystallizes out, which is then thoroughly purified by recrystallization. From its solution the insoluble lead salt is precipitated by means of lead acetate, this decomposed with sulphuretted hydrogen, and the filtrate evaporated.

Properties. Crystallizable with difficulty; deliquescent in the air; has a very acid taste; fusing point, 100°; at 180° it is decomposed, forming water, fumaric acid, maleïc acid, and maleïc anhydride. The aqueous solution of the acid occurring in nature rotates the plane of polarization to the left. On being heated with hydrobromic acid, it is transformed into bromsuccinic acid; heated with hydriodic acid, into succinic acid. It is also reduced to succinic acid in the animal organism.

The *neutral salts of the alkalies* are deliquescent. The *neutral calcium salt* $C^4H^4O^5Ca$ forms large, shining, easily soluble, folious prisms. When a solution of this salt is boiled, an almost insoluble salt with 1 molecule water of crystallization is thrown down. The *acid calcium salt* $(H.C^4H^4O^5)^2Ca + 8H^2O$ forms large, transparent crystals, difficultly soluble in cold water, more easily in hot water.

Ethyl malate, $C^2H^3(OH)\left\{\begin{matrix} CO.O.C^2H^5 \\ CO.O.C^2H^5, \end{matrix}\right.$ is produced by saturating a solution of malic acid in absolute alcohol with hydrochloric acid gas.—Liquid, not distillable without decomposition.

Ethyl acetylmalate, $C^2H^3(O.C^2H^3O)\left\{\begin{matrix} CO.O.C^2H^5, \\ CO.O.C^2H^5, \end{matrix}\right.$ is formed by the action of acetyl chloride on the ether, without the aid of heat.—A liquid, boiling at 258°.

When treated with caustic potassa, it is resolved into acetic and malic acids and alcohol.

---

3. *Oxypyrotartaric acid*, $C^5H^8O^5 = C^3H^5(OH)\left\{\begin{matrix}CO.OH\\CO.OH.\end{matrix}\right.$
Is formed when dicyanhydrine (p. 170) is boiled with caustic potassa.—Colorless crystals, easily soluble in water, alcohol, and ether. Fusing point, 135°.

The following acids are isomeric with this:—

Itamalic acid, $C^5H^8O^5$, is produced by heating itamonochlorpyrotartaric acid (p. 163) with water or carbonates.—Long, white needles. Deliquescent, easily soluble in alcohol and ether. Fusing point, 60–65°. By heating, it is resolved into water, itaconic acid, and citraconic anhydride.

Citramalic acid, $C^5H^8O^5$. Citraconic acid combines directly with hypochlorous acid, forming *chlorcitramalic acid* $C^5H^7ClO^5$, a solid amorphous substance which, when heated with zinc in an aqueous solution, is converted into citramalic acid. Chlorcitramalic acid is also formed by the action of chlorine on citraconic acid.—Amorphous deliquescent mass.

Mesamalic acid, $C^5H^8O^5$. Is obtained from mesamonochlorpyrotartaric acid, like itamalic acid. Deliquescent mass, that fuses at about 60°.

Glutaric acid, $C^5H^8O^5$. Is produced by the action of nitrous acid on amidopyrotartaric acid (p. 163).—Syrupy mass, difficultly crystallizable.

---

4. *Adipomalic acid*, $C^6H^{10}O^5$. From monobromadipic acid by decomposing it with potassa-ley.—Easily soluble, sticky, gradually crystallizing mass.

D. Tribasic, Triatomic Acids, $C^nH^{2n-4}O^6$.

Tricarballylic Acid.

$$C^6H^8O^6 = \begin{matrix} CH^2.CO.OH \\ CH.CO.OH. \\ CH^2.CO.OH. \end{matrix}$$

Formation. By heating tricyanhydrine with caustic potassa and by the action of nascent hydrogen (sodium-amalgam and water) on aconitic acid.

Properties. Colorless, transparent, rhombic prisms. Fusing point, 157–158°. Easily soluble in water, alcohol, and ether.

The **calcium salt,** $(C^6H^5O^6)^2Ca^3 + 4H^2O$, is a white, amorphous powder, difficultly soluble in water.

Ethyl tricarballylate, $C^3H^5(CO.O.C^2H^5)^3$. Colorless liquid, but slightly soluble in water, boiling at about 300°.

E. Tribasic, Triatomic Acids, $C^nH^{2n-6}O^6$.

Aconitic Acid.

$$C^6H^6O^6 = C^3H^3(CO.OH)^3.$$

Occurrence. As calcium salt in *Aconitum napellus*, in *Delphinium consolida*, and *Equisetum fluviatile*.

Formation. By rapidly heating citric acid until oily streaks appear in the neck of the retort. It is extracted from the residue with ether.

Properties. Colorless laminæ or verrucose crystalline mass, easily soluble in water, alcohol, and ether. Fusing point, 140°. It is resolved at a high temperature into carbonic anhydride, water, itaconic acid, and citraconic anhydride. It combines with hydrogen, forming tricarballylic acid.

The **calcium salt,** $(C^6H^3O^6)^2Ca^3 + 6H^2O$, forms colorless, difficultly soluble prisms.

An acid, isomeric with aconitic acid, *aceconitic acid*, is produced in the form of its ethyl ether by the action of sodium on ethyl bromacetate.

SIXTH GROUP.

A. Tetratomic Alcohols, $C^nH^{2n+2}O^4$.

Erythrite (Erythroglucin, Phycite).
$C^4H^{10}O^4 = C^4H^6(OH)^4$.

Occurrence and preparation. Exists ready formed in *Protococcus vulgaris.* Is formed by the decomposition of erythrine (a substance contained in a number of lichens, for example *Roccella Montagnei*) by means of alkalies or alkaline earths.—The lichens are exhausted with milk of lime, the extract evaporated to one-quarter its volume, and the lime then precipitated by means of carbonic acid, and the filtrate evaporated to syrupy consistence. After the addition of alcohol, and after standing for some time, erythrite crystallizes out, and can then be purified by recrystallization.

Properties. Large, clear crystals, easily soluble in water, difficultly soluble in cold alcohol, insoluble in ether. Tastes sweet; fuses at 120°, and volatilizes at 300°, undergoing partial decomposition.—When heated with caustic potassa to 240°, it yields oxalic acid; when heated with concentrated hydriodic acid, the iodide of secondary butyl alcohol (p. 69) is formed.

Erythrite nitrate (nitroerythrite), $C^4H^6(O.NO^2)^4$, is produced by treating erythrite with a mixture of nitric and sulphuric acids.—Large, shining, lamellar crystals; fusing point, 61°. Detonates under the hammer.

B. Monobasic, Tetratomic Acids, $C^nH^{2n}O^5$.

Erythroglucic Acid.

$$C^4H^8O^5 = C^3H^4(OH)^3CO.OH.$$

Is produced when a solution of erythrite, to which is added platinum black, is allowed to stand in contact with the air for a long time; and by the action of nitric acid on erythrite.—Crystalline, very deliquescent mass.

C. Bibasic, Tetratomic Acids, $C^nH^{2n-2}O^6$.

1. *Tartaric Acid.*

$$C^4H^6O^6 = C^2H^2(OH)^2 \begin{cases} CO.OH \\ CO.OH. \end{cases}$$

Occurrence and formation. Most particularly in grape-juice. The *crude tartar*, which is deposited from new wine, is potassium bitartrate. It is formed by boiling several salts of dibromsuccinic acid, especially the silver salt, with water; and, in small quantity, by the oxidation of the carbohydrates with nitric acid, in company with saccharic and mucic acids.

Preparation. Purified tartar, in the form of a fine powder, is mixed with one-quarter its weight of finely pulverized chalk, and the mixture gradually thrown into boiling water in small portions. By this means the tartar is decomposed, forming neutral potassium tartrate, which remains in solution, and in calcium tartrate, which is thrown down as a white insoluble powder. By means of a solution of calcium chloride the neutral potassium salt is, in its turn, converted into calcium tartrate. From the calcium salt the tartaric acid is separated by digesting with dilute sulphuric acid, the calcium sulphate filtered off, and the solution of the acid evaporated to crystallization.

Properties. It crystallizes in clear, oblique, rhombic prisms, of a strongly acid taste; it is inodorous and easily soluble in water. Its solution rotates the plane of polarization towards the right; the solution of the acid, obtained from succinic acid, is, however, optically inactive. It differs from similar acids in the fact that it gives a powdery precipitate of potassium bitartrate with a saturated solution of saltpetre and potassium chloride. Heated in the air, it, as well as its salts, diffuses an odor of burnt sugar. It melts at 135°, and, at this temperature, without loss of water, is converted into two isomeric, deliquescent acids, *metatartaric* and *isotartaric acids.* When more strongly heated, water is given off, and it is transformed into tartaric anhydride $C^4H^4O^5$, a white powder, which in contact with water or bases, is gradually reconverted into tartaric acid. Subjected to dry distillation, it is resolved into pyruvic (p. 175) and pyrotartaric acids (p. 162), secondary products being formed at the same time. Hydriodic acid (phosphorus iodide and water) reduces it to malic and succinic acids.

Potasssium tartrate. The *neutral salt*, $C^4H^4O^6K^2$, forms large, very soluble crystals. The *acid salt*, $C^4H^4O^6.HK$ (tartar), is deposited in an impure condition in wine casks, in the form of gray or dirty red crusts. The pure salt forms small, transparent, weakly acid-tasting, very difficultly (in 240 parts cold water) soluble crystals, or white crystalline crusts.

Sodium tartrate. The *neutral salt*, $C^4H^4O^6Na^2 + 2H^2O$, and the *acid salt*, $C^4H^4O^6HNa + H^2O$, are both crystallizable, the latter much more soluble than tartar.

Potassium-sodium tartrate (Seignette salt), $C^4H^4O^6KNa + 4H^2O$, results from saturating cream of tartar with sodium carbonate. Large crystals, easily soluble, stable in the air.

Calcium tartrate, $C^4H^4O^6Ca + 4H^2O$, occurs in a number of plants, also in grape-juice, hence often found in crystals on crude tartar. Formed by double decomposition: a white crystalline powder, scarcely soluble in water, easily soluble in alkalies, acetic acid, and ammonium chloride.—Is deposited from a mixture of lime-water, with an excess of tartaric acid, in the form of large shining crystals.

Lead tartrate, $C^4H^4O^6Pb$, white voluminous precipitate, insoluble in water. Soluble in ammonia. When this solution is boiled, there is thrown down a salt $C^4H^2O^6Pb^2 = C^2H^2\begin{pmatrix}O \\ O\end{pmatrix}Pb\Big) \left\{ \begin{matrix} CO.O \\ CO.O \end{matrix} \right. Pb.$

Antimonyl-potassium tartrate (Tartar emetic), $C^4H^4O^6(SbO)K + \frac{1}{2}H^2O$, is prepared by digesting antimony oxide with cream of tartar and water. Crystallizes in shining octahedral or tetrahedral crystals, which lose their water of crystallization at 108°, and become white and opaque. Soluble in 14 parts of water of the ordinary temperature. Acids precipitate from its solution insoluble antimony compounds. Barium, lead, and silver salts precipitate salts which are analogous in composition to tartar emetic. At 200° the anhydrous salt loses another molecule of water, and is converted into a salt $C^4H^2O^6SbK$ (analogous to the lead salt $C^4H^2O^6Pb^2$), from which with the aid of water tartar emetic is regenerated.

Ethyl tartrate, $C^2H^2\begin{pmatrix}OH \\ OH\end{pmatrix} \left\{ \begin{matrix} CO.O.C^2H^5 \\ CO.O.C^2H^5, \end{matrix} \right.$ is formed, when hydrochloric acid gas is conducted into an alcoholic solution of tartaric acid.—Liquid; mixes with water; not volatile without decomposition.—*Ethyl-tartaric acid* $C^2H^2\begin{pmatrix}OH \\ OH\end{pmatrix} \left\{ \begin{matrix} CO.O.C^2H^5 \\ CO.OH. \end{matrix} \right.$ When a solution of tartaric acid in absolute alcohol is evaporated, this compound is left behind.—Crystalline, very deliquescent, easily decomposable acid.

Ethyl acetyltartrate, $C^2H^2\begin{pmatrix}O.C^2H^3O \\ OH\end{pmatrix} \left\{ \begin{matrix} CO.O.C^2H^5 \\ CO.O.C^2H^5 \end{matrix} \right.$ and **ethyl diacetyltartrate** $C^2H^2\begin{pmatrix}O.C^2H^3O \\ O.C^2H^3O\end{pmatrix}$ $\left\{ \begin{matrix} CO.O.C^2H^5 \\ CO.O.C^2H^5 \end{matrix} \right.$ result from the action of acetyl chloride on cooled ethyl tartrate. The former is a colorless oil, not volatile without decomposition; the latter forms large clear crystals, which fuse at 67° and boil at 288–290° without undergoing decomposition, and can be crystallized from boiling water.

Nitrotartaric acid, $C^2H^2\begin{pmatrix}O.NO^2\\O.NO^2\end{pmatrix}\begin{cases}CO.OH\\CO.OH.\end{cases}$ When finely pulverized tartaric acid is dissolved in very concentrated nitric acid, and concentrated sulphuric acid gradually added to the solution, there is formed a pasty mass, which, when pressed between porous stones, leaves behind a white shiny mass of nitrotartaric acid. This is soluble in water, but the solution decomposes very rapidly, and yields tartronic acid (p. 176) by spontaneous evaporation.

Ethyl nitrotartrate, $C^2H^2(O.NO^2)^2(CO.O.C^2H^5)^2$. Is produced when ethyl tartrate and a mixture of concentrated nitric and sulphuric acids are brought together.—Colorless prisms; fusing point, 45–46°.

The following acid is isomeric with and very similar to tartaric acid:—

2. *Racemic Acid.*

$C^4H^6O^6$.

This occurs, together with tartaric acid, in a number of varieties of grapes. It can be prepared artificially by heating cinchonin tartrate to 170°; by oxidizing several of the carbohydrates. The acid formed from dibromsuccinic acid with water and silver oxide appears to be racemic acid. It forms clear, rhombic prisms with 1 molecule of water of crystallization, which is given off at 100°. It is less soluble in water than tartaric acid. At an elevated temperature and towards reagents, it conducts itself almost precisely like the latter. It causes, however, a precipitate in a solution of calcium chloride and even of gypsum, whereas free tartaric acid does not precipitate these solutions. Precipitated calcium racemate is insoluble in acetic acid and ammonium chloride.

When acid sodium racemate is saturated with ammonia, there are formed, when the crystallization takes place slowly, two different salts of the same composition and appearance $C^4H^4O^6NaNH^4$. These have, however, such dissimilar and unsymmetrical hemihedral faces, that the forms are not congruent, the form of the one being a reflection of that of the other. The crystals of the one salt rotate the plane of polarization towards the right, those of the other towards the left. The free acids, separated from the salts, show the same difference in their form and their conduct towards polarized light. The acid, which rotates the plane towards the right is ordinary tartaric acid, the other is a peculiar acid, isomeric with this, *antitartaric acid.* With bases they form two series of salts, which differ from each other in the same way. When both acids are mixed together in a solution, there are formed crystals of racemic acid, which are optically inactive, an evolution of heat accompanying the formation.

The acids homologous with tartaric acid are only very imperfectly known.

D. TRIBASIC, TETRATOMIC ACIDS, $C^nH^{2n-4}O^7$.

Citric Acid.

$$C^6H^8O^7 = C^3H^4.OH(CO.OH)^3$$

Occurrence. Particularly in lemon-juice, in a free state. Further, in currants and gooseberries, and many other acidulous sweet fruits; in beet-juice; in the bark of *Pyrus malus* and *Æsculus hippocastanum.*

Preparation. Has not as yet been prepared artificially. Lemon-juice, clarified by boiling with albumen, is saturated while warm with chalk and milk of lime; the insoluble calcium citrate, which separates, filtered off, washed, and decomposed by means of dilute sulphuric acid.—Direct evaporation of the juice does not yield the acid in a crystalline form, on account of the presence of other substances.

Properties. Colorless, transparent, rhombic prisms, of a strong, agreeable, acid taste; easily soluble in water. Fuses at 100° in its water of crystallization, in an anhydrous condition at 153–154°. Its solution is not precipitated at the ordinary temperature by lime-water, but the precipitation ensues on heating the solution.

Potassium citrate (neutral), $C^6H^5O^7K^3 + H^2O$, clear deliquescent needles, insoluble in alcohol.—*Monacid salt*, $C^6H^5O^7HK^2$, white, amorphous, easily soluble mass, is produced by evaporating the mixed solutions of 2 molecules of the neutral potassium salt with 1 molecule of free citric acid.—*Diacid salt*, $C^6H^5O^7H^2K + 2H^2O$, large, transparent prisms, soluble in water and alcohol, is formed, when citric acid is added to a solution of the neutral salt, in double the quantity in which it is already present in the salt, and then the whole evaporated.

Calcium citrate, $(C^6H^5O^7)^2Ca^3 + 4H^2O$. Fine, crystalline powder, difficultly soluble in water; is precipitated by heating its solution.

Silver citrate, $C^6H^5O^7Ag^3$. White flocculent precipitate, insoluble in water, is decomposed by boiling with water.

Methyl citrate, $C^3H^4(OH)\{(CO.O.CH^3)^3$, and **Ethyl citrate,** $C^3H^4(OH)\{(CO.O.C^2H^5)^3$, are formed by conducting hydrochloric acid into solutions of citric acid in methyl or ethyl alcohol.—Compounds not volatile without decomposition. The methyl ether crystallizes; the ethyl ether is oleaginous.

Ethyl acetylcitrate, $C^3H^4(O.C^2H^3O)\{(CO.O.C^2H^5)^3$, is formed by the action of acetyl chloride on ethyl citrate.—A liquid, insoluble in water, boiling at 288°.

Decompositions. Heated to 175°, citric acid is converted into aconitic acid, water being eliminated. When distilled it yields citraconic anhydride. Oxidizing agents decompose it, yielding carbonic anhydride, oxalic, acetic, and formic acids. Chlorine and bromine decompose the free acid and its alkaline salts in aqueous solution, forming chlorine or bromine substitution-products of methyl acetate, together with chloroform or bromoform.

SEVENTH GROUP.

Alcohols which are derived from the hydrocarbons by the replacement of five hydrogen atoms by five hydroxyl groups are not known. Only one pentatomic acid is known. It is bibasic.

Aposorbic Acid, $C^5H^8O^7 = C^3H^3(OH)^3 \left\{ \begin{array}{l} CO.OH \\ CO.OH. \end{array} \right.$ Is formed, together with tartaric and racemic acids, by the oxidation of sorbine (p. 197) with nitric acid.—Colorless laminæ or pointed rhombohedral crystals; easily soluble in water; fuses at 110°, at the same time giving off water; decomposes completely at 170°.

EIGHTH GROUP.

A. Hexatomic Alcohols, $C^nH^{2n+2}O^6$.

Only such hexatomic alcohols are known that are derived from hexyl hydrides. Of these there can exist four isomeric modifications, of which two, mannite and dulcite, are known. Mannite is undoubtedly a derivative of normal hexyl hydride, and has the constitution expressed by the formula $CH^2.OH.CH.OH.CH.OH.CH.OH.CH.OH.CH^2.OH$. Dulcite appears to be a derivative of ethyl-isobutyl (p. 30); and has, hence, the constitution expressed by the formula $CH^2.OH.CH.OH.CH.OH.C.OH \begin{cases} CH^2.OH \\ CH^2.OH. \end{cases}$

1. *Mannite.*

$$C^6H^{14}O^6 = C^6H^8(OH)^6.$$

Occurrence. Pretty widely distributed. In celery, in sponges, in sea grasses, in the wood of *Pinus larix*, in the bark of *Canella alba;* most particularly, however, in manna, the dried juice of *Fraxinus ornus.*

Formation. During the fermentation of sugar under certain circumstances. By the action of nascent hydrogen (sodium-amalgam) on grape-sugar.

Preparation. Manna is dissolved in boiling alcohol, and, when the solution cools, mannite crystallizes out; by pressing and recrystallization, it is purified.

Properties. Crystallizes from alcohol in fine, colorless crystals; from water in large, clear prisms. Of very sweet taste, easily soluble in water. Fusing point, 166°. It does not reduce an alkaline solution of copper subchloride nor the solutions of the noble metals.—In contact with cheese and chalk it is resolved, at 40°, into hydrogen, carbonic anhydride, alcohol, and lactic, butyric, and acetic acids.—In aqueous solution, in contact with platinum-black or with testicle-substance for some time, it is converted into a syrupy sugar, mannitose $C^6H^{12}O^6$, which is capable of undergoing fermentation.—When heated with concentrated hydriodic acid, it yields β-hexyl iodide (p. 72).—It combines with bases forming amorphous, easily decomposable compounds.

Mannite nitrate (nitromannite), $C^6H^8(O.NO^2)^6$, is formed by treating mannite with a mixture of concentrated sulphuric and nitric acids.—Fine prisms, easily soluble in hot alcohol. Fuses at 72°; and detonates at 120°, or under the hammer, with a sharp report.

Mannite acetate, $C^6H^8(O.C^2H^3O)^6$. Is produced by heating mannite for a long time with acetic anhydride.—White, granular, crystalline mass, but slightly soluble in cold water, more easily in hot water and in alcohol. Fuses at about 100°.

Mannitan, $C^6H^{12}O^5$, is formed from mannite by heating to 200°, and by continued boiling with concentrated hydrochloric acid.—Sweet tasting syrup, easily soluble in water and alcohol, insoluble in ether. Is reconverted into mannite, slowly in moist air, rapidly by boiling with barium hydroxide.—Ethers of mannitan are formed by heating mannite with organic acids.

Several bodies, which are isomeric with mannitan, occur in nature.

Quercite, $C^6H^{12}O^5$. In acorns.—Colorless, monoclinic crystals of a sweet taste; fusing point, 235°.

Pinite, $C^6H^{12}O^5$. In the sap of the California pine (*Pinus lambertiana*).—Colorless, nodular crystals, easily soluble in water.

Isodulcite, $C^6H^{12}O^5 + H^2O$. By decomposition of quercitrine (see Glucosides) with dilute sulphuric acid.—Large, colorless, transparent crystals, easily soluble in water. Fuses at 105–110°, at the same time losing its water of crystallization.

Hesperidine sugar, $C^6H^{12}O^5 + H^2O$. By the decomposition of hesperidine.—Colorless, easily soluble crystals. Fuses at 71–76°, and loses its water at 100°.

2. *Dulcite* (*Melampyrin*).

$$C^6H^{14}O^6 = C^6H^8(OH)^6.$$

Occurrence. In *Melampyrum nemorosum*, *Scrophularia nodosa*, *Rhinanthus Christa Galli*, *Evonymus Europaeus*, and other plants. In the largest quantity in dulcite-manna, a variety of manna of unknown origin, coming from Madagascar.

Formation. It appears to be formed by the action of sodium-amalgam on sugar of milk.

Properties. Large, well-formed, monoclinic crystals. Easily soluble in water, but slightly in alcohol; fusing point, 182°; like mannite, it gives β-hexyl iodide with concentrated hydriodic acid.

Dulcite acetate, $C^6H^8(O.C^2H^3O)^6$. Is formed, together with the diacetate and other acetyl derivatives, by the action of glacial acetic acid, acetic anhydride, or acetyl chloride on dulcite.—Small white crystals, that fuse at 171°; but slightly soluble even in boiling water, moderately easily soluble in alcohol.

The following substance is isomeric with mannite and dulcite:—

Sorbite, $C^6H^{14}O^6$. Is contained in the berries of *Sorbus acuparia.*—Small, fine needles, which contain ½ molecule of water of crystallization, and fuse at 110–111° in an anhydrous condition. Almost insoluble in cold water, soluble in water in all proportions.

B. MONOBASIC, HEXATOMIC ACIDS, $C^nH^{2n}O^7$.

Glucic Acid.

$$C^6H^{12}O^7 = C^5H^6(OH)^5.CO.OH.$$

Formation. From grape-sugar. The dilute aqueous solution of sugar is saturated with chlorine, the excess of chlorine removed by means of a current of air, silver oxide then added until the acid reaction disappears, and the filtrate from silver chloride evaporated, after the removal of the dissolved silver by means of sulphuretted hydrogen.

Properties. Almost colorless syrup, very easily soluble in water, insoluble in absolute alcohol; monobasic acid.

The *calcium salt,* $(C^6H^{11}O^7)^2Ca + 2H^2O$, and the *barium salt,* $(C^6H^{11}O^7)^2Ba + 3H^2O$, crystallize in prisms.

The following acids are isomeric with glucic acid:—

Mannitic acid, $C^6H^{12}O^7$, is produced, together with mannitose, by the oxidation of mannite, under the influence of platinum-black.—Amorphous, gummy mass, easily soluble in water and alcohol. Apparently bibasic.

Dextrinic acid, $C^6H^{12}O^7$. Is obtained from dextrin, the same as glucic acid from grape-sugar.—Colorless syrup, which does not crystallize.

Lactonic acid, $C^6H^{10}O^6$. Is obtained from sugar of milk, or gum Arabic, in the same way as glucic acid from grape-sugar.

C. Bibasic, Hexatomic Acids, $C^nH^{2n-2}O^8$.

1. *Saccharic Acid.*

$$C^6H^{10}O^8 = C^6H^4(OH)^4 \begin{cases} CO.OH \\ CO.OH. \end{cases}$$

Formation. By careful oxidation of mannite, cane-sugar, grape-sugar, fruit-sugar, starch, and other carbo-hydrates with nitric acid.

Properties. Gummy, uncrystallizable, very deliquescent mass, also very easily soluble in alcohol. By further oxidation, it is converted into tartaric acid and then into oxalic acid.

The acid potassium salt, $C^6H^8O^8.HK$, and the *acid ammonium salt*, $C^6H^8O^8.HNH^4$, are difficultly soluble in cold water (in from 80 to 90 parts), easily crystallizable, and, hence, are well adapted to the separation of saccharic acid from the excess of nitric acid in the preparation of the former.—The neutral alkaline salts are deliquescent; the remaining salts are mostly insoluble or difficultly soluble in water.

2. *Mucic Acid.*

$$C^6H^{10}O^8 = C^4H^4(OH)^4 \begin{cases} CO.OH \\ CO.OH. \end{cases}$$

(Isomeric with saccharic acid.) Is formed, generally together with saccharic acid, by careful oxidation of dulcite, gum Arabic, mucilage, sugar of milk, and melitose, with nitric acid.—White, crystalline powder, very slightly soluble in cold water, more easily in boiling (in 50 parts), insoluble in alcohol.

The *neutral potassium salt*, $C^6H^8O^8K^2$, crystallizes readily, is almost insoluble in cold water, easily soluble in hot water; the acid salt, $C^6H^8O^8HK$, forms more easily soluble crystals.—The *neutral ammonium salt*, $C^6H^8O^8(NH^4)^2$, crystallizes in prisms, difficultly soluble in cold water. Is decomposed, by heating, into ammonia, water, *pyrrol* (C^4H^5N, a colorless, liquid base, boiling at 133°), and other products.

Ethyl mucate, $C^4H^4(OH)^4\left\{\begin{matrix}CO.O.C^2H^5\\CO.O.C^2H^5\end{matrix}\right.$, is obtained by heating mucic acid with sulphuric acid and alcohol.

Four-sided columns, easily soluble in boiling water and boiling alcohol, but slightly in cold. Fuses at 158°. When treated with $C^2H^3O.Cl$, it yields *ethyl tetracetylmucate* $C^4H^4(O.C^2H^3O)^4(CO.O.C^2H^5)^2$. Colorless needles of a vitreous lustre, but slightly soluble in water, cold alcohol, and ether, easily soluble in hot alcohol. Fuses at 177°, and sublimes even at 150°.

Phosphorus chloride converts mucic acid into a crystallizing chloride $C^6H^2Cl^2O^2Cl^2$, which is decomposed by water into hydrochloric acid and *chlormuconic acid* $C^6H^4Cl^2O^4$. Crystals, which are difficultly soluble in cold water, easily in boiling. Bibasic. Treated with sodium-amalgam and water, it yields *muconic acid* $C^6H^8O^4$. Prisms, sometimes an inch in length, difficultly soluble in cold water, easily soluble in hot water and alcohol. Bibasic acid. By continued action of hydrogen, adipic acid (p. 164) is produced.

Pyromucic acid, $C^5H^4O^3$, is produced by the dry distillation of mucic acid. The distillate is supersaturated with sodium carbonate, filtered, and evaporated down to a small volume; and, after acidifying with sulphuric acid, the pyromucic acid extracted by means of ether. Its potassium salt is thrown down, when alcoholic potassa is added to a solution of furfurol in alcohol.—Colorless laminæ or needles, easily soluble in water, especially in hot water and in alcohol. Fuses at 134°, and sublimes even at 100°. Monobasic acid. Is converted into a well crystallizing acid, *mucobromic acid* $C^4H^2Br^2O^3$, by bromine in the presence of water; carbonic anhydride and hydrobromic acid being formed at the same time.

Furfurol (Pyromucic aldehyde), $C^5H^4O^2$, is formed by the distillation of sugar and by the distillation of bran with dilute sulphuric acid.—Colorless oil, of a peculiar odor, soluble in 12 parts water, easily in alcohol; specific gravity, 1.165; boiling point, 162°. It turns dark in contact with the air, and is converted into a pitchy mass.—It is oxidized by boiling with water and silver oxide, forming pyromucic acid. Combines with alkaline bisulphites, and yields with ammonia, *furfuramide* $(C^5H^4O)^3N^2$, water being eliminated. Colorless crystals, insoluble in water, soluble in alcohol; without basic properties. Turns brown under the influence of light; and, when heated to 120°, or when boiled with caustic potassa, it is converted into a base of the same composition *furfurin*, which crytallizes in small colorless prisms, but slightly soluble in water, easily in alcohol, fusing under 100°.

D. Carbohydrates.

The so-called carbohydrates are derivatives of the hexatomic alcohols $C^6H^8(OH)^6$. They may be divided into three groups, according to their composition.

1. Grape-sugar group, $C^6H^{12}O^6$. The bodies belonging to this group are in all probability the first aldehydes of the hexatomic alcohols, formed by the oxidation of one group $CH^2.OH$ to CHO. They still contain five alcoholic groups.

2. Cane-sugar group, $C^{12}H^{22}O^{11}$. The bodies of this group are formed by the combination of two molecules of the aldehydes, water being eliminated. They contain eight OH groups, and are readily converted into the bodies of the first group by assimilation of water.

3. Cellulose group. The bodies of this group have the composition expressed by the empirical formula $C^6H^{10}O^5$; most of them, however, appear to have a higher (double or triple) molecular weight. They assimilate water readily, and are thus converted into the aldehydes of the first group.

1. *Grape-Sugar* (*Glucose*).

$$C^6H^{12}O^6 = CH^2.OH(CH.OH)^4.CHO.$$

Occurrence. In the juice of grapes, plums, cherries, figs, and other sweet fruits; in honey; in the urine of diabetic patients, and in other animal fluids.

Formation. It is formed, together with fruit, sugar, from cane-sugar by the action of acids or ferments; from starch or cellulose by boiling them with dilute sulphuric acid, or by the action of diastase; from amygdalin, salacin, tannic acid, etc. (See Glucosides.)

Preparation. The juice of grapes is neutralized with chalk, clarified by the white of eggs, and evaporated to the point of crystallization.—50 parts starch are added gradually to a boiling mixture of 100 parts water and 5 parts sulphuric acid; the starch dissolves, is at first converted into dextrin, and, after being boiled for several hours, into sugar. The acid is then saturated with calcium carbonate, the saccharine solution filtered from the gypsum, treated with animal charcoal, and evaporated to a syrupy consistence.—400 parts water are heated to 60–62° with 8 parts barley malt,* and 100 parts starch stirred into the mixture in small portions. The starch is soon dissolved, and, by the action of the diastase in the malt, is con-

* Germinated and then thoroughly dried barley.

verted, first, into dextrin, and, after continued digestion, into sugar.

Properties. Obtained from the aqueous solution, it forms colorless, not very hard, granular, crystalline masses, with 1 molecule of water of crystallization; crystallizes from alcohol in fine anhydrous prisms, united in compact nodular masses; easily soluble in water (in its own weight), difficultly soluble in absolute alcohol; fuses below 100°, the water of crystallization being given off.

A solution of grape-sugar rotates the plane of polarization towards the right; reduces the noble metals from their solutions; on the addition of alkalies turns bismuth nitrate a dark color; and, with the aid of heat, separates copper suboxide from an alkaline copper solution;* from an alkaline solution of mercury cyanide, metallic mercury.

Grape-sugar can be mixed with concentrated sulphuric acid without discoloration, and forms with it a saccharo-sulphuric acid, which gives a soluble barium salt.

It combines with bases, suffers a change, however, very rapidly, particularly with an excess of alkali and access of air, the solution becoming brown, and humus-substances being formed. A solution of grape-sugar, saturated with lime or baryta, and allowed to stand for a long time without access of air, becomes neutral, the sugar being converted into *glucic acid* $C^{12}H^{18}O^{9}$. This is amorphous, deliquescent; tastes and reacts acid. Sodium chloride and grape-sugar, dissolved together, combine when the solution is allowed to evaporate spontaneously, forming a very regular crystallizing compound $NaCl + C^{6}H^{12}O^{6} + 1\frac{1}{2}H^{2}O$. It crystallizes sometimes from evaporated diabetic urine. It can

* *Fehling's Solution.* Can be used for the quantitative estimation of sugar. 1 molecule grape-sugar completely reduces 5 molecules copper sulphate. For the preparation of this solution, 34.64 grm. of crystallized copper sulphate, which is not effloresced, are dissolved in water, a solution of 200 grm. Seignette salt (p. 182) or 160 grm. neutral potassium tartrate and 600–700 grm. caustic soda of specific gravity 1.12 added, and the whole now diluted so as to make 1000 cc. 10 cc. of this solution are completely reduced by 0.05 grm. grape-sugar.

always be obtained from the latter when sodium chloride is added to it after a sufficient concentration has been reached.

Grape-sugar is resolved into alcohol and carbonic anhydride by the process of fermentation (p. 42). Nitric acid oxidizes it, forming saccharic, tartaric, and oxalic acids.

Diacetyl-grapesugar, $C^5H^6\left(\begin{matrix}(O.C^2H^3O)^2\\(OH)^3\end{matrix}\right).CHO$. Is formed when 1 part of grape-sugar is heated with $2\frac{1}{2}$ parts acetic anhydride to the boiling point of the latter. —Colorless, amorphous, bitter-tasting mass. Fuses below 100°. Easily soluble in water, alcohol, and ether. When heated with water to 160°, it is resolved into acetic acid and grape-sugar.

Triacetyl-grapesugar, $C^5H^6\left(\begin{matrix}(O.C^2H^3O)^3\\(OH)^2\end{matrix}\right).CHO$. Is formed when the preceding compound is heated at 140° with twice its weight of acetic anhydride.—Solid, white mass, but slightly soluble in pure water, soluble in alcohol, ether, and dilute acetic acid.

2. *Fruit-Sugar.*
$C^6H^{12}O^6$.

Occurrence and formation. Together with grape-sugar in honey and the juices of ripe fruits. Cane-sugar is decomposed, by heating with dilute acids, into equal parts by weight of fruit-sugar and grape-sugar. A similar process appears to go on in connection with the ripening of fruits, for in unripe fruits cane-sugar is contained.

Properties. Not crystallizable; forms, when dried at 100°, a gummy, deliquescent mass. Easily soluble in water and alcohol; insoluble in absolute alcohol and ether. Its solution rotates the plane of polarization towards the left. Conducts itself towards an alkaline copper solution, and, in connection with fermentation, like grape-sugar. By oxidation with nitric acid there result saccharic, racemic, and oxalic acids.

3. *Lactose.*
$C^6H^{12}O^6$.

Formation. From sugar of milk by heating with dilute acids; together with another variety of sugar, that appears to be grape-sugar.

Properties. Easily soluble, microscopic crystals, united in nodules. Does not combine with sodium-chloride. Its solution rotates the plane of polarization towards the right, more strongly than that of grape-sugar. Exhibits the same conduct towards an alkaline copper solution, and in connection with fermentation, as grape-sugar; yields, however, mucic acid, on being heated with nitric acid.

---

4. *Sorbine*, $C^6H^{12}O^6$, in the juice of the berries of the mountain-ash.—Large, colorless, easily soluble crystals. Not capable of fermentation with yeast. When oxidized, it yields aposorbic acid (p. 187) together with tartaric and racemic acids.

5. *Inosite* (phaseomannite), $C^6H^{12}O^6 + 2H^2O$. Occurs in the animal organism, particularly in the muscular substance of the heart; is, however, also contained in the lungs, kidneys, liver, spleen, in the brain; and, in certain diseases (Morbus Brightii), it has also been detected in the urine. Occurs, further, pretty widely distributed in the vegetable kingdom, particularly in the unripe fruits of many papilinaces (beans, peas, lentils, acacias), in cabbage, in *Digitalis purpurea*, *Taraxacum officinale*, in the shoots of potatoes, in the green portions and the unripe berries of asparagus, in the leaves of *Fraxinus excelsior*, in grape-juice, etc.—Large, colorless, rhombic crystals of a sweet taste, losing their water of crystallization in the air and becoming white and opaque. Easily soluble in water, insoluble in absolute alcohol and in ether. Fuses at 210°. Not capable of fermentation with yeast.—Evaporated nearly to dryness with nitric acid, then mixed with an ammoniacal solution ot calcium chloride and

again carefully evaporated, it yields a beautiful rose color.

6. *Cane-Sugar.*

$$C^{12}H^{22}O^{11} = C^{12}H^{14}(OH)^{8}O^{3}.$$

Occurrence. Widely distributed. Particularly in the juice of sugar-cane, beets, madder roots, sugar-maple, and several palms. In small quantity in nearly all sweet fruits.

Extraction. Carefully cleansed beets are pressed, mixed with half per cent. of lime, heated to 100°, in order to neutralize any free acids and precipitate albuminous and mucous substances, then freed of lime by means of carbonic acid, filtered through thick layers of bone-black for the purpose of decolorization, and evaporated in a vacuum. On cooling, a granular and more or less discolored mass, *raw sugar* or *Muscovado sugar*, is deposited, which is separated from the uncrystalline portion (molasses).—The extraction from sugar-cane takes place in a similar manner.

Raw Sugar, an article still rendered impure by the presence of syrupy sugar and other substances, is now refined for the purpose of removing these impurities. It is dissolved in a little water, the solution again filtered through bone-black, and again evaporated in a vacuum until crystals separate, while it is still hot.—The mass is then brought into iron or clay moulds of a conical form and allowed to cool. The uncrystalline syrup which still remains is removed by the process of bottoming. From the broad end of the loaf a layer two inches thick is removed; this is stirred with water and formed into a thick pasty mass and then replaced in the mould. The concentrated solution of sugar in percolating through the loaf carries the molasses with it towards the apex, where it flows out through an opening in the mould. The loaves are afterwards dried in warm air and by passing a current of air through the moulds.

Properties. As loaf-sugar, it forms a perfectly colorless aggregate of small granular crystals; as sugar-candy, well developed regular crystals. Soluble in one-third part of water; in alcohol the less soluble the less water it contains. Fusible at 160°. The solution rotates the plane of polarization to the right, and does not reduce an alkaline solution of copper.

Sugar combines with bases. A saccharine solution dissolves a large amount of calcium and barium hydroxides, and loses by this means its sweet taste. Further, it dissolves lead oxide, forming a soluble saccharate, which has an alkaline reaction. All of these compounds, however, are decomposed even by carbonic acid.

Decompositions. Heated to the fusing point, sugar becomes amorphous without losing water, and becomes crystalline again only after standing for a long time. It suffers the same alteration when its solution is boiled for a long time. Heated to 190–200°, it is converted into a brown, uncrystalline mass called *caramel.* Distilled with a large excess of caustic lime, sugar is decomposed into water, carbonic anhydride, acetone, and metacetone $C^6H^{10}O$, a colorless, agreeably smelling liquid, which boils at 84°, floats on water, and, when heated with potassium bichromate and sulphuric acid, yields carbonic, acetic, and propionic acids.—Melted carefully with an excess of potassium hydroxide, sugar forms potassium carbonate, oxalate, formate, acetate, and propionate, hydrogen being evolved.

Concentrated sulphuric acid converts cane-sugar into a black mass, the action being accompanied by an elevation of temperature and formation of formic acid. Boiled with dilute sulphuric acid or with hydrochloric acid, it breaks up into equal molecules of grape-sugar and uncrystalline fruit-sugar, the elements of water being taken up.—It is not capable of fermentation as such; in contact with yeast, however, it is decomposed in the same manner as by means of dilute acids, and transformed into the two varieties of fermentable sugar. When boiled for a long time with acids, it is converted into brown bodies.

Gently heated with nitric acid, it yields saccharic,

tartaric, and racemic acids; boiled with it, it yields oxalic acid.

Octacetyl-canesugar, $C^{12}H^{14}(O.C^2H^3O)^8O^3$. Is produced when cane-sugar is heated with an excess of acetic anhydride to 160°.—White, amorphous mass, insoluble in water and acetic acid.—A very similar compound of the same composition is also produced when grape-sugar is heated for a long time at 160°, with a large excess of acetic anhydride.

7. *Sugar of Milk.*
$C^{12}H^{22}O^{11} + H^2O$.

Occurrence. Only in the milk of animals.

Preparation. The casein of the milk is precipitated by heating with a dilute acid, or, better, with rennet (calves' stomachs), and the yellow liquid (whey) evaporated to syrupy consistence. On standing for a length of time in a cool place, the sugar of milk crystallizes out. It is purified by repeated recrystallization.

Properties. Colorless, translucent, four-sided prisms, possessing but a slightly sweet taste; is slowly soluble in water, and crystallizes even from a saturated solution but slowly. But slightly soluble in alcohol. The aqueous solution rotates the plane of polarization to the right, and reduces alkaline solutions of copper and silver.—When heated with dilute acids, or left in contact with yeast, it is resolved into lactose and another variety of sugar, probably grape-sugar.—When heated with nitric acid, it yields mucic, saccharic, tartaric, racemic acids, and, as final product, oxalic acid.—Heated with acetic anhydride, it yields an ether similar to that formed from cane-sugar under the same circumstances.

8. *Mycose (Trehalose.)*
$C^{12}H^{22}O^{11} + 2H^2O$.

In ergot of rye, several other fungi and trehala-manna, an article of food used in the East.—Lustrous, rhombic crystals. Easily soluble in water and boiling alcohol.

9. *Melezitose.*

$C^{12}H^{22}O^{11} + H^2O$.

In the manna of Briançon (from *Pinus larix*). Small crystals, easily soluble in water, but slightly soluble in alcohol.

10. *Melitose.*

$C^{12}H^{22}O^{11} + 3H^2O$.

In Australian Eucalyptus-manna.—Thin, needly crystals. Is resolved into glucose and an unfermentable syrupy sugar, *eucalyn* $C^6H^{12}O^6$, when heated with dilute acids or when placed in contact with yeast.

11. *Synanthrose.*

$C^{12}H^{22}O^{11}$.

In the bulbs of composites, together with inulin. Most easily isolated from *Dahlia variabilis* and *Helianthus tuberosus*.—White, very light, amorphous, deliquescent mass. Treated with dilute sulphuric acid, it is resolved into fruit-sugar and another variety of sugar.

12. *Cellulose.*

$(C^6H^{10}O^5)^x$.

Occurrence. Is the most widely distributed substance in the vegetable kingdom; is present in the organs of all plants. In a chemical point of view, the material of the cell-membranes of all plants and parts of plants is the same. In those cases, in which it shows certain varieties in the chemical properties, it is to be assumed that these are caused by the presence of substances from which it can be separated only with great difficulty, if at all. In the latter respect, it exhibits the most varied conditions, as can be seen from the dissimilar mechanical constitution of vegetable germs and the young organs of plants, of pith, of the soft, pulpy mass of juicy fruits and roots, and, on the other hand, of the tough tissue-fibres of cotton, flax, and hemp. These differences are produced chiefly by the deposit of woody substance (incrusting substance, xylon, lignin) on the cell-walls, which forms the principal mass of wood and the woody portions of fruit kernels, but which appears to be nothing but cellulose in a very compact state of aggregation, thoroughly penetrated by or combined with other substances, simultaneously secreted.

Purification. From the tender parts of plants, crushed as thoroughly as possible, pure cellulose is obtained by successive digestions with dilute potassa, dilute sulphuric acid, water, alcohol, and ether. It can be also obtained from fine white paper, already almost pure cellulose, which, during the process of preparation, has been thoroughly disorganized.

Properties. Pure cellulose is an amorphous, white body, insoluble in water, alcohol, ether, dilute alkalies, and acids. It is soluble in a solution of copper hydroxide in ammonia, and is precipitated from these solutions in an amorphous condition by acids, alkaline salts, solutions of gum and sugar.

Transformations. When heated with potassium hydroxide and a little water to 200°, without access of air, it forms hydrogen, methyl alcohol, potassium oxalate, acetate, propionate, and carbonate, without separation of carbon.

Cellulose, or broken-up wood, straw, linen, paper, cotton, etc., when gradually ground together with concentrated sulphuric acid, so that no elevation of temperature or discoloration takes place, are transformed into an homogeneous, pasty mass. If water is then quickly added, a white amorphous body, *amyloid*, is separated, which, with iodine, gives a blue color. When paper is dipped for a few moments into moderately concentrated sulphuric acid, and then washed with water and ammonia, it is changed superficially into amyloid, and forms a substance similar to parchment (vegetable parchment), which is translucent, of great firmness, and swells up in water, forming a lubricous mass.

By long-continued action of sulphuric acid on cellulose, it is completely dissolved by subsequent addition of water, and this solution contains *dextrin* (wood dextrin), which is converted into grape-sugar by boiling the solution, water being assimilated.

Cellulose-nitrate (Pyroxilin, gun-cotton). When cotton, which has been cleansed by means of dilute caustic soda, washing and drying, is inserted for five minutes into a mixture of 1 volume concentrated sulphuric acid and 2 volumes fuming nitric acid, then thoroughly washed with water and dried, it is changed into gun-cotton, although its external appearance remains the same. It burns up instantaneously in contact with a burning body, and acts exactly like gunpowder, but more violently. It is insoluble in alcohol, acetic, hydrochloric, and nitric acids. Ordinary gun-cotton consists of a mixture of several compounds. Hence, according to the method of preparation, it possesses somewhat different properties. When prepared by adding 1 part of cotton to a warm concentrated solution of 20 parts of dry, finely-pulverized saltpetre in 31 parts of concentrated sulphuric acid, allowing to stand for twenty-four hours, and afterwards thoroughly washing and drying, it has the property of dissolving in a mixture of ether with a little alcohol. This solution (collodion), on the evaporation of the ether, leaves the compound behind in the form of a transparent, flexible coating, impervious to water. If, on the other hand, very concentrated nitric and sulphuric acids are employed in the preparation, there is obtained an exceedingly explosive compound of the composition $C^6H^7(O.NO^2)^3O^2$ (trinitrocellulose), entirely insoluble in ether and alcohol.—Wood, flax, tow, and paper conduct themselves in a similar manner towards the acid mixture.

Triacetyl-cellulose, $C^6H^7(O.C^2H^3O)^3O^2$. Is obtained by heating cotton or Swedish filter paper with six to eight times its weight of acetic anhydride in sealed tubes to 180°, and treating the syrupy product with water.—White, flocculent mass, insoluble in water, alcohol, and ether; soluble in concentrated acetic acid.

13. *Starch (Amylum).*
$(C^6H^{10}O^5)^x$.

Occurrence. Very widely distributed; in large quantity in the seeds of the different varieties of grain; in leguminous plants, chestnuts, potatoes; in the trunks of a number of pines, etc.; further, in most roots, in a great many kinds of bark, even in fruits, for example in apples. Always deposited in plant cells, in the form of microscopic grains.

Extraction. Technically it is prepared from wheat and potatoes by washing. They are ground, and the starch grains washed out from the cellular substance in a fine wire sieve. The starch settles from the milky water as a white, solid sediment, which is repeatedly stirred up with water, washed out and finally dried in the air.

Sago, from the pith of the sago palm, *cassava* and *tapioca*, from the poisonous root (containing hydrocyanic acid) of *Jatropha Manihot*, and *arrowroot*, from the root of *Maranta arundinacea*, consist of the same kind of starch.

Properties. Perfectly white powder, glistening in sunlight, consisting of small, shining, transparent grains, recognizable under the microscope. These are formed of layers, arranged upon each other, surrounded by a more delicate and compact envelope, which is, perhaps, cellulose. The grains are of various sizes and forms, sometimes spherical, sometimes spheroidal, according to the plant from which they take their origin. Tasteless, inodorous, insoluble in cold water. Insoluble in alcohol and ether; these, however, usually extract from most starch small quantities of wax and fat.

Heated with water to 60°, the envelopes are burst and the starch forms a gelatinous, translucent mass.

When dried, starch, which has been treated with warm water, forms a colorless, transparent, hard mass.

Compounds. In the form of powder as well as of jelly, starch combines with iodine, forming a body of a deep blue color, which, when heated with water, becomes colorless and on cooling turns blue again. With bromine it turns orange-red.

Starch combines with bases. Insoluble white amylates of calcium and barium are produced by precipitating baryta and lime-water with a hot solution of starch; lead amylate from a solution of starch with basic lead acetate.

Transformations. Heated to about 200°, starch is converted into dextrin. A solution of starch heated with a solution of diastase, or boiled for some time with water containing sulphuric acid, is changed first into an isomeric modification, *soluble starch*, which is soluble in hot and cold water and can be precipitated from its solution by alcohol. Iodine colors the solution blue, and baryta water gives a heavy precipitate. If the action of the diastase or acids is continued further, it is first converted into dextrin and then into grape-sugar. Starch dissolves in very concentrated nitric acid. If the solution is immediately mixed with water, all the starch, employed at first, is precipitated as a white, powdery body, *xyloidine*. This is a compound similar to gun-cotton $C^6H^7\begin{pmatrix}O.NO^2\\(OH)^2\end{pmatrix}O^2$. After being washed and dried, it takes fire even at 180°, and burns with violence. If to the solution in nitric acid are added first sulphuric acid and then water, a similar body $C^6H^7\begin{pmatrix}(O.NO^2)^2\\OH\end{pmatrix}O^2$ separates.—By heating with nitric acid, starch yields the same products as grape-sugar.

Triacetyl-amylum, $C^6H^7(O.C^2H^3O)^3O^2$. Is formed by heating starch with an excess of acetic anhydride to 140°.—Amorphous mass, insoluble in water, alcohol, and acetic acid. Heated with alkalies, it is reconverted into starch and potassium acetate.

14. *Inulin*, $C^6H^{10}O^5$, occurs chiefly in the roots of *Inula Helenium*, *Georgina purpurea*, *Helianthus tuberosus*, *Leontodon taraxacum*, etc., and is prepared from them by a similar process to that described in connection with starch.

Very fine, white powder, tasteless and inodorous. But slightly soluble in cold water; very easily soluble in hot water, forming a mucous liquid (not a jelly), from which it separates again as a powder.—It turns yellow with iodine, not blue. When boiled for a long time it loses the property of separating in a powdery form, being finally completely converted into non-crystallizing fruit-sugar.

15. *Glycogen* (animal starch), $C^6H^{10}O^5$, is a constant ingredient of the liver; occurs in the tissues which surround the fœtus in the uterus, and also in the fœtus itself, but only during the period of the fœtal life; is also contained in the yolk of eggs and in mollusks.—For the purpose of preparing it, liver, as fresh as possible, is chopped up, immersed in boiling water, boiled an hour, filtered, and alcohol added to the watery solution. The precipitate is boiled with concentrated caustic potassa, as long as ammonia is evolved, for the purpose of destroying albuminous substances, and the diluted solution again precipitated with alcohol. By repeated dissolving in acetic acid and precipitating with alcohol, it is finally obtained pure.—A white, amorphous powder; forms with water an opalescent solution; is insoluble in alcohol; and turns a brownish-red color with iodine. When heated with dilute acids, or when in contact with diastase, blood, saliva, etc., it is rapidly converted into grape-sugar. It suffers the same change in the liver very quickly after the death of the animal.

16. *Moss-starch*, $C^6H^{10}O^5$, occurs very generally in lichens, particularly in Iceland moss. Moss is broken

up, and, for the purpose of removing other substances, at first washed successively with ether, alcohol, a dilute solution of sodium carbonate, with water containing hydrochloric acid, and pure water, then dissolved by boiling with water. The strained solution coagulates, forming a jelly, which, on being dried, leaves the starch behind as a colorless, gummy mass, that swells up with water again to a jelly. Tasteless and inodorous. Becomes brown with iodine.

17. *Dextrin*, $C^6H^{10}O^5$, occurs in the vegetable kingdom, although not very widely distributed; and is also contained in muscular tissue. Is formed from starch by heating to 180°; heating with water to 150°; by boiling with dilute acids; and by warming with water and diastase to 65–70°.—Amorphous, gummy mass; attracts moisture from the air. Very easily soluble in water, insoluble in absolute alcohol and ether. The aqueous solution rotates the plane of polarization to the right, and does not reduce an alkaline solution of copper. By further action of dilute acids or diastase, it is converted into grape-sugar. It conducts itself towards nitric acid the same as starch.

Triacetyl-dextrin, $C^6H^7(O.C^2H^3O)^3O^2$. Is obtained by heating dextrin with acetic anhydride and is also formed when, in the preparation of triacetyl-amylum, the temperature rises to 160°.—Amorphous mass, insoluble in water and alcohol; soluble in acetic acid.

18. *Gum* (Arabin), $C^6H^{10}O^5$, exudes spontaneously as a concentrated solution from a great many trees, and solidifies in large, transparent drops; as, for instance, gum Arabic and gum Senegal, of various species of acacia, cherry and plum-tree gum.—Colorless, transparent, vitreous mass, with a conchoidal, shining fracture, completely uncrystallizable; without taste and odor. Easily soluble in water, forming a thick, sticky, tasteless liquid (mucilage).

Pure gum (gummic acid, arabin) combines with bases. Gum Arabic consists essentially of the calcium and potassium salts of gummic acid. The latter is obtained from this pure, by precipitating its solution, containing hydrochloric acid, with alcohol. A white, amorphous, easily soluble mass separates, which is vitreous after being dried at 100°; has the composition $C^6H^{10}O^5 + \frac{1}{2}H^2O$, and does not lose its water under 120–130°.

A solution of gum rotates the plane of polarization to the left. When boiled for a long time with dilute sulphuric acid, it is converted into sugar (grape-sugar?). Nitric acid oxidizes it, forming mucic, saccharic, tartaric, and oxalic acids.

19. *Vegetable Mucus* (*Bassorin*), $C^6H^{10}O^5$. Is very widely distributed in the vegetable kingdom, as a solid mass deposited on the cell-walls, or in a condition similar to solution as a glairy mass. The following substances are richly supplied with vegetable mucus: Tragacanth (the spontaneously exuded sap of *Astragalus* varieties); gum of Bassora, cherry-tree gum, plum-tree gum, salep (from different *Orchis* varieties), carageen moss; further, linseed, *Semen psyllii*, quince seeds, the root of *Althæa officinalis*, and *Symphytum officinale*, etc.—Colorless or yellowish, translucent thick mass, inodorous and tasteless. Swells up with water, forming an exceedingly voluminous mucus, without dissolving.—Yields, with nitric acid, the same products as gum.

NINTH GROUP.

Cyanogen Compounds.

Cyanogen.

$$C^2N^2 = (CN)^2.$$

Formation. Cyanogen compounds are produced from carbon and nitrogen, which unite at a high temperature in the presence of metals; by the distillation of a great many organic compounds (sugar, fat) with dilute nitric acid, or by their explosion with saltpetre; by heating ammonium formate and oxalate or oxamide.

Preparation. By heating mercury cyanide. A portion of the cyanogen remains behind in the reaction as a black, amorphous substance, *paracyanogen*, which has the same percentages of its constituents as cyanogen, but a higher molecular weight, and is transformed into the latter at a very high temperature.

Properties. Colorless gas, of a peculiar pungent odor; density, 1.801. Easily condensable by pressure (4 atmos. at 17°) and cooling (to —25°); more strongly cooled it becomes solid. Water absorbs four and a half times its volume, alcohol twenty-three times. The aqueous, as well as the alcoholic, solution decomposes when kept, a brown body (azulmic acid) being thrown down and hydrocyanic acid, ammonium oxalate and carbonate, urea, etc., being formed; if a very small quantity of aldehyde be present, oxamide is almost the exclusive product.

Cyanhydric Acid, Prussic Acid.
CNH.

Formation and preparation. By the passage of electrical sparks through a mixture of acetylene and nitrogen; by the action of nitric acid on a great many organic bodies. Most particularly prepared by the distillation of 10 parts of potassium ferrocyanide with a mixture of 7 parts of sulphuric acid and 14 parts of water. The vapors are passed through an U-shaped tube filled with pieces of calcium chloride and standing in water at 30°, for the purpose of drying, and then in a vessel surrounded by ice. For the preparation of the aqueous acid, the vapors are conducted directly into cold water.—A very dilute solution is obtained by distilling with water parts of plants containing amygdalin. (See Glucosides, Amygdalin.)

Properties. Clear liquid, of a peculiar narcotic odor, like that of bitter almonds; 0.7, specific gravity; boiling point, 27°. Congeals in a crystalline form at —15°.

Miscible with water in all proportions. Exceedingly poisonous.

It can be easily detected by means of the following reactions: If to a solution containing it potassa be added, then a ferroso-ferric salt, and the whole acidified, Prussian blue is precipitated; if to a solution containing it yellow ammonium sulphide be added and the excess of the sulphide evaporated, the residue exhibits an intensely blood-red color on the addition of a drop of iron sesquichloride.

In an anhydrous condition it combines with dry hydrochloric, -bromic, and -iodic acids, forming white, crystalline, but very unstable compounds, which, brought in contact with water, are instantaneously resolved into ammonium chloride (bromide, iodide) and formic acid.

Decompositions. When kept for any length of time it is decomposed, a brown body being deposited in the vessel; very small quantities of other acids prevent this. Boiled with acids or alkalies, it is decomposed, forming formic acid and ammonia, water being taken up. Nascent hydrogen (zinc and sulphuric acid) converts it into methylamine (p. 40).

Forms metallic cyanides with bases. Combines also with several chlorides.

The compounds of cyanogen with alcohol radicles (nitriles) are described in connection with the alcohols.

Cyanogen chloride. There are two polymeric compounds of cyanogen with chlorine known.

Liquid cyanogen chloride, CNCl, is produced by the action of chlorine on metallic cyanides or dilute cyanhydric acid.—Colorless, very mobile liquid; boils at +15,5°; and congeals at —5° to —6°. Heavier than water. Preserved in sealed tubes, it is transformed into solid cyanogen chloride. Combines with several metallic chlorides. Alkalies decompose it, chlorides and cyanates being formed. The so-called gaseous *cyanogen chloride* does not appear to have a separate existence, but is probably the vapor of the liquid variety.

Solid cyanogen chloride, $(CN)^3Cl^3$, is formed from anhydrous cyanhydric acid and chlorine, in the direct light of the sun; and by conducting chlorine in a solution of 1 part of mercury cyanide and 4 parts of ether; is also formed by distilling cyanuric acid with phosphorus chloride.—Shining needles or laminæ, which fuse at 145°, and boil at 190°. But slightly soluble in cold water, easily in alcohol and ether. Decomposed by boiling with water or alkalies into cyanuric acid and hydrochloric acid. It suffers the same decomposition even at the ordinary temperature, when its solution in dilute alcohol is allowed to stand.

Cyanogen bromide, CNBr, and **Cyanogen iodide,** CNI, are formed by heating potassium or mercury cyanide with bromine or iodine. They are crystallizing compounds, easily volatile, soluble in water and alcohol. Cyanogen bromide, when heated alone or with anhydrous ether to 130–140°, is converted into a polymeric compound $(CN)^3Br^3$. This body forms a white, amorphous powder, fusing above 300°, which is decomposed by boiling with water and even in contact with moist air, yielding cyanuric and hydrobromic acids.

Cyanic Acid.
CN.OH.

Formation. The cyanates of the alkalies are produced by heating metallic cyanides in the air or in contact with easily reducible metallic oxides. Cyanic acid cannot be separated from these salts, inasmuch as it breaks up with water, the moment it becomes free, forming carbonic acid and ammonia. The free acid can only be obtained by the dry distillation of cyanuric acid, or by heating urea with phosphoric anhydride.

Properties. Colorless liquid, of a penetrating, pungent, acid smell. Only stable below 0°. Removed from the freezing mixture, it becomes turbid, and is rapidly converted into a white, amorphous mass, *cyamelide*, a spontaneous elevation of the temperature and an explosive boiling accompanying the action. *Cyamelide* is polymeric with cyanic acid, and is again transformed into it by means of distillation. It is decomposed by water into ammonia and carbonic acid.

Potassium cyanate, CN.OK. A mixture of 8 parts of previously dehydrated iron ferrocyanate and 3 parts of potassium carbonate is heated to fusing, and 15 parts of red lead added gradually to the somewhat cooled, but still liquid mass. After the reduced lead has been separated, the salt-mass is poured off, and the potassium cyanate extracted by means of alcohol.—Lamellæ, similar to potassium chloride; easily soluble in water, but decomposes very rapidly with water, yielding potassium carbonate and ammonia.

Ammonium cyanate, $CN.ONH^4$, is produced when the vapors of cyanic acid meet with dry ammonia: white, solid mass, soluble in water. When heated or dissolved in water, it is rapidly converted into urea, with which it is isomeric.

Silver cyanate, CN.OAg, white precipitate, produced by adding silver nitrate to a solution of potassium cyanate; is also formed, together with ammonium nitrate, by evaporating a solution of urea with silver nitrate. When treated with dry hydrochloric acid, it yields a volatile, easily decomposable compound CO.NH + HCl.

Ethyl cyanate (Cyanetholin), $CN.O.C^2H^5$, is produced by the action of cyanogen chloride on sodium ethylate.—Liquid, not distillable without decomposition; insoluble in water; specific gravity, 1.127. Caustic potassa decomposes it, yielding ethyl alcohol, carbonic anhydride, and ammonia.

A compound $CO:N.C^2H^5$, isomeric with the preceding, usually called *ethyl cyanate*, is produced, together with ethyl cyanurate, by the distillation of a mixture of potassium cyanate and ethyl sulphate.—Colorless liquid, boiling at 60°, of a strong odor.—With water it decomposes immediately, forming diethylurea (which see), an evolution of carbonic anhydride taking place at the same time; heated with potassa, it yields ethylamine (p. 56) and potassium carbonate. It combines directly with one molecule of dry hydrochloric or -bromic acid, forming liquid compounds, which are decomposed by water, yielding ethylamine hydrobromate (or hydrochlorate) and carbonic acid.

Sulphocyanic Acid (Rhodanic Acid).
CN.SH.

The alkaline salts of this acid are produced by the direct combination of the cyanides with sulphur. For the purpose of preparing the free acid, the potassium salt is distilled with dilute sulphuric acid. It is obtained in an anhydrous state by decomposing the mercury salt with hydrochloric acid.—Colorless oil, mixes in all proportions with water, congeals at —12.5°. Monobasic acid. The solutions of the free acid, as well as of its salts, are turned intensely red on the addition of a solution containing iron oxide.—Decomposes very easily, particularly in an anhydrous condition, into potassium cyanide and yellow crystalline *persulphocyanic acid* (hydroxanthanic acid) $(CN)^2H^2S^3$; but very slightly soluble in water.

Potassium sulphocyanate, CN.SK, can be most readily prepared by fusing together 46 parts dehydrated potassium ferrocyanide, 17 parts potassium carbonate, and 32 parts sulphur. The melted mass is poured off, allowed to cool, and then extracted with alcohol.—Large, colorless prisms, very easily soluble in water. When air is excluded it fuses without decomposition.

Ammonium sulphocyanate, $CN.S(NH^4)$, is obtained by digesting mercury cyanide with ammonium sulphide; or ammonium cyanide with sulphur. Can be prepared most easily by treating carbon bisulphide with a solution of ammonia in dilute alcohol.—Colorless, very easily soluble crystals; fuses at 147°, and is transformed at 170° into the isomeric compound, sulphocarbamide (see Urea).

Mercury sulphocyanate, $(CNS)^2Hg$. Produced by adding potassium sulphocyanate to a solution of mercury nitrate.—Amorphous, colorless, heavy precipitate, insoluble in water; burns on being heated, increasing greatly in volume (Pharaoh's serpents).

Ethyl sulphocyanate, $CN.S.C^2H^5$, is formed by the distillation of a mixture of potassium suphocyanate and potassium ethylsulphate.—Colorless oil, boiling at 146°; insoluble in water; specific gravity, 1.033 at 0°. Combines directly with the simple hydrogen acids, like ethyl cyanate. When boiled with nitric acid it is oxidized, forming ethylsulphurous acid (p. 51); it is decomposed by treatment with alcoholic potassium sulphide, forming potassium sulphocyanate and ethyl sulphide (p. 54); when treated with aqueous ammonia, it yields urea, ethyl bisulphide, and ammonium cyanide.

Ethylene sulphocyanate, $(CNS)^2C^2H^4$, is produced when ethylene chloride (p. 113) is heated with an alcoholic solution of potassium sulphocyanate.—Crystallizes from boiling water in stellate needles, from alcohol in rhombic plates. But slightly soluble in alcohol, easily in water. Fuses at 90°, and is not volatile without decomposition.

The following compounds, called *mustard-oils*, are isomeric with the sulphocyanic ethers.

Ethyl mustard-oil, $CS:N.C^2H^5$. When carbon bisulphide is brought in contact with ethylamine, ethylamine ethylsulphocarbamate (p. 227) is formed by the direct combination of both bodies. When to the solution of this salt silver nitrate or mercury chloride is added, white precipitates are formed, which, when distilled with water, yield ethyl mustard-oil, sulphuretted hydrogen, and sulphides. Or, to the alcoholic solution of ethylamine ethylsulphocarbamate, iodine may be added until the solution gives a reaction with starch; the liquid is then distilled off, and the ethyl mustard-oil precipitated from the distillate by water. Diethylamine, treated in a similar manner, also yields ethyl mustard-oil.—Colorless liquid, of a penetrating odor, exciting to tears; specific gravity, 1.019 at 0°; boiling point, 133°. Does not mix with water. Causes a burning pain when brought in contact with the skin. When heated with water to 200°, or with hydrochloric acid to 100°, it is decomposed into ethylamine, carbonic anhydride, and sulphurretted hydrogen. With nascent hydrogen (zinc and hydrochloric acid) it yields ethylamine, formylsulphaldehyde (p. 102), and sulphuretted hydrogen.

Methyl mustard-oil, $CS:N.CH^3$, and **Amyl mustard-oil,** $CS:N.C^5H^{11}$, are prepared like the ethyl compound, and are very similar to this. The first is solid, crystalline, fuses at 34°, and boils at 119°; the latter is liquid, and boils at 183–184°.

Butyl mustard-oil, $CS:N.C^4H^9$, is contained in oil of spoon-wort (from *Cochlearia officinalis*).—Boiling point, 159–160°.

Allyl mustard-oil, $CS:N.C^3H^5$. When black mustard-seed (from *Sinapis nigra*) is freed of fatty oils as thoroughly as possible by pressure, digested with water, and then distilled, the *potassium myronate* (see Glucosides) contained in the seed is decomposed, yielding sugar, potassium bisulphite, and allyl mustard-oil, and the latter distills over with the water vapors.—It can be prepared by the decomposition of allyl bromide or iodide, by means of an alcoholic solution of potassium sulphocyanate (in this point differing from the other mustard-oils).—Colorless liquid, of an exceedingly strong odor; specific gravity, 1.01; boiling point, 148°; but slightly soluble in water. Raises blisters on the skin very rapidly.

Cyanogen sulphide (sulphocyanic anhydride), $(CN)^2S$, is produced by mixing silver sulphocyanide intimately with an ethereal solution of cyanogen iodide; and is extracted from the mass, which remains behind after evaporation, by means of carbon bisulphide; is also formed when mercury cyanide and sulphur subchloride (S^2Cl^2) are shaken together.—Clear, rhombic plates, of a strong odor, similar to that of cyanogen iodide; sublimes at 30–40°; fuses at 65°; and is decomposed at a higher temperature; soluble in water, alcohol, and ether. In aqueous solution, and even by the moisture of the air, it is quickly decomposed, a bright-yellow powder being thrown down. Hydrogen, sulphuretted hydrogen, and potassium sulphide decompose it into hydrocyanic and sulphocyanic acids.

Cyanuric Acid.

$$C^3N^3H^3O^3 + 2H^2O = (CN)^3(OH)^3 + 2H^2O.$$

Formation. From solid cyanogen chloride or the corresponding cyanogen bromide, by boiling with water or alkalies. More practically by heating urea until the evolution of ammonia ceases and the mass becomes solid again; or by passing chlorine over melted urea at 140°, and afterwards removing the ammonium chloride by means of cold water.

Properties. Colorless, rhombic prisms with a slightly acid taste; difficultly soluble in cold water (40 parts), more easily in hot water and in alcohol.

Tribasic acid. Yields three series of salts. The *neutral sodium salt* $C^3N^3O^3Na^3$ is almost insoluble in concentrated hot caustic soda. The *cuprammonium salt* is particularly characteristic.—An amethyst-colored precipitate, difficultly soluble in water and ammonia, which is formed by the addition of an ammoniacal solution of copper sulphate to an aqueous solution of cyanuric acid. —The alkaline salts are converted into cyanates by heat.

Methyl cyanurate, $C^3N^3O^3(CH^3)^3$, is produced by the distillation of a dry mixture of potassium cyanate and methyl sulphate.—Prismatic crystals, fusing at 175°, and boiling at 295°. Boiled with caustic potassa, it yields methylamine and carbonic anhydride.

Ethyl cyanurate, $C^3N^3O^3(C^2H^5)^3$, is prepared like the methyl ether.—Colorless, rhombic crystals, which fuse at 85° and boil at 276°. Soluble in boiling water, alcohol and ether. When boiled with caustic potassa, it conducts itself like the methyl ether.

Diethylcyanuric acid, $C^3N^3O^3(C^2H^5)^2H$, can be best obtained from the alcoholic mother-liquor from the crystallization of crude ethyl cyanurate. This is boiled with baryta water, the barium oxide removed by means of sulphuric acid, and the filtrate evaporated.—Rhombohedral crystals, soluble in hot water, alcohol and ether. Fuses at 173°, volatile at a higher temperature. Weak acid; dissolves easily in alkalies; on being evaporated, however, these solutions yield the acid in the free state.

Cyanamides.

Cyanamide, $CH^2N^2 = CN.NH^2$, is formed by mixing gaseous or liquid cyanogen chloride with dry ammonia gas; and by the action of carbonic anhydride on sodiumamide. Is most easily obtained by conducting cyanogen chloride into an anhydrous ethereal solution of ammonia, filtering from the sal-ammoniac, which is thrown down, and evaporating.—Colorless crystals, easily soluble in water, alcohol, and ether; fuses at 40°. Its solution gives a yellow precipitate, CN^2Ag^2, with silver nitrate and a little ammonia; when nitric acid is added to its solution, water is taken up, and it is converted into urea.

Dicyano-diamide, $C^2H^4N^4 = (CN)^2H^4N^2$. When cyanamide is left to itself for some time, it is spontaneously converted into dicyano-diamide; or when its aqueous solution is evaporated, the same change takes place, especially when a little ammonia is previously added; is further formed, when a solution of sulpho-carbamide is digested with silver oxide, lead oxide, or mercury oxide on a water-bath.—Transparent, thin rhombic plates, pretty easily soluble in water and alcohol, sparingly soluble in ether. Fuses at 205°, and decomposes at a higher temperature. On the addition of silver nitrate to its solution, long, colorless crystals, $C^2H^3AgN.HNO^3$, of a silken lustre, separate, from which the nitric acid can be extracted by means of ammonia.

When boiled with baryta water, dicyano-diamide yields the barium salt of *dicyan-amidic acid* $C^2H^3N^3O$, together with cyanic acid and cyanamide. This acid crystallizes in long lancet-shaped needles; is monobasic. Its potassium salt is also formed by the direct combination of cyanamide and potassium cyanate, when a dry mixture of both bodies is heated to 60°, or the solution of the mixture allowed to stand for twenty-four hours.

When a solution of dicyano-diamide in dilute acids is evaporated, it is converted into a strongly alkaline base *dicyano-diamine*, $C^2H^6N^4O$, water being assimilated. It is easily soluble in water, but sparingly in alcohol. Its salts, which crystallize well, remain behind on evaporation, and from these it can be separated by means of stronger bases.

Cyanuramide (Melamine), $C^3H^6N^6 = (CN)^3H^6N^3$. When cyanamide is heated to 150°, it is transformed into this polymeric compound, the change being accompanied by a violent reaction.—Large, shining, rhombic, octahedral crystals; but slightly soluble in cold water, more easily in hot water; insoluble in ether and alcohol. Combines with acids, forming salts. When boiled with hydrochloric acid, however, it is converted into *ammeline*, $C^3H^5N^5O$ (a white powder, insoluble in water), which combines with nitric acid, forming a salt that crystallizes well.

Triethylmelamine, $C^3H^3(C^2H^5)^3N^6$. Is produced, when a solution of ethylsulphocarbamide is digested with lead oxide or mercury oxide.—Crystalline, easily soluble mass. Boiled for a short time with hydrochloric acid, it is converted into *triethylammeline* $C^3H^2(C^2H^5)^3N^6O$; when digested for several hours with it, ethyl cyanurate is formed.

Guanidine, $CH^5N^3 = C\left\{\begin{matrix}(NH^2)^2\\NH.\end{matrix}\right.$ The hydriodate is formed by heating cyanogen iodide with alcoholic ammonia to 100°; the hydrochlorate by heating an alcoholic solution of cyanamide with ammonium chloride to 100°; by heating chloropicrin (p. 37) or orthocarbamic ether with aqueous ammonia to 150–160°, or with alcoholic ammonia to 100°; by oxidizing guanine with hydrochloric acid and potassium chlorate; by conducting dry hydrochloric acid over biuret (p. 221).—The free base, separated from the sulphate by means of baryta water, is a crystalline, strongly alkaline, caustic tasting mass, which takes up moisture and carbonic anhydride from the air. Strong monatomic base.

Guanidine hydrochlorate, $CH^5N^3.HCl$, is easily soluble in water and alcohol. With platinum chloride it yields a double salt $(CH^5N^3.HCl)^2PtCl^4$, which crystallizes in yellow needles. With gold chloride, long needles $CH^5N^3.HCl.AuCl^3$.

Guanidine nitrate, $CH^5N^3.HNO^3$, forms large lamellar crystals, difficultly soluble in water.

Methylguanidine (Methyluramine), $C^2H^7N^3 = CH^4(CH^3)N^3$. Is formed by boiling a solution of creatine with mercury oxide; by the action of potassium hypermanganate on a warm solution of creatinine, which contains caustic potassa; and by heating cyanamide with methylamine in alcoholic solution.—Colorless, very deliquescent mass. Yields salts that crystallize well.

Triethylguanidine, $CH^2(C^2H^5)^3N^3$. Is obtained by digesting an alcoholic solution of diethylsulphocarbamide, containing ethylamine, with mercury oxide.—

Strongly alkaline liquid, which attracts carbonic anhydride from the air, and then crystallizes.

Fulminic acid, $C^2H^2N^2O^2$.* In a free condition, unknown. Cannot be isolated from its salts.

Mercury fulminate (Fulminating mercury), $C^2N^2O^2Hg+H^2O$, is formed by the spontaneous heating of a mixture of mercury nitrate, excess of nitric acid and alcohol. 1 part mercury is dissolved in 12 parts nitric acid of specific gravity 1.36, this solution poured into 5½ parts of alcohol (90° Tralles), and the vessel shaken violently. In a short time reaction begins, which becomes more violent very rapidly, and is moderated by the gradual addition of 6 parts alcohol of the same strength as that first employed. The black color, which at first makes its appearance from the presence of metallic mercury, soon disappears, and crystalline flocks of mercury fulminate separate, which are purified by recrystallization from hot water.—White, silky needles, sparingly soluble in cold water, easily in hot. Heated or forcibly struck, it detonates with a loud report. Copper and zinc, boiled in water with fulminating mercury, decompose the latter substance, yielding metallic mercury and copper and zinc fulminate. Sulphuretted hydrogen precipitates the mercury from the solution, but the liberated acid is immediately decomposed by sulphuretted hydrogen into carbonic anhydride and ammonium sulphocyanate. Free bromine forms mercury bromide and *methyl dibromnitrocyanide* $C^2Br^2N^2O^2 = C(NO^2)Br.^2CN$ (large crystals, fusing at 50°), together with cyanogen bromide, hydrobromic acid, and other products.

Silver fulminate (Fulminating silver), $C^2N^2O^2Ag^2$, is obtained in the same manner as fulminating mercury, and is very similar to the latter in all its properties. Potassium chloride precipitates only half the silver from a boiling solution, and, on the evaporation

* Probably nitrocyanmethyl $C(NO^2)(CN)H^2$.

of the filtered solution, a white, easily-detonating salt $C^2N^2O^2.KAg$ crystallizes out.

Fulminuric acid (Isocyanuric acid), $C^3H^3N^3O^3$. The alkaline salts are produced, together with mercury chloride or iodide, by boiling fulminating mercury with chlorides or iodides of the alkalies.—The free acid, separated from the lead salt by means of sulphuretted hydrogen, crystallizes in small prisms; soluble in water, alcohol, and ether; explodes when heated to 145°. Monobasic acid.—When added to a cooled mixture of very concentrated nitric and sulphuric acids, it yields carbonic anhydride, ammonia, and *trinitroacetonitrile* $C^2N^4O^6 = C(NO^2)^3.CN$. Colorless, crystalline, camphoraceous mass, which fuses at 41.5°, explodes when heated above 200°, and is decomposed by water and alcohol even at the ordinary temperature. (See Nitroform, p. 36.)

Allophanic acid, $C^2H^4N^2O^3$. Not known in a free state. Is decomposed, at the instant of its liberation, into urea and carbonic anhydride. Its ethers are formed by conducting the vapor of cyanic acid into anhydrous alcohols.

Ethyl allophanate, $C^2H^3N^2O^3.C^2H^5$. Colorless prisms, soluble in alcohol and hot water. Is resolved by distillation into alcohol and cyanuric acid. When mixed with barium hydroxide and water, it yields *barium allophanate* $(C^2H^3N^2O^3)^2Ba$.

Biuret (Allophanamide), $C^2H^5N^3O^3 + H^2O$. Is formed by heating ethyl allophanate with aqueous ammonia to 100°; and by heating urea to 150–160°.—Long, colorless needles; fusing point, 190°; pretty difficultly soluble in cold water, easily soluble in hot water and in alcohol. Heated above its melting point, it is decomposed, yielding ammonia and cyanuric acid. When copper salts or caustic potassa are added to its solution, it turns red.

Trigenic acid, $C^4H^7N^3O^2$, is produced when cyanic acid vapor is conducted into well-cooled aldehyde, a lively evolution of carbonic anhydride taking place at the same time.—Small, colorless prisms, difficultly soluble in water. Not fusible without decomposition.

TENTH GROUP.

DERIVATIVES OF CARBONIC ACID.

The isolated bivalent radicle of these compounds *carbonyl* is carbonic oxide CO. Carbonic acid is only known in the form of the anhydride $CO^2 = CO.O$; the true acid $CO(OH)^2$ cannot be prepared.

Carbonyl chloride (Phosgene), $COCl^2$, is produced by the direct union of chlorine and carbonic oxide in sunlight; or, in the presence of heated spongy platinum, by heating carbon tetrachloride with sulphuric anhydride. Can be prepared most readily by heating 20 parts chloroform with 50 parts potassium bichromate and 400 parts concentrated sulphuric acid in a flask, connected with an inverted condensing apparatus.—Colorless liquid, of a suffocating odor; boiling point, +8°; specific gravity, 1.432 at 0°. Is decomposed by water, yielding hydrochloric acid and carbonic anhydride; by alcohols, yielding hydrochloric acid and the ethers of *chlorcarbonic (chlorformic) acid* COCl.OH, a body which cannot be prepared in a free state.

Ethyl carbonate, $CO(O.C^2H^5)^2$, is formed, together with carbonic oxide, by heating ethyl oxalate with sodium or sodium ethylate at 80°; by the action of bromine on orthoformic and orthocarbonic ethers (p. 35 and 37).—Colorless, thin liquid, of a pleasant odor; specific gravity, 0.975; boiling point, 126°.

Ethyl carbonic acid, $CO\left\{\begin{matrix}O.C^2H^5\\OH.\end{matrix}\right.$ Cannot be prepared in a free state. The *potassium salt* $CO^3.C^2H^5.K$ is formed by saturating a solution of freshly-melted potassa in alcohol with carbonic acid.—Laminæ, of a mother-of-pearl lustre, which are decomposed by water into alcohol and potassium bicarbonate.

Ethyl chlorcarbonate, $COCl.O.C^2H^5$. Is produced by bringing carbonyl chloride together with well-cooled absolute alcohol.—Colorless liquid of a suffocating odor, exciting to tears; specific gravity, 1.13; boiling point, 94°. Heated with absolute alcohol it yields ethyl carbonate.

Carbon sulphoxide, COS. Occurs apparently in a number of mineral springs. Is produced when carbonic oxide and sulphur are conducted together through a red-hot tube; together with sulphur and sulphurous anhydride, by the action of sulphuric anhydride on carbon bisulphide, slowly at the ordinary temperature, quickly by heating; together with ethylamine, allylamine, etc., by shaking the mustard-oils with concentrated sulphuric acid; by heating carbon bisulphide with urea, oxamide, or acetamide; by conducting dry sulphuretted hydrogen into ethyl cyanate; by heating thiacetic acid to 300°. Can be obtained most readily by pouring moderately concentrated sulphuric acid on potassium sulphocyanate.—Colorless, easily inflammable gas, of peculiar odor. Yields an explosive gas-mixture with oxygen. Water absorbs about an equal volume of the gas; alkalies absorb it easily, forming sulphides and carbonates. At a red heat it is partially decomposed, yielding carbonic oxide and sulphur.

Carbon bisulphide, CS^2. Is formed by direct combination of carbon and sulphur at a high temperature.—Colorless, strongly refracting liquid. In a pure state (obtained by shaking the commercial substance with metallic mercury or mercury chloride, and then rectifying) it has a pleasant ethereal odor; boiling point, 47°; specific gravity, 1.27; easily inflammable, but sparingly soluble in water; mixes with alcohol and ether in all proportions. Excellent solvent for many substances, for example iodine, phosphorus, sulphur, fats, resins.—Combines with water at a low temperature, forming a crystalline hydrate, which is decomposed again at —3°.

Conducted over red-hot metallic oxides, carbon bisulphide forms sulphides of the metals, carbonic anhydride being evolved. Dry chlorine resolves it into sulphur chloride and *sulphocarbonyl chloride* $CSCl^2$, a liquid, boiling at 70°. Treated with a chlorine mixture, there is further produced a chloride, $CSCl^4$, a liquid, boiling at 146–147°; and this, when oxidized, yields *trichlormethyl sulphochloride* $CCl^3.SO^2Cl$ (p. 39). Phosphorus chloride and antimony chloride convert it into carbon tetrachloride CCl^4 (p. 35). The same compound is formed, together with sulphur chloride, when the vapor of carbon bisulphide, mixed with chlorine, is passed through a red-hot tube. By the action of iodine chloride on carbon bisulphide, there are produced sulphur chloride, carbon tetrachloride, and a crystalline compound $2(CS^2) + ICl^3$.—Carbon bisulphide combines directly with $(C^2H^5)^2Zn$, forming a brown, amorphous compound $C^5H^{10}S^2Zn$, which, when treated with hydrochloric acid, or subjected to dry distillation, yields a liquid $C^5H^{10}S$, boiling at 130–150°.—It combines directly with triethylphosphine, forming a red crystalline body, a violent reaction taking place.

Sulphocarbonic acid, $CH^2S^3 = CS(SH)^2$. Carbon bisulphide dissolves in alkaline sulphides, forming the alkaline salts of sulphocarbonic acid. From these the free acid can be separated by hydrochloric acid and rapid addition of water.—Reddish-brown, oily liquid.

The *sodium salt* $CS(SNa)^2$ can be precipitated from a concentrated solution of sodium sulphide, to which is added carbon bisulphide, by means of alcohol or ether and alcohol.—Thick red liquid, soluble in water.

Ethyl sulphocarbonate, $CS(S.C^2H^5)^2$, is produced by pouring an alcoholic solution of ethyl iodide on the sodium compound.—Yellow oil, boiling at 240°; insoluble in water; combines directly with two atoms of bromine, forming red crystals, which in the air or when treated with water or caustic potassa, give up the bromine and regenerate the ether. Heated with alcoholic ammonia, the ether is decomposed into mercaptan and ammonium sulphocyanate. When oxidized with nitric acid, it yields ethylsulphurous acid (p. 51).

Ethylene sulphocarbonate, $CS.S^2.C^2H^4$, is formed from the sodium compound and ethylene bromide or chloride like the ethyl ether.—Large, yellow crystals, which fuse at 36.5°. Slowly soluble in alcohol, easily in ether and ether-alcohol. By oxidation with dilute nitric acid, it is transformed into

Ethylene oxysulphocarbonate, $COS^2.C^2H^4$. Thin plates fusing at 31°, easily soluble in alcohol and ether. Yields, when further oxidized with concentrated nitric acid, disulphetholic acid (p. 141).

Xanthogenic acid, $C^3H^6OS^2 = CS\left\{\begin{matrix} O.C^2H^5 \\ SH \end{matrix}\right.$ (Ethyldisulphocarbonic acid). The potassium salt of this acid separates in colorless, silky needles when carbon bisulphide is added to a solution of caustic potassa in alcohol. The free acid, separated from the potassium salt by means of sulphuric acid at the ordinary temperature, is oily, insoluble in water; decomposes at 24° into alcohol and carbon bisulphide.

Ethyl xanthogenate, $CS\left\{\begin{matrix} O.C^2H^5 \\ S.C^2H^5, \end{matrix}\right.$ is obtained by the action of ethyl chloride on potassium xanthogenate.—Colorless oil, boiling at 200°. With ammonia, it yields *xanthogenamide* $C^3H^7NOS = CS\left\{\begin{matrix} O.C^2H^5 \\ NH^2, \end{matrix}\right.$ a substance which forms large crystals and fuses at 36°. At the same time there are formed ethyl sulphide and ammonium sulphide; when mixed with an alcoholic solution of potassa, it gives the potassium salt of *ethylmonosulphocarbonic acid*, $CS\left\{\begin{matrix} O.C^2H^5 \\ OK. \end{matrix}\right.$ This salt can also be prepared by direct union of carbonic anhydride and potassium mercaptide (p. 54); and by conducting carbon sulphoxide into a cold alcoholic solution of potassa. The free acid cannot be extracted from this salt, because it is resolved into carbonic anhydride, alcohol, and sulphuretted hydrogen, the instant it is liberated.

Carbonyldisulphethyl, $CO \begin{cases} S.C^2H^5 \\ S.C^2H^5 \end{cases}$ (isomeric with ethyl xanthogenate), is produced by the action of sulphuric acid or hydriodic acid on ethyl sulphocyanate (p. 214).—Colorless liquid, boiling at 196–197°. With alcoholic potassa, it yields mercaptan and potassium carbonate; with alcoholic ammonia—mercaptan and urea.

Carbamic acid, $CH^3NO^2 = CO \begin{cases} NH^2 \\ OH. \end{cases}$ Cannot be isolated. The ethers of this acid are formed by the action of ammonia on the ethers of carbonic acid at the ordinary temperature (above 100° urea is formed); by conducting cyanogen chloride into alcohol; and by boiling urea for a long time with the alcohols. The *methyl ether* (urethylan) and the *ethyl ether* (urethan) are crystalline compounds; easily soluble in water, alcohol, and ether; volatile without decomposition. The former fuses at 52–55°, the latter at 100°. Alkalies decompose them, forming carbonic anhydride, ammonia, and the respective alcohols.

Ammonium carbamate, $CO \begin{cases} NH^2 \\ O.NH^4 \end{cases}$ (so-called anhydrous ammonium carbonate), is formed by bringing ammonia and carbonic anhydride together, best in the presence of absolute alcohol.—White, loose powder, subliming at 60°; or thin laminæ. With water, it rapidly forms ammonium carbonate.

Sulphocarbamic acid, $CS \begin{cases} NH^2 \\ SH. \end{cases}$ The ammonium salt is formed by the action of alcoholic ammonia on carbon bisulphide. It crystallizes in long, pale-yellow needles. The free acid, separated from this salt by means of hydrochloric acid, is solid, easily soluble in water, alcohol, and ether. Exceedingly unstable.

Ethylsulphocarbamic acid, $CS \left\{ \begin{array}{l} NH.C^2H^5 \\ SH. \end{array} \right.$ The crystalline ethylamine salt (fusing point, 103°) of this acid is formed by adding carbon bisulphide to an alcoholic solution of ethylamine. It is easily soluble in water. From its solution metallic salts precipitate the salts of ethylsulphocarbamic acid. When these are boiled with water, they are resolved into sulphides, sulphuretted hydrogen, and ethyl mustard-oil (p. 214).

Carbamide (Urea).

$$CH^4N^2O = CO(NH^2)^2.$$

Occurrence. In many of the animal fluids, especially in urine. (See Animal Chemistry—Urine).

Formation. From ammonium cyanate by the evaporation of its aqueous solution; from cyanamide by assimilation of the elements of water (p. 217); by the action of ammonia on phosgene or carbonic ether; from oxamide by means of heating with mercury oxide; by heating ammonium carbamate or commercial ammonium carbonate to 130–140°.

Preparation. 1. *Extraction from urine.* Urine is evaporated to syrupy consistence, and, when cool, mixed with an excess of strong nitric acid. Urea nitrate separates in the form of dark brown crystalline masses. It is now filtered off, pressed, and purified by recrystallization from moderately strong nitric acid. It is most easily obtained colorless, but not without loss, by gradually adding small quantities of finely powdered potassium chlorate to the hot concentrated solution in nitric acid, then allowing to cool and recrystallizing the almost colorless crystals which now separate, either from water or nitric acid. The urea nitrate, purified in this manner, is now decomposed by heating with water and barium carbonate, the filtrate evaporated to dryness and the urea extracted from barium nitrate by means of cold alcohol. It crystallizes from the solution, when concentrated by distilling off a portion of the alcohol. 2. *Artificial preparation.* Crude potassium cyanate (prepared according to p. 212) is dissolved in water without the aid of heat, and to the solution as much ammonium sulphate is added as potassium ferrocyanide was employed; the liquid is evaporated down to a small volume, the potassium sulphate, that crystallizes out on cooling, filtered off, and the filtrate evaporated to dryness. The urea is extracted from the residue by means of alcohol.

Properties. Colorless, four-sided prisms, without odor, of a cooling taste; fuses at 130°. Easily soluble in water and alcohol.

Heated above its fusing point, it is decomposed, ammonia is given off, and, according to the duration of the heating, the residue consists either of *biuret* (p. 221) or *cyanuric acid* (p. 216).—By heating with water in fused tubes above 100°; by boiling with alkalies; by heating with concentrated sulphuric acid; by evaporation of the solution, to which is added lead acetate, urea is resolved into carbonic anhydride and ammonia, water being assimilated. It suffers the same change in foul urine. When heated for some time with alcoholic carbon bisulphide, ammonium sulphocyanate and carbonic anhydride are formed.

Urea combines with bases, acids, and salts, forming crystallizing compounds.

Urea nitrate, $CH^4N^2O.HNO^3$, crystallizes from a solution of urea on the addition of nitric acid. A salt which is but slightly soluble in water, alcohol, and concentrated nitric acid.

Urea hydrochlorate, $CH^4N^2O.HCl$, is produced by the action of hydrochloric acid gas on urea. Yellow oil, which soon congeals. An elevation of temperature accompanies the action. Is decomposed by water, even by lying in contact with moist air; and, when heated, it yields cyanuric acid and ammonium chloride.

Urea phosphate, $CH^4N^2O.H^3PO^4$, crystallizes occasionally from evaporated urine (of swine); is always produced when phosphoric acid is added to urea in the proportion required by the formula of the salt, and the solution evaporated down to a small volume.—Large, well-formed, rhombic crystals, easily soluble in water, but not deliquescent.

Urea-mercury oxide, $CH^4N^2O.2HgO$, white precipitate, which a solution of mercury nitrate produces in a solution of urea mixed with potassa. When a solution of mercury chloride is employed, there is formed a gelatinous, snowy-white precipitate $2(CH^4N^2O).3HgO$, which, when washed with boiling water, becomes yellow.

Urea and sodium chloride, $CH^4N^2O.NaCl + H^2O$. Shining crystals, which separate on the evaporation of a solution of urea containing sodium chloride.

Urea also unites with other chlorine compounds and a great many nitrates. Mercury nitrate precipitates from its solution insoluble compounds of varying composition. By mixing very dilute solutions, there is produced a heavy, white powder of the composition $2(CH^4N^2O) + Hg(NO^3)^2 + 3HgO$.*

* This reaction is employed for the purpose of estimating urea quantitatively. For this object a solution of mercury nitrate—prepared by dissolving 77.2 grms. pure mercury oxide in nitric acid, evaporating to dryness, and diluting with water so as to make 1000 cc.—is added to the solution of urea until an addition of sodium carbonate to a small portion, removed each time for the purpose, commences to give a yellow color. Every cc. of the mercury solution employed corresponds to 0.01 grms. of urea.—Before the estimation of urea in urine, all sulphuric and phosphoric acids must be removed. This is accomplished best by means of a mixture of two volumes of a solution of barium hydroxide (saturated at the ordinary temperature) with one volume of a similarly prepared solution of barium nitrate. Two volumes of urine are mixed with one volume of this mixture, filtered, and the urea precipitated exactly from 15 cc. of the filtrate (corresponding to 10 cc. urine) by means of the mercury solution.—In the case of very exact estimations, another correction of the result is necessary in consequence of the presence of sodium chloride in the urine.

COMPOUND UREAS.

In urea, one, two, or three atoms of hydrogen can still be replaced by means of alcohol or acid radicles. If, for instance, a solution of potasssium cyanate is evaporated with methylamine sulphate, instead of with ammonium sulphate, an urea is produced, which, together with the radicle CO, contains the radicle methyl. A large number of such compounds are known, of which only a few will be described here. They all show the greatest analogy with urea in their conduct towards reagents and in their decompositions.

Ethyl-urea, $CO\left\{\begin{matrix} NH.C^2H^5 \\ NH^2, \end{matrix}\right.$ is produced by the decomposition of potassium cyanate by means of ethylamine sulphate; and by the action of ethyl cyanate on aqueous ammonia.—Large prisms, easily soluble in water and alcohol; fuses at 92°; not volatile without decomposition. Is not precipitated from its solution by means of nitric acid; when evaporated with nitric acid, however, it yields a crystalline nitrate.

Diethyl-urea, $CO\left\{\begin{matrix} NH.C^2H^5 \\ NH.C^2H^5, \end{matrix}\right.$ is formed by the action of ethyl cyanate on ethylamine; and by the decomposition of the former by means of water or sulphydric acid, in which case ethylamine always results at first.—Long prisms, which fuse at 112.5°, and boil undecomposed at 263°. Easily soluble in water and alcohol. Is decomposed by boiling with water, yielding ethylamine and carbonic acid.—A diethyl-urea $CO\left\{\begin{matrix} N(C^2H^5)^2 \\ NH^2, \end{matrix}\right.$ isomeric with the one described, is produced by the decomposition of potassium cyanate with diethylamine sulphate. It is resolved into carbonic acid, ammonia, and diethylamine by boiling with alkalies.

Triethyl-urea, $CO\left\{\begin{matrix} N(C^2H^5)^2 \\ NH.C^2H^5, \end{matrix}\right.$ is produced by the action of ethyl cyanate on diethylamine.—Crystalline;

fuses at 63°; and boils without decomposition at 223°; soluble in water, alcohol, and ether.

Ureas with bivalent alcohol radicles are also known. In the formation of these, however, two or more molecules of urea usually unite.

Acetyl-urea, $CO\left\{\begin{matrix} NH.C^2H^3O \\ NH^2, \end{matrix}\right.$ is produced, when acetyl chloride is poured upon urea (a spontaneous elevation of temperature takes place, and hydrochloric acid is evolved); and by heating urea with acetic anhydride.—Long, silky needles; but slightly soluble in cold water, more easily in hot water and alcohol; fuses at 112°, and is decomposed at a higher temperature, forming cyanuric acid and acetamide. Does not combine with acids.—*Bromacetyl-urea*, $CO.N^2H^3(C^2H^2BrO)$, is formed by the action of bromacetyl bromide on urea.—Colorless needles, difficultly soluble in cold water, more easily but with decomposition in hot.—*Tribromacetyl-urea* (see Barbituric Acid, p. 239).

Similar compounds, ureas, in which hydrogen atoms are replaced by acid radicles, are formed by the action of various agents on uric acid.

Sulphocarbamide (Sulpho-urea), $CS(NH^2)^2$. Is produced by heating dry ammonium sulphocyanate to 170°.—Long, silky needles, or thick, short rhombic prisms; fusing point, 149°; easily soluble in water and alcohol. Heated with water to 140°, it is reconverted into ammonium sulphocyanate. Combines with acids, oxides, and salts, like urea.

The *hydriodate* and *hydrochlorate* are produced by treating persulphocyanic acid with hydriodic acid, or tin and hydrochloric acid.

When its solution is digested with silver, lead, or mercury oxides, it is converted into dicyano-diamide (p. 217).

Ethyl-sulphocarbamide, $CS\left\{\begin{matrix} NH.C^2H^5 \\ NH^2, \end{matrix}\right.$ is produced by dissolving ethyl mustard-oil (p. 214) in alcoholic ammonia.—Colorless needles, pretty easily soluble in water; fusing point, 106°.

Diethyl-sulphocarbamide, $CS(NH.C^2H^5)^2$. Is prepared in the same way from ethyl mustard-oil and ethylamine; or by boiling ethylamine ethylsulphocarbamate (p. 227) on a water bath.—Crystals, that fuse at 77°, which, under the influence of phosphoric anhydride or hydrochloric acid gas, yield ethyl mustard-oil.

Uric Acid.
$C^5H^4N^4O^3$.

Occurrence. In urine, urinary calculi, and urinary sediments (compare Animal Chemistry). In small quantity in the blood and in the muscular fluid. In the form of sodium urate in the concretions found in the joints of gouty patients. In the excrements of birds, amphibious animals, insects, these excrements often consisting entirely of sodium urate.

Preparation. Calculi, consisting of uric acid, or, better, the excrement of serpents (ammonium urate, with various foreign substances), are dissolved in dilute caustic potassa or soda at the boiling temperature, the solution filtered and poured boiling hot into an excess of dilute hot sulphuric acid. The precipitated uric acid is washed out and dried. If it is not white, it is redissolved and again precipitated.—Or a current of carbonic anhydride is conducted into the solution of uric acid in potassa, by means of which, white acid potassium urate is precipitated; or the solution is mixed with a solution of ammonium chloride, when acid ammonium urate is precipitated. In both cases the precipitated salts are washed out, and decomposed by adding them to boiling dilute hydrochloric acid.

In order to prepare uric acid from guano, the latter is boiled with a solution of borax (1 part borax in 120 parts water), filtered, and the uric acid precipitated with hydrochloric acid. Or, dried and finely pulverized guano is added to an equal weight of concentrated hydrochloric acid, heated over the water-bath, allowed to stand on the water-bath until the smell of hydrochloric acid has disappeared, and the uric acid then precipitated by the addition of from twelve to fifteen times the volume of water. The crude acid obtained in this way is dissolved in alkali and purified, as above directed.

Properties. Light, white powder, consisting of fine crystalline scales, without taste and odor, exceedingly sparingly soluble in water, insoluble in alcohol and ether. Soluble in concentrated sulphuric acid without decomposition; from this solution there crystallizes on cooling a very deliquescent compound of uric acid with sulphuric acid $C^5H^4N^4O^3 + 2SO^4H^2$, which is decomposed by water. Easily soluble in nitric acid, undergoing decomposition; on evaporating the solution, there remains a mass, which turns purple with ammonium carbonate, and violet with potassa. By distillation, it yields, among other products, a great deal of hydrocyanic acid, a sublimate, consisting of urea, cyanuric acid, and ammonium cyanide, and leaves behind nitrogenous carbon. When heated with concentrated hydriodic acid to 160–170°, it yields glycocol (p. 84), ammonium iodide, and carbonic acid.

Uric acid is a very weak, bibasic acid. The neutral alkaline urates are white, granular, crystalline, difficultly soluble in water, but easily soluble in an excess of potassa. Carbonic anhydride precipitates from the solution the acid salt, in the form of a translucent jelly, which soon becomes powdery. The acid ammonium salt separates in the same form, when the dissolved potassium salt is mixed with sal-ammoniac. It afterwards shrinks up, forming a white powder.

From uric acid, a long series of transformation products can be produced. The most remarkable are the following:—

Uroxanic acid, $C^5H^8N^4O^6$. When a solution of uric acid, in an excess of concectrated potassa, is allowed to stand for a long time in contact with the air, *potassium uroxanate*, $C^5H^6N^4O^6K^2 + 3H^2O$, separates in the form of laminæ of a mother-of-pearl lustre. From a solution of the salt, the free acid is precipitated as a crystalline powder by means of hydrochloric acid. It is soluble in hot water, but only with decomposition and evolution of carbonic anhydride.

Alloxan (Mesoxalylurea), $C^4H^2N^2O^4 =$
$CO\left\{\begin{matrix} NH.CO \\ NH.CO \end{matrix}\right\}CO$. This is produced by the action of concentrated, cold nitric acid on uric acid, urea being formed at the same time. When uric acid is added to nitric acid, alloxan is thrown down immediately as a white crystalline powder, which, when perfectly freed of acid, need only be recrystallized from water. It is prepared most practically from alloxantine. The latter is moistened with very concentrated nitric acid, so as to form a thick, pasty mass, and then allowed to stand for a few days, until, as may be tested with a small portion, it dissolves readily and completely in cold water. The mass is then allowed to dry completely in the air, spread out on bricks, and, after the removal of the last trace of nitric acid, by heating over the water-bath, recrystallized from hot water.

On the cooling of the warm aqueous solution, it crystallizes out with four molecules of water of crystallization; if the solution is evaporated by the aid of heat, it crystallizes with only one molecule of water. The former consists of large, shining, transparent, rhombic crystals of the form of heavy spar; effloresces in the air and loses three molecules of water; crystallized with one molecule, it forms smaller, hârder, monoclinic crystals, which do not effloresce. It is easily soluble in water; the solution imparts to the skin a repulsive odor, and colors it purple; it tastes unpleasantly sour and saltish; shows an acid reaction; is decomposed by heating. It gives an indigo-blue color with salts of iron suboxide.—It unites with alkaline bisulphites by heating, forming salts which can be obtained in large crystals.

The aqueous solution decomposes slowly at the ordinary temperature, rapidly by boiling, into alloxantine, parabanic acid, and carbonic anhydride. When boiled with baryta water or lead acetate, it is resolved into mesoxalic acid (p. 158) and urea. Heated with dilute acids, an evolution of carbonic anhydride takes place, and alloxantine, urea, and oxalic acid are formed. Nitric acid converts it into parabanic acid and carbonic anhydride; lead peroxide into urea, oxalic acid, and carbonic anhydride; reducing agents into alloxantine, and dialuric acid.

Alloxanic acid, $C^4H^4N^2O^5 = CO\left\{\begin{matrix} NH^2 \\ NH.CO.CO.CO.OH. \end{matrix}\right.$ Is produced by combination of alloxan with alkalies. When baryta water is dropped into a solution of alloxan at 60°, until the white precipitate, which is formed on the addition of each drop, is no longer redissolved, *barium alloxanate*, $C^4H^2N^2O^5Ba + 4H^2O$, crystallizes out on cooling, in small, very difficultly soluble crystals. This salt, after being washed, is decomposed by sulphuric acid. The filtrate, on evaporation, at a temperature below 40°, yields a syrup, which solidifies after a time, forming a crystalline mass.

Alloxanic acid forms a radiated, crystalline mass; very easily soluble; very acid; dissolves zinc with an evolution of hydrogen; is not changed by sulphuretted hydrogen; and cannot be reconverted into alloxan. Its salts are decomposed by boiling with water, yielding urea and mesoxalates (p. 158).

Parabanic acid (Oxalylurea), $C^3H^2N^2O^3 =$ $CO\left\{\begin{matrix} NH.CO \\ NH.CO, \end{matrix}\right.$ is formed when uric acid or alloxan is dissolved in moderately strong nitric acid, and the solution evaporated to a syrupy consistence; or when uric acid is treated with manganese peroxide and sulphuric acid. Occasionally it is obtained, instead of alloxan, in the preparation of the latter.—Crystallizes in colorless, broad, very thin prisms; is permanent in the air; tastes very acid; and is easily soluble in water. It gives a white precipitate, $C^3Ag^2N^2O^3$, with silver nitrate; and this, when heated with methyl iodide, yields silver iodide, and *dimethylparabanic acid* (cholestrophane), $C^3(CH^3)^2N^2O^3$, a compound, crystallizing in broad laminæ, of a silvery lustre, easily fusible and sublimable.

Parabanic acid is decomposed when boiled with dilute acids, yielding urea and oxalic acid. Nascent hydrogen (zinc and hydrochloric acid) converts it into oxalantine.

Oxaluric acid, $C^3H^4N^2O^4 = CO\left\{\begin{matrix} NH^2.CO.OH \\ NH.CO. \end{matrix}\right.$

Bears the same relation to parabanic acid that alloxanic acid bears to alloxan. The real salts of parabanic acid do not appear to be capable of existence. Strong bases immediately cause an assimilation of the elements of water, with which parabanic acid is transformed into oxaluric acid. The ethyl ether is produced by the action of ethyloxy-oxalyl chloride (p. 155) on urea.

When parabanic acid is dissolved in ammonia and heated, the solution turns into a white mass of fine crystals of a silky lustre. These are ammonium oxalurate, $C^3H^3N^2O^4.NH^4$. This salt is also contained in small quantity in urine. It is difficultly soluble in water. When its solution in hot water is mixed with an acid, oxaluric acid separates as a white crystalline powder. It is very difficultly soluble, though the solution tastes and reacts acid. Its *silver salt*, $C^3H^3N^2O^4Ag$, formed by double decomposition from the ammonium salt, separates in thick, white flocks. It is soluble in hot water, from which it crystallizes, on cooling, in fine needles of a silky lustre. When a solution of oxaluric acid is heated for some time to boiling, it is converted into urea oxalate and oxalic acid.

Oxaluramide (Oxalan), $C^3H^5N^3O^3 = CO\left\{\begin{matrix} NH^2. \\ NH. \end{matrix}\right. \begin{matrix} CO.NH^2 \\ CO, \end{matrix}$ is formed, as a white precipitate, when a solution of alloxan is mixed with hydrocyanic acid and then with ammonia. Alloxan is thus resolved into oxaluramide, dialuric acid, and carbonic anhydride—water and ammonia being assimilated.—White, crystalline powder, but slightly soluble in water;

soluble in concentrated sulphuric acid without decomposition; is reprecipitated by water; is resolved, by boiling with water, into oxalic acid, urea, and ammonia.

If methylamine, ethylamine, or analogous bases are employed instead of ammonia in the preparation of oxaluramide, crystalline precipitates, similar to oxaluramide, consisting of methyl- or ethyl-oxaluramide, are formed.

Oxalantine, $C^6H^4N^4O^5 + H^2O$. Parabanic acid, in contact with zinc and hydrochloric acid, yields a crystalline powder, containing zinc, which dissolves with great difficulty in boiling water. If water is added to it, and it is then treated with sulphuretted hydrogen, zinc sulphide is precipitated, and the aqueous solution contains oxalantine, which it deposits in crystals on being evaporated. It is produced, together with other substances, by boiling a concentrated solution of alloxanic acid.—White, crystalline crusts, difficultly soluble in water, is not decomposed by hot concentrated nitric acid; but it reduces the metals from silver or mercury salts after an addition of ammonia.

Alloxantine, $C^8H^4N^4O^7 + 3H^2O$. Is produced by spontaneous decomposition of alloxan, when left to itself; by the action of dilute nitric acid on uric acid; or of reducing agents on alloxan. It is prepared most readily by dissolving uric acid in warm dilute nitric acid (1 part nitric acid of specific gravity 1.42 and 8–10 parts of water of 60–70°), and adding carefully a solution of tin chloride, containing concentrated hydrochloric acid. It is obtained from alloxan by conducting concentrated hydrogen into the aqueous solution of the latter. It is hereby thrown down mixed with sulphur, from which it may be separated by dissolving in boiling water.—Small, colorless, hard prisms; becomes red and purple in an ammoniacal atmosphere; is very difficultly soluble in cold water, easily in hot. The solution gives, with baryta water, a beautiful violet precipitate, which, when heated, becomes white, and is resolved into barium alloxanate and dialurate; with silver nitrate it gives a grayish-black precipitate of metallic silver. It is converted into alloxan by nitric acid; by boiling with hydrochloric acid into alloxan, parabanic acid, and a difficultly soluble crystalline acid, *allituric acid* $C^6H^6N^4O^4$. Heated with water to 180–190° it is decomposed, forming oxalic acid, ammonia, carbonic anhydride, and carbonic oxide.

Alloxantine, dissolved in water, at the ordinary temperature, is transformed, in the air, into ammonium oxalurate, oxygen being absorbed and water formed.

Dialuric acid (Tartronyl- or Oxymalonylurea), $C^4H^4N^2O^4 = CO\left\{\begin{matrix} NH.CO \\ NH.CO \end{matrix}\right\}CH.OH$, is produced by the reduction of alloxantine, particularly when sulphuretted hydrogen is conducted for a long time into its boiling solution. The solution, filtered from the sulphur, gives, with ammonium carbonate, a fine crystalline precipitate of *ammonium dialurate*, $C^4H^3N^2O^4.NH^4$, which is difficultly soluble in cold water, more easily in hot: becomes red in the air. It is decomposed when dissolved in warmed hydrochloric acid, and, on cooling, the free acid separates in crystals. Further, it is produced from alloxantine by the action of sodium-amalgam; and by mixing a solution of alloxantine with a hot solution of tin subchloride and an excess of hydrochloric acid. It is prepared most easily by the last method.—Long needles; moderately easily soluble in water; the crystals turn red in the air, and are gradually converted into alloxantine.

When a solution of alloxan is added to a solution of dialuric acid, there is formed a precipitate of alloxantine.

Hydurilic acid, $C^8H^6N^4O^6$. Is produced, together with glycocol and pseudoxanthine (p. 246), by heating uric acid with double its weight of concentrated sulphuric acid to 110–130°, and then adding the mass to a great deal of water.—When a solution of dialuric acid in glycerin is heated to 140–150°, it is resolved into carbonic anhydride, formic acid, and acid ammonium hydurilate. The same salt is produced by boiling alloxantine with very concentrated sulphuric acid. In order to separate the acid, the salt is dissolved in boiling water with ammonia, and to the filtered solution, copper sulphate added. From the dark-green solution are deposited black, anhydrous crystals, if the solution was still hot on the addition of the copper sulphate; if the solution was cold, red crystals, containing water, are deposited; in both cases the crystals consist of the neutral copper salt. This is thrown into hot hydrochloric acid, the crystalline hydurilic acid, which separates, washed with dilute hydrochloric acid, and recrystallized from water.—It crystallizes from water in small four-sided columns with two molecules of water of crystallization; from its salt, on the addition of hydrochloric acid, with the aid of heat, it separates in small, rhombohedric plates with one molecule of water. Difficultly soluble in water and alcohol. Strong, bibasic acid. The solutions of the acids and its salts become colored a beautiful dark green on the addition of a solution of iron sesquichloride.—A mixture of hydrochloric acid and potassium chlorate converts it into *dichlorhydurilic acid*, $C^8H^4Cl^2N^4O^6$. Fuming nitric acid converts it into alloxan; ordinary nitric acid yields, in addition to this, violuric acid, violantine, and dilituric acid. When heat is employed only the last acid is produced.

Barbituric acid (Malonylurea), $C^4H^4N^2O^3 = CO\left\{\begin{matrix} NH.CO \\ NH.CO \end{matrix}\right\}CH^2$. By heating a solution of alloxantine in three to four parts concentrated sulphuric acid on a water-bath, until the evolution of sulphurous anhydride has ceased, there is obtained a honey-colored solution, which becomes thick on cooling. When this is diluted with an equal volume of water, an abundant precipitate of a difficultly soluble body is obtained, which is completely dissolved by boiling with water. On the cooling of this solution barbituric acid crystallizes out and parabanic acid remains in the mother liquor.—Large, colorless prisms, but sparingly soluble in cold, easily in hot water. Is resolved, by heating with alkalies, into malonic acid (p. 157) and urea or its decomposition products, carbonic anhydride and ammonia.

Dibrombarbituric acid, $C^4H^2Br^2N^2O^3$, is produced by the action of bromine on barbituric, violuric, or dilituric acids.—Crystallizes in laminæ or prisms; is difficultly soluble in cold water, easily in hot, in ether and alcohol; it is decomposed by boiling with water, alloxan being formed. When treated with reducing agents, it yields different products, according to the nature of the agent employed; metallic zinc converts it into monobrombarbituric acid; sulphuretted hydrogen into dialuric acid; hydriodic acid, in small quantity, into hydurilic acid. If a solution of the acid, saturated by the aid of heat and then cooled, is allowed to stand with bromine, or if chlorine is conducted into the solution, carbonic anhydride is evolved and *tribromacetyl-urea*, $CO.N^2H^3(C^2Br^3O)$, is formed, which crystallizes in needles or laminæ; fuses at 148°; is difficultly soluble in water; and is decomposed by boiling with it, yielding bromoform.

Monobrombarbituric acid, $C^4H^3BrN^2O^3$, is produced together with cyanogen bromide by the action of aqueous hydrocyanic acid on dibrombarbituric acid.—White crusts, which consist of small needles, difficultly soluble in cold water. Its salts are formed by the action of metals, hydroxides, or acetates on dibrombarbituric acid. By the action of baryta water at the ordinary temperature, tribromacetyl-urea is formed at the same time.

Nitro-barbituric acid (Dilituric acid), $C^4H^3(NO^2)N^2O^3+3H^2O$, is formed by treating barbituric acid with fuming nitric acid and by heating hydurilic acid with ordinary nitric acid.—Colorless, quadratic prisms or laminæ, which effloresce in the air; easily soluble in hot water, with more difficulty in cold water, the solution being of an intense yellow color; but slightly soluble in alcohol, and insoluble in ether. Tribasic acid; usually forms salts, however, with one atom of metal, and these are so stable that mineral acids cannot liberate the dilituric acid from them. By heating with bromine and a little water to 100° it is converted into dibrombarbituric acid.

Nitrosobarbituric acid (Violuric acid), $C^4H^3(NO)N^2O^3+H^2O$, is produced by the action of nitric or nitrous acids on hydurilic acid. The potassium salt can be most easily obtained by treating hydurilic acid with potassium nitrate and alcohol with the employment of heat; it is also formed by the action of potassium nitrite on barbituric acid. From the potassium salt is prepared the insoluble red barium salt by precipitating with barium chloride, and this, suspended in hot water, is exactly decomposed by means of sulphuric acid.—It crystallizes in rhombic octahedrons; is moderately soluble in cold water, easily in hot. Warmed with caustic potassa, it is resolved into nitrosomalonic acid (p. 158) and urea; by heating with nitric acid, it is converted into dilituric acid. Monobasic acid.

Potassium violurate, $C^4H^2N^3O^4K + 2H^2O$. Deep blue prisms or laminæ; much more soluble in hot water than in cold, the solution having a violet color. Caustic potassa colors the solution red. The *ammonium salt* resembles the *potassium salt.* The *sodium salt* and the salts of the alkaline earths form intensely red colored crystals. The free acid produces a deep dark-blue color in a solution of ferrous acetate, and on the addition of alcohol the ferrous salt is deposited in six-sided plates with a red metallic lustre.

Amidobarbituric acid (Uramile, Murexan), $C^4H^3(NH^2)N^2O^3$, is formed by the action of hydriodic acid on violuric acid or dilituric acid; by conducting sulphuretted hydrogen into a solution of violuric acid. Can be prepared most easily by mixing a solution of alloxantine with a boiled solution (freed of air) of sal-ammoniac; it is then deposited in the form of fine white crystals.—Colorless, white needles, which become red in the air; insoluble in cold water, somewhat soluble in boiling. Nitric acid converts it into alloxan. It dissolves without change in ammonia; if this solution is boiled, however, it is converted into murexide. If boiled with water, and mercury oxide be gradually added, it is converted into murexide, metallic mecury being thrown down.

Thionuric acid, $C^4H^5N^3SO^6$. If a solution of alloxan be saturated at the ordinary temperature with sulphuric acid and afterwards with ammonia and then heated to boiling, it deposits, on cooling, a difficultly soluble salt, crystallizing in thin scales of a mother-of-pearl lustre, *ammonium thionurate*, $C^4H^3N^3SO^6(NH^4)^2 + H^2O$. The same salt is produced by warming violuric acid with ammonium sulphite. The free acid separated from this salt is a white, crystalline, easily soluble, acid mass. The ammonium salt precipitates metallic silver from dissolved silver salts. By boiling its aqueous solution, it is decomposed, forming uramile and sulphuric acid.

Purpuric acid, $C^8H^5N^5O^6$. Unknown in the free state.

Acid ammonium purpurate (Murexide), $C^8H^4N^5O^6.NH^4 + H^2O$, is produced, when ammonia gas is conducted for a long time over dried alloxantine at 100°; or when a solution of alloxantine and alloxan is mixed with ammonia and diluted with half a volume of hot water. The formation of murexide is the cause of the reaction of uric acid mentioned above (p. 233).—Prepared most practically by heating slowly to boiling 4 parts uramile and 3 parts mercury oxide with water. The boiling hot, filtered solution yields crystals of murexide.—Crystallizes in four-sided columns or plates of an exceedingly beautiful green color of a metallic lustre, greatly resembling the color of the wings of the gold-beetle. By transmitted light they are red, and they give a red powder. Difficultly soluble in cold water, more easily in hot, the solution having a purple color.

Acid potassium purpurate, $C^8H^4N^5O^6.K$, is produced by boiling a solution of murexide with saltpetre. It resembles the ammonium salt.

Purpuric acid cannot be isolated from its salts, as water resolves it into uramile and alloxan, at the moment of its liberation.

Pseudo-uric acid, $C^5H^6N^4O^4$. The potassium salt of this acid, $C^5H^5N^4O^4K + H^2O$, is produced, when uramile or murexide is heated with a concentrated solution of potassium cyanate, until the solution loses the property of turning red in the air. If the potassium salt, which has been separated, is now dissolved in caustic potassa and decomposed with hydrochloric acid, the free acid is thrown down as a white, crystalline powder, consisting of small prisms. Inodorous, tasteless, but sparingly soluble in water, easily soluble in free alkalies. It differs from uric acid only by containing the elements of water more, and yields, like this, alloxan, when treated with nitric acid; when treated with lead peroxide, however, it yields no allantoine, but carbonic anhydride, oxalic, and oxaluric acids and urea. Monobasic acid.

Allantoine, $C^4H^6N^4O^3$. Contained in the urine of sucking calves, in the urine of dogs whose respiration is disturbed, in human urine, especially after large quantities of tannic acid have been taken internally. The allantoic fluid of the cow (*i. e.* the urine of the fœtus) is particularly rich in it. It can be obtained from this liquid in crystalline form by concentrating it.—Uric aid, heated with water and lead oxide gradually added, is converted into allantoine, urea, oxalic acid, and carbonic anhydride—oxygen and water being assimilated. The hot filtrate from lead oxalate (and urate), after the removal of the dissolved lead by means of sulphuretted hydrogen, deposits allantoine in crystals on evaporation. In the last mother-liquor urea remains. It is also produced, when uric or dialuric acid is boiled with potassium nitrate and acetic acid; further, by the oxidation of uric acid with manganese peroxide or potassium ferricyanide, and by the action of ozone upon it.—Colorless rhombohedric prisms, but slightly soluble in cold water, more easily in boiling. It combines with several metallic oxides. When a hot solution of it is mixed with silver nitrate and ammonia, a white precipitate of *allantoine-silver*, $C^4H^5N^4O^3Ag$, is deposited.—Heated with sulphuric acid, it is resolved into ammonia, carbonic anhydride, and carbonic oxide. —When allantoine is dissolved in ordinary nitric acid, *allanic acid*, $C^4H^5N^5O^5+H^2O$, is formed. This acid crystallizes from a small amount of water in stellate groups of needles. It is decomposed at 210–220°, without previously being fused.—When a solution of allantoine in an excess of potassa-ley is left to itself for several days, the potassium salt of a new acid, *allantoic acid*, $C^4H^3N^4O^4$, crystallizes out on the addition of acetic acid and a little alcohol.

Glycolurile, $C^4H^6N^4O^2$, is formed by the action of sodium-amalgam on a warm solution of allantoine.—Small octahedral crystals or lance-shaped needles. More difficultly soluble in water than allantoine.

Hydantoine (Glycolylurea), $C^3H^4N^2O^2 =$

$CO\left\{\begin{array}{l}NH.CH^2\\NH.CO,\end{array}\right.$ is produced, together with urea, by heating allantoine with hydriodic acid; and by boiling glycolurile with acids; together with carbonic anhydride, water, and free iodine, by heating alloxanic acid with hydriodic acid. Is further formed by the action of an excess of alcoholic ammonia on monobromacetyl-urea (p. 231).—Colorless crystals, easily soluble in hot water, moderately in cold; fuse at 206°; do not react on litmus paper; taste sweetish.

Methylhydantoine, $C^4H^6N^2O^2 = CO \left\{ \begin{matrix} N(CH^3).CH^2 \\ NH \quad .CO, \end{matrix} \right.$
is produced by prolonged heating of creatinine (p. 249) with baryta water to 100°.—Colorless crystals, easily soluble in water and alcohol. Fuses at 145°, and when carefully heated sublimes at 145° in small lustrous crystalline flakes. When boiled with silver or mercury oxides, it yields salts.

Ethylhydantoine, $C^3H^3(C^2H^5)N^2O^2$, is formed by heating ethylglycocol (p. 86) with urea to 120–125°.—Large, colorless rhombic prisms, appearing of a plate form, easily soluble in water and alcohol, in ether more difficultly soluble. Fuses below 100°.

Hydantoic acid (Glycoluric acid), $C^3H^6N^2O^3 =$
$CO \left\{ \begin{matrix} NH \ .CH^2 \\ NH^2.CO.OH, \end{matrix} \right.$ is formed by boiling allantoine, hydantoine or glycolurile with barium hydroxide; and by heating glycocol with urea.—Large rhombic prisms, easily soluble in hot water, rather difficultly soluble in cold water. Monobasic acid. The *barium salt* is precipitated from an aqueous solution by means of alcohol, as an amorphous, flocky, or syrupy mass. It is soluble in water in all proportions, and does not crystallize. Heated with concentrated hydriodic acid, hydantoic acid is resolved into glycocol, ammonia, and carbonic anhydride. It bears the same relation to hydantoine as alloxanic acid to alloxan, and oxaluric acid to parabanic acid.

Allanturic acid (Glyoxylurea ?), $C^3H^4N^2O^3$. Is produced at first, together with urea, by boiling allantoine with baryta water, and by treating allantoine with lead peroxide or nitric acid.—Non-crystalline, deliquescent acid.—But slightly known.—Is resolved, by further action of baryta water, into hydantoic and parabanic acids, and the latter acid (oxalylurea) is immediately decomposed, forming oxalic acid, carbonic anhydride, and ammonia. Hence hydantoic acid, oxalic acid, carbonic anhydride, and ammonia are the final products of the action of barium hydroxide on allantoine.

Xanthine (Xanthic oxide), $C^5H^4N^4O^2$. Is formed by the reduction of uric acid by means of sodium-amalgam. Occurs in urine, in muscular flesh, and in a number of glandular organs. In large quantity in certain rare urinary calculi which are found in human bladders and often consist entirely of it. Further, it occurs in some varieties of guano from which it can be extracted by caustic soda and afterward precipitated by carbonic anhydride. After the use of sulphur baths, it appears in larger quantity in the urine.—In the form of calculi, it has a brownish flesh-color. In a purified state it forms a white, amorphous mass, or small scales. Insoluble in cold water, very difficultly soluble in hot; sublimable, but only with partial decomposition. Combines with acids and bases. It dissolves in ammonia and the boiling saturated solution deposits crystals of *xanthine-ammonia* on cooling. Silver nitrate gives a precipitate, $C^5H^4N^4O^2 + Ag^2O$, in an ammoniacal solution, which is insoluble in ammonia, soluble in hot nitric acid.

A compound isomeric with xanthine, *pseudoxanthine*, is produced together with hydurilic acid and glycocol by heating uric acid with concentrated sulphuric acid.

Sarcine (Hypoxanthine), $C^5H^4N^4O$, is formed by long-continued action of sodium-amalgam on uric acid or xanthine. Occurs in a great many of the animal organs and fluids; in muscular flesh, particularly in the cardiac muscles of the horse; in the liver and the spleen of the ox; in the human liver in certain diseases of this organ; in urine; in blood, etc. It is always accompanied by xanthine.—Microscopic, needly crystals, difficultly soluble in cold water, more easily in hot, but sparingly in alcohol. It is readily dissolved by alkalies, as well as by diluted acids, crystalline compounds being formed. It combines also with salts. On the addition of silver nitrate to its ammoniacal solution, there is formed a precipitate, $C^5H^4N^4O + Ag^2O$, which is insoluble in ammonia. This dissolves in hot dilute nitric acid, and, on cooling, white needles appear in the solution. These are a compound of sarcine with silver nitrate, $C^5H^4N^4O + NO^3Ag$. This compound is completely insoluble in water, and can be employed for the purpose of estimating sarcine quantitatively.

Guanine, *glycocyamine* and *glycocyamidine*, *creatine* and *creatinine* bear a close relation to uric acid, but have not yet been prepared from it.

Guanine, $C^5H^5N^5O$. Is contained in guano, the changed excrements of sea-birds; and in the excrements of garden-spiders. It has besides been shown to be present in small quantity in the liver and pancreas and in the scales of the bleak. In pork, in a certain disease (guanine-gout), concretions of guanine occur. In order to prepare it, guano is suspended in water and gradually milk of lime added to it, boiled and filtered. This is repeated until the filtrate is no longer colored. The residue, which consists essentially of guanine and uric acid, is boiled repeatedly with sodium carbonate, until the solution no longer gives a precipitate on the addition of hydrochloric acid. The combined extracts are then mixed with sodium acetate and hydrochloric acid, until the solution shows a strongly acid reaction. After washing the precipitate thus obtained, the guanine may be taken up by boiling with dilute hydrochloric acid, while the uric acid remains for the greater part undissolved. On evaporating the solution in hydrochloric acid, guanine hydrochlorate crystallizes out, from which the guanine can be separated by means of ammonia. Prepared in this way, it is still impure from the presence of some uric acid, which must be decomposed by dissolving in concentrated nitric acid. From the guanine nitrate, which crystallizes out on evaporating, the base is obtained pure by decomposing with ammonia.

White, amorphous powder, insoluble in water; combines with acids, with bases, and also with salts, forming compounds, that crystallize well.

Nitrous acid decomposes guanine, with an evolution of nitrogen, forming xanthine and a nitro-compound, as yet but little known, which, when treated with reducing substances, yields xanthine. When moderately concentrated hydrochloric acid is poured on guanine, and potassium chlorate then added gradually, until all is dissolved, water and oxygen are assimilated, and the guanine is resolved into *guanidine* (p. 219), parabanic acid, and carbonic anhydride; at the same time, however, xanthine, oxaluric acid, urea, and oxalic acid are formed as secondary products.

Glycocyamine, $C^3H^7N^3O^2$, is formed by direct combination of glycocol with cyanamide, and separates, when the aqueous solution of both bodies, containing ammonia, is allowed to stand for several days.—Colorless crystals, difficultly soluble in cold water, more easily in hot water, insoluble in alcohol; yields salts with acids and with bases.

Glycocyamidine, $C^3H^5N^3O$. The hydrochlorate is formed by heating glycocyamine hydrochlorate to 160°. The free base, separated from the salt by means of lead hydroxide, forms easily soluble laminæ, which have an alkaline reaction.

Creatine (Methylglycocyamine), $C^4H^9N^3O^2 + H^2O$. Occurs in the juice of flesh of all classes of animals; in the blood and brain in small and varying quantities; not in the urine.—To prepare it, chopped meat is pressed out with cold water, the liquid boiled in order to coagulate the albumen, the filtrate mixed with baryta water to remove phosphoric acid, and the filtered liquid evaporated to one-twentieth its volume. On cooling, the creatine gradually crystallizes out, and is then purified by recrystallization. In the same manner, it can be readily prepared from commercial extract of meat. It is obtained artificially by the direct combination of methylglycocol (sarcosine, p. 85) with cyanamide.

Colorless, lustrous prisms, of a slightly bitter taste, very difficultly soluble in cold water, more easily in hot, insoluble in absolute alcohol.

It combines with acids, forming crystallizable salts. By heating with baryta water, it yields urea, sarcosine and methylhydantoine (p. 245). By boiling with mercury oxide, oxalic acid, carbonic anhydride, and methyluramine are formed.

Creatinine, $C^4H^7N^3O$, does not occur in muscular flesh; is, however, contained in urine in considerable quantity (in normal urine of twenty-four hours, 1–1.3 grms. creatinine). It is a decomposition-product of creatine. Even by evaporating its aqueous solution with the aid of heat, creatine is partially converted into creatinine; if acetic acid is previously added the conversion is complete.

In order to prepare it from urine, this is quickly evaporated to from one-eighth to one-tenth its volume, precipitated by calcium chloride and milk of lime, and the filtrate evaporated until the sodium chloride crystallizes out. From the mother liquid the creatinine is precipitated by means of a thick solution of zinc chloride, free of hydrochloric acid. In a few days, a thick, crystalline pulp is deposited which is washed with cold water, then dissolved in boiling water, and decomposed by boiling with lead hydroxide. On evaporating this solution creatinine remains behind, still impure from the presence of some creatine, from which it can be separated by dissolving in absolute alcohol.

Colorless prisms, much more soluble in alcohol and water than creatine. Strong base, reacts alkaline and drives out ammonia from its salts. Unites also with acids and a few salts.—*Creatinine-zincochloride*, $2(C^4H^7N^3O) + ZnCl^2$, is particularly characteristic. It is a granular, crystalline powder, difficultly soluble in water, insoluble in alcohol. It dissolves in hydrochloric acid, forming *creatinine-zincochloride hydrochlo-*

rate, $2(C^4H^7N^3O.HCl)+ZnCl^2$, a crystalline compound, easily soluble in water, from which solution creatinine, zincochloride is precipitated by a concentrated solution of sodium acetate.

In contact with bases, creatinine, even at ordinary temperatures, is generally converted into creatine. Heated for a length of time with barium hydroxide, it is resolved into ammonia and methylhydantoine (p. 245), water being assimilated. By boiling with mercury oxide and water, it yields the same products as creatine.

Ethyl iodide combines directly with creatinine, forming crystalline *ethylcreatinine iodide*, $C^4H^7N^3O.C^2H^5I$, from which, by the action of silver oxide, *ethylcreatinine hydroxide*, $C^4H^7N^3O.C^2H^5.OH$, is obtained. This latter compound does not combine with ethyl iodide, but is decomposed by it, yielding alcohol and ethylcreatinine iodide.

II. BENZENE DERIVATIVES (AROMATIC COMPOUNDS).

THE aromatic compounds are derived from benzene, C^6H^6, and the hydrocarbons homologous with it, just as the fatty bodies are derived from marsh-gas and its homologues. Benzene is the common nucleus of all these bodies. The carbon-atoms in benzene are combined in the form of a ring, in such a manner that they are united alternately with first one and then two affinities with each other, and the fourth affinity of each carbon-atom is saturated with a hydrogen-atom:—

$$\begin{array}{c}(6)HC:CH(1)\\ (5)H\dot{C}\quad \dot{C}H(2)\\ (4)H\ddot{C}\,.\,\ddot{C}H(3).\end{array}$$

By the displacement of one or more hydrogen-atoms by methyl CH^3, ethyl C^2H^5, etc., the homologues of benzene are formed; by the displacement of hydrogen-atoms by hydroxyl groups, the real aromatic alcohols (phenols); by the displacement of hydrogen atoms by CHO, the aldehydes; and, finally, by the displacement of hydrogen-atoms by CO.OH, the aromatic acids. All facts lead us to the conviction, that, just as we have observed in the case of marsh-gas that all the hydrogen-atoms had the same relative value, the six hydrogen-atoms in benzene also have the same value. Hence, of all the compounds, formed by the displacement of one hydrogen-atom in benzene by a monovalent element or monovalent group of atoms, only one modification can exist. When two or more hydrogen-atoms are displaced, the properties of the compounds are materially dependent upon the relative position of the displacing atoms or atomic groups to each other. Of every disubstitution-product (for instance, $C^6H^4Cl^2$, $C^6H^4(OH^2)$, $C^6H^4(CO.OH)^2$) three isomeric varieties can exist, viz.:—

(1) Those in which the hydrogen-atoms of two neighboring carbon-atoms are displaced (1 : 2 or 1 : 6, 2 : 3, 3 : 4, etc.). These compounds have been designated by means of the prefix *ortho;*

(2) Those in which the hydrogen-atoms of two carbon-atoms are displaced, which are separated from each other by one group CH (1 : 3 or 1 : 5, 2 : 4, etc.). *Meta-compounds;*

(3) Those in which the hydrogen-atoms of two carbon-atoms are displaced, which are separated by two groups CH from each other (1 : 4, 2 : 5 or 3 : 6). *Para-compounds.*

A similar method of consideration shows that by the displacement of three hydrogen-atoms by one and the same kind of atoms or atomic groups, also three isomeric compounds (1 : 2 : 3, 1 : 2 : 4 and 1 : 2 : 5) can be formed, and that the number of the possible cases of isomerism becomes much greater when the three atoms or atomic groups are unlike.

In the homologues of benzene, cases of isomerism may also be caused by the substitution taking place in the group CH^3, and not in the benzene-residue. From methylbenzene (toluene) $C^6H^5.CH^3$ are derived thus the two classes of isomeric compounds:—

$$C^6H^4Cl.CH^3 \quad \ldots \quad C^6H^5.CH^2Cl$$

$$C^6H^4\left\{\begin{matrix}OH\\CH^3\end{matrix}\right. \quad \ldots \quad C^6H^5.CH^2.OH$$

$$C^6H^4\left\{\begin{matrix}CO.OH\\CH^3\end{matrix}\right. \quad \ldots \quad C^6H^5.CH^2.CO.OH$$

$$C^6H^4\left\{\begin{matrix}CH^3\\CH^3\end{matrix}\right. \quad \ldots \quad C^6H^5.CH^2.CH^3.$$

FIRST GROUP.

A. Hydrocarbons, C^nH^{2n-6}.

In the preparation of coal-gas by the destructive distillation of coal, a secondary product is obtained in the form of a tar (coal-tar), which, subjected to distillation, yields a large number of bodies of various character. At first an oil distills over, which is lighter than water (*light oil*), and which consists mainly of benzene, toluene, dimethylbenzene, and trimethylbenzene. At a later stage of the process an oil passes over, which sinks under water (*creosote oil*, *dead oil*). This contains particularly two alcoholic bodies, *phenol* and *cresol*; and, besides these, volatile bases, *anilin*, *pyridine bases* (p. 130), and several hydrocarbons, partially liquid, partially solid.

In order to prepare the hydrocarbons from the light oil, this is first shaken successively with sulphuric acid, alkalies, and water, for the purpose of removing foreign substances, and then that portion of the oil, which remains undissolved, separated into its constituents by means of long-continued partial distillation. The different isomeric modifications of dimethyl- and trimethylbenzene, however, cannot be separated from each other in this manner.

1. *Benzene* (*Benzol*).
C^6H^6.

Preparation. That portion of the purified light oil that boils at 80–85°, congeals almost completely when cooled down to —5° to —10°. That which remains liquid is poured off and the crystals pressed between layers of filtering-paper below 0°. Can be most readily prepared in a pure condition by the distillation of an intimate mixture of 1 part benzoic acid with 3 parts quicklime.

It is also produced, together with other hydrocarbons of higher boiling points, when acetylene (p. 131) is heated to a temperature at which glass begins to soften. Three molecules of acetylene combine to form one molecule of benzene.

Properties. Colorless liquid of a peculiar odor, specific gravity at 0°, 0.899. Boils at 81–82°, and congeals at 0°. Burns with a luminous flame. Excellent solvent for resins, fats, etc.

When chlorine acts on benzene, products of substitution and of addition are formed simultaneously.

Benzene hexachloride, $C^6H^6Cl^6$, and **Benzene hexabromide,** $C^6H^6Br^6$, are formed by the action of an excess of chlorine or bromine on benzene in direct sunlight. The former compound is also produced by passing chlorine into boiling benzene. Both compounds are solid and crystallizable (the chloride fuses at 157°), and are decomposed partially when heated alone, completely when heated with bases, into hydrochloric acid or hydrobromic acid and trichlor- or tribrombenzene.

Benzene hypochlorite, $C^6H^6(ClOH)^3$, is produced by bringing benzene and an aqueous solution of hypochloric acid together.—Colorless laminæ, which fuse at 10°. But slightly soluble in water, easily in alcohol, ether, and benzene. Decomposes when kept in contact with the air, and, when heated with a very dilute solution of sodium carbonate, yields *phenose* $C^6H^6(OH)^6$, an amorphous, deliquescent mass. Heated with hydriodic acid to 120°, benzene hypochlorite and phenose both yield β-hexyl iodide $C^6H^{13}I$ (p. 72).

Chlorine substitution-products are obtained by conducting chlorine into benzene containing iodine. In this way are produced: *monochlorbenzene*, C^6H^5Cl, also formed by the action of phosphorus pentachloride on phenol. Liquid fusing at 135°; congeals only below —40°.—*Paradichlorbenzene*, $C^6H^4Cl^2$. Colorless crystals; fusing point, 53°; boiling point, 172°.—*Trichlorbenzene*, $C^6H^3Cl^3$. Fusing point, 17°; boiling point, 206–210.°—*Tetrachlorbenzene*, $C^6H^2Cl^4$. Fine needles; fusing point, 139°; boiling point, 240°.—*Pentachlorbenzene*, C^6HCl^5. Fine needles; fusing point, 85°; boiling point, 270°.—*Perchlorbenzene* (Julin's carbon chloride), C^6Cl^6, is easily formed by the action of antimony chloride on benzene; and is also produced when the vapor of chloroform or carbon chloride (C^2Cl^4) is passed through a tube heated to red-heat, or acetylene tetrachloride (p. 132) is heated for some time at 360°.—Long, colorless, thin prisms. Fusing point, 222–226°; boiling point, 332°.

Compounds isomeric with these ($C^6H^4Cl^2$, liquid, boiling point, 175°: $C^6H^3Cl^3$, fusing point, 60°: $C^6H^2Cl^4$, fusing point, 35°; boiling point, 253°: C^6HCl^5, fusing point, 198–199°) are produced, when the products, obtained from benzene with chlorine without iodine, or from sulphobenzide with chlorine, are treated with alcoholic potassa.

Bromine substitution-products. When bromine is allowed to act upon benzene at the ordinary temperature, monobrombenzene and a trace of dibrombenzene are formed very slowly; at a higher temperature substitution-products containing more bromine are produced. These compounds are also obtained by the action of phosphorus bromide on phenol and its substitution-products. *Monobrombenzene*, C^6H^5Br, is a liquid, boiling at 154°.—*Paradibrombenzene*, $C^6H^4Br^2$, crystals, fusing point, 89°; boiling point, 219°. Together with this is produced, in small quantity, an isomeric compound, that fuses at —1°, and boils at 214°. By replacing the amide group in dibromanilin (brominated bromanilin) by hydrogen, a third modification of dibrombenzene is obtained, which boils at 215°, and does not congeal at —28°.—*Tribrombenzene*, $C^6H^3Br^3$. Needles of a silky lustre, fusing point, 44°, sublimes without decomposing. A second variety of tribrombenzene is formed as a secondary product, in the preparation of dibrombenzene from dibromanilin, and by the replacement of the amide group of tribromanilin by hydrogen.—White, broad needles; fusing point, 118.5°.—*Tetrabrombenzene*, $C^6H^2Br^4$, needles, which fuse at 137–140°. The tetrabrombenzene, obtained from tribromphenol by the action of phosphorus bromide, appears to be different from this compound, obtained from benzene. It fuses at 98°.—*Pentabrombenzene* C^6HBr^5 needles sublime without decomposing, fuse above 240°.

Iodine substitution-products. These are produced by heating benzene with iodine and iodic acid at 200–240°; by the action of iodine and phosphorus on phenol; and by treating silver benzoate with iodine chloride. *Monoiodobenzene* C^6H^5I, colorless liquid, boiling at 185°.—*Diiodobenzene* $C^6H^4I^2$, laminæ, fusing point, 127°; boiling point, 277°.—*Triiodobenzene* $C^6H^3I^3$, needles, fusing at 76°, sublimable without decomposition.

Fluorbenzene, C^6H^5Fl. By heating a mixture of calcium fluorbenzoate and calcium hydroxide.—Crystalline mass. Fusing point, 40°; boiling point, 180–183°.

Cyanbenzene (Benzonitrile), $C^6H^5.CN$, is formed by the distillation of a mixture of potassium sulphobenzolate with potassium cyanide; by the distillation of ammonium benzoate or hippuric acid; by heating benzamide with benzoyl chloride or benzoic anhydride; and in small quantities in a great many other ways.—Colorless oil, boiling at 191°. Combines directly with hydrogen, bromine, hydrobromic and hydriodic acids; and yields benzoic acid and ammonia when heated with alkalies.

Paradicyanbenzene, $C^6H^4(CN)^2$. By distilling a mixture of potassium parabromsulphobenzolate or paradisulphobenzolate with potassium cyanide.—Colorless prisms, sublimable without decomposition. Heated with alkalies, it yields terephtalic acid.

A compound isomeric with benzonitrile is

Phenylcarbylamine (Phenyl cyanide), $C^6H^5.CN$. Is produced by distilling a mixture of anilin, chloroform, and alcoholic potassa.—Liquid, not distillable without decomposition. Combines with other cyanides, especially silver cyanide, forming crystallizing compounds. Is scarcely acted upon by alkalies; with acids, however, it decomposes easily, yielding formic acid and anilin.

Nitrosubstitution-products, *Nitrobenzene*, $C^6H^5(NO^2)$, is formed, when benzene is added gradually to very concentrated nitric acid, which is kept cool.—Bright yellow liquid, with an odor similar to that of oil of bitter almonds. Boils at 205°, and congeals at 3°.—*Paradinitrobenzene*, $C^6H^4(NO^2)^2$, is produced by heating the preceding compound for a long time with very concentrated nitric acid; more readily by dropping benzene into a mixture of two volumes concentrated sulphuric acid, and one volume very concentrated nitric acid. Crystallizes from alcohol in very long, shiny, nearly colorless needles, that fuse at 86°.

By the action of nitric acid or nitric-sulphuric acid on chlorine, bromine, and iodine substitution-products of benzene, mono- and dinitro-derivatives of these compounds are formed. Usually several isomeric modifications are produced at the same time, the constitution of which is as yet unknown. Nearly all of these compounds are solid and crystallize well. All these modifications of *nitrochlorbenzene* $C^6H^4(NO^2)Cl$ are known; two of them are solid and have the fusing points 85° and 46°; the third is fluid, and boils at 240°; also of *nitrobrombenzene* $C^6H^4(NO^2)Br$, all three modifications are known; they all crystallize in yellowish prisms, and have the fusing points, 125°, 50°, and 37°.—*Nitroiodobenzene* $C^6H^4(NO^2)I$. Of this, two modifications are known (fusing points, 171°.5 and 34°); also of *nitrobenzonitrile* $C^6H^4(NO^2)CN$, (metanitrobenzonitrile, from benzonitrile with nitric acid and from metanitrobenzamide with phosphorus pentachloride. Fusing point, 117–118°.—Paranitrobenzonitrile from paranitrobenzamide with phosphoric anhydride: laminæ; fusing point, 139°).

Dinitrochlorbenzene, $C^6H^3(NO^2)^2Cl$, from chlorbenzene with nitric-sulphuric acid; and from dinitrophenol with phosphorus pentachloride. Prisms, fusing point, 48–49°.

Trinitrochlorbenzene, $C^6H^2(NO^2)^3Cl$, from trinitrophenol with phosphorus pentachloride. Needles, fusing point, 83°.

Anilin (Amidobenzene), $C^6H^5.NH^2$, is obtained by treating nitrobenzene with reducing substances (by heating with tin and hydrochloric acid; by conducting sulphuretted hydrogen in its solution containing ammonia; by heating with arsenious acid and caustic soda; by treating with grape-sugar and caustic soda; by heating gently with zinc-dust); and prepared on the large scale by heating a mixture of 1 part of nitrobenzene, 1 part of concentrated acetic acid, and 1.2 parts of iron-filings. It is produced further by the distillation of indigo and several of its derivatives, either alone or with caustic potassa; and in small quantity by heating phenol with ammonia to 200°.—From dead oil (p. 253) it can be extracted with dilute acids, but in this way it can only with difficulty be prepared in a pure condition and free from other bases contained in the oil.—Colorless, clear liquid: specific gravity, 1.036. Boils at 184.5°, is difficultly (in 31 parts at 12°) soluble in water, in all proportions in alcohol and ether. Congeals at —8°. In contact with the air it turns brown, and becomes resinous. Its aqueous solution turns purple on the addition of a solution of chloride of lime. When to its solution in concentrated sulphuric acid, a few drops of a solution of potassium bichromate are added, it becomes at first red, and then deep blue.—Strong base; yields well-characterized salts with acids; and combines with aldehydes like ammonia. Combines also with metallic salts like ammonia.

Anilin hydrochlorate, $C^6H^7N.HCl$. Colorless needles, very easily soluble in water and alcohol, sublimes without undergoing decomposition. Combines with a number of metallic chlorides Platinum chloride precipitates from its alcoholic solution fine, yellow, needly crystals of *anilin platinum chloride*, $(C^6H^7N.HCl)^2PtCl^4$.—*Anilin oxalate*, $2(C^6H^7N)H^2C^2O^4$, crystallizes from water in thick, hard prisms, is easily soluble in hot water and hot alcohol, much less easily soluble in the cold solvents.

For the so-called anilin colors, see Toluidin.

Substitution-products of anilin. In the benzene residue of anilin one or more hydrogen atoms may be replaced by the halogenes. These compounds are formed by the action of chlorine, bromine, or iodine on anilin; by the decomposition of the substitution-products of acetanilide and other anilides by means of caustic potassa; by treating the mononitrochlorine, bromine, or iodine substitution-products of benzene with reducing agents; by heating nitrobenzene with concentrated hydrochloric or hydrobromic acids to a high temperature; and by the distillation of the substitution-products of isatine (see Indigo), with caustic potassa. The basic properties of anilin are lessened by the entrance of the halogenes. Trichlor- and tribromanilin do not combine with acids. By heating anilin with methyl alcohol under pressure at 300°, the hydrogen of the benzene-residue can be replaced by methyl CH^3.

Monochloranilin, $C^6H^4Cl.NH^2$. Is known in three modifications. One of these (probably ortho), which is obtained from the nitrochlorbenzene of fusing point 85°, and from monochloracetanilide, forms shiny octahedrons, insoluble in cold water, difficultly soluble in boiling, easily in alcohol and ether. Fusing point, 64°. Distills almost without decomposition.—Both the other modifications are liquid.—*Dichloranilin* $C^6H^3Cl^2.NH^2$, crystallizes in needles. Fusing point, 50°.—*Trichloranilin* $C^6H^2Cl^3.NH^2$. Long, colorless needles. Fusing point, 96.5°; boiling point, 270°. Somewhat soluble in boiling water, easily in alcohol and ether.—*Tetrachloranilin* $C^6HCl^4.NH^2$. Fine needles. Fusing point, 90°.

Monobromanilin, $C^6H^4Br.NH^2$. There are three modifications known, which are obtained from the three nitrobrombenzenes. One of these (orthobromanilin) forms octahedrons, that fuse at 63–64°; another is liquid; and the third crystallizes in needles, which fuse at 31°.—*Dibromanilin* $C^6H^3Br^2.NH^2$. Flat needles. Fusing point, 79.5°.—*Tribromanilin* $C^6H^2Br^3.NH^2$. Long, colorless needles. Fusing point, 117°; boiling point, 300°.

Orthoiodanilin, $C^6H^4I.NH^2$. Colorless needles, which fuse at 60°. The hydriodate is produced when anilin is mixed with powdered iodine. *Iodanilin hydrochlorate* $C^6H^4I(NH^2).HCl$, prepared from this with hydrochloric acid, crystallizes in laminæ, which are difficultly soluble in water, and still more difficultly in hydrochloric acid.—A second iodanilin, prepared from the nitroiodobenzene of fusing point 34°, forms laminæ of a silvery lustre. Fusing point, 25°.

Orthonitranilin, $C^6H^4(NO^2)NH^2$. Cannot be prepared directly from anilin. Is produced by boiling nitracetanilide and other anilides with caustic potassa.—Yellow needles or plates. In water very difficultly, in alcohol easily soluble. Fusing point, 141°. Sublimable. Weak base.—*Metanitranilin* is formed by the action of alcoholic ammonia on metabromnitrobenzene.—Dark, yellow, long, fine needles; fusing point, 66°. Weak base.—*Paranitranilin* is formed by conducting sulphuretted hydrogen into a warm alcoholic solution of paradinitrobenzene, to which has been added ammonia.—Long, yellow needles, which fuse at 108°. Easily sublimable. But slightly soluble in water, but more readily than the preceding compound; in alcohol easily soluble. Weak base. The hydrochlorate is decomposed even by water.

Dinitranilin, $C^6H^3(NO^2)^2.NH^2$. By heating dinitrochlorbenzene with alcoholic ammonia.—Bright-yellow prisms. Fusing point, 175°.—*Trinitranilin* $C^6H^2(NO^2)^3.NH^2$. From trinitrochlorbenzene with aqueous or alcoholic ammonia.—Long, furrowed needles. Fusing point, 179–180°.

Ethylanilin, $C^6H^5.NH.C^2H^5$. Anilin combines directly with ethyl bromide, slowly at the ordinary temperature, rapidly by the aid of heat, forming ethylanilin hydrobromate, a crystalline substance, from which the base can be separated by means of potassa.—Colorless liquid; becomes brown in contact with the air; boils at 240°; and forms with acids, crystallizing, easily soluble salts.

Diethylanilin, $C^6H^5.N(C^2H^5)^2$. The hydrobromate of this base is produced by direct combination of ethylanilin with ethyl bromide.—Colorless oil; does not turn brown; boils at 213°.

Triethylphenylammonium. The iodide, $C^6H^5.N(C^2H^5)^3I$, is formed by heating diethylanilin with ethyl-iodide at 100°, for a long time. Silver oxide produces from this the hydroxide $C^6H^5.N(C^2H^5)^3.OH$, which is strongly alkaline; easily soluble in water; and is resolved into diethylanilin, ethylene, and water by distillation.

Ethylenediphenyldiamine, $(C^6H^5)^2N^2H^2.C^2H^4$, is produced, together with anilin hydrobromate, by boiling ethylene bromide with an excess of anilin.—Crystalline base, fusing at 57°.

A large number of analogous bases can be prepared in the same manner, by allowing the bromides of other alcoholic radicles to act on anilin. When anilin is heated with amyl bromide, for instance, amylanilin $C^6H^5.NH.C^5H^{11}$, is produced; this yields, with ethyl bromide, amylethylanilin $C^6H^5.N\left\{\begin{matrix}C^5H^{11}\\C^2H^5\end{matrix}\right.$, and, finally, if methyl iodide is allowed to act upon this, the iodide of an ammonium is produced, in which each hydrogen atom is displaced by a different radicle (phenylamylethylmethylammonium iodide $C^6H^5N\left\{\begin{matrix}C^5H^{11}\\C^2H^5I\\CH^3\end{matrix}\right.$). A great many such compounds have been prepared and carefully investigated.

Diphenylamine, $(C^6H^5)^2NH$. Is produced by heating anilin with anilin hydrochlorate, and by distilling anilin-blue (see Anilin Colors).—Crystals, that fuse at 45°, and distill, without decomposition, at 310°. Is colored deep blue by nitric acid. Weak base. The salts are decomposed by water.

The hydrogen of the NH^2 in anilin can also be displaced by acid radicles. The compounds formed in this way, which are called *anilides*, may also be considered as the amides of the acids, in which hydrogen is displaced by phenyl. There are a great many such compounds known. The following may serve as examples:—

Formanilide (Phenylformamide), $C^6H^5.NH.CHO$, is produced by digesting ethyl formate with anilin, and by heating equal molecules of oxalic acid and anilin rapidly; in the latter cases secondary products are formed.—Prisms, fusing point, 46°; easily soluble in hot water, alcohol, and ether. In an aqueous solution it gives a precipitate of *sodiumformanilide*, $C^6H^5.NNa.CHO$, with concentrated soda-ley, which is again resolved into formanilide and sodium hydroxide by means of water. When distilled with concentrated hydrochloric acid, formanilide yields benzonitrile (p. 256).

Acetanilide (Phenylacetamide), $C^6H^5.NH.C^2H^3O$, is produced by mixing anilin with acetic anhydride or acetyl chloride, and also by heating equal molecules of glacial acetic acid and anilin together for an hour.—Colorless, shiny, lamellar crystals, that fuse at 112–113°, and volatilize without decomposition at 295°. But slightly soluble in cold water, more readily in hot water and in alcohol. Treated with soda-ley it yields acetic acid and anilin.

Oxanilide (Diphenyloxamide), $\begin{matrix} CO.NH.C^6H^5 \\ \dot{C}O.NH.C^6H^5 \end{matrix}$, is formed by heating anilin oxalate to 160–180°, and, together with *monophenyloxamide*, $\begin{array}{l}CO.NH.C^6H^5\\ \dot{C}O.NH^2,\end{array}$ by evaporating a solution of anilin cyanide (p. 265) with hydrochloric acid.—Shiny crystals, that fuse at 245°, and are sublimable.

Oxanilic acid (Phenyloxamic acid), $\begin{array}{l}CO.NH.C^6H^5\\ \dot{C}O.OH,\end{array}$ is produced by heating anilin with an excess of oxalic acid.—Crystalline scales, easily soluble in hot water, but slightly in cold. Has a strong acid reaction. Monobasic acid.

Phenylcarbamide (Phenylurea), $CO\left\{\begin{array}{l}NH.C^6H^5\\ NH^2,\end{array}\right.$ is produced like ethylurea (p. 230) by the decomposition of phenol cyanate with ammonia; by the mixing of potassium cyanate with anilin sulphate; by the slow action of the vapor of cyanic acid on anilin, etc.—Colorless, needle-shaped crystals, difficultly soluble in cold water, easily in hot water, in alcohol and ether. Is decomposed by heat, yielding ammonia, cyanuric acid, and

Diphenylurea (Carbanilide), $CO(NH.C^6H^5)^2$, which is also produced by bringing together phenol cyanate with water or anilin; by heating 1 part urea with 3 parts anilin at 150–170°; and, together with formanilide, by heating oxanilide.—Needles of a silken lustre, sparingly soluble in water, easily soluble in alcohol; fuses at 205°. Volatile without decomposition.

Phenylcarbamic acid (Carbanilic acid), $CO\left\{\begin{array}{l}NH.C^6H^5\\ OH.\end{array}\right.$ Not known in an isolated condition. Its ethyl ether is produced by the action of ethyl chlorcarbonate on anilin. It forms colorless needles, that fuse at 52°. Treated with concentrated potassa-ley, or heated with anilin, it yields diphenylurea.

Phenylsulphocarbamide, $CS\left\{\begin{array}{l}NH.C^6H^5\\ NH^2.\end{array}\right.$ Is produced by the action of ammonia on phenyl mustard-oil;

and by heating ammonium sulphocyanate with anilin for a long time.

Diphenylsulphocarbamide (Sulphocarbanilide), $CS(NH.C^6H^5)^2$. Is produced by bringing together carbon bisulphide with anilin, slowly at the ordinary temperature, rapidly by heating a mixture of carbon bisulphide, anilin, and alcohol.—Colorless laminæ. Fusing point, 140°. Insoluble in water, easily soluble in alcohol and ether.

Phenyl Mustard-oil, $CS:N.C^6H^5$. Is produced from diphenylsulphocarbamide by distilling it with phosphoric anhydride; by heating with concentrated hydrochloric acid, the vessel being connected with an inverted condenser; and by mixing its alcoholic solution with an alcoholic solution of iodine.—Colorless liquid, with an odor very similar to that of mustard-oil. Boiling point, 222°. Combines directly with ammonia, forming phenylsulphocarbamide; with anilin forming diphenylsulphocarbamide; with alcohol at 110–115°, forming *phenylxanthogenamide*, $CS\left\{\begin{matrix} NH.C^6H^5 \\ O.C^2H^5, \end{matrix}\right.$ which is also formed by heating phenylsulphocarbamide for a long time with alcohol at 140–150°. Colorless crystals; fusing point, 65°.

Phenylcyanamide (Cyananilide), $CN.NH.C^6H^5$, is produced by conducting dry cyanogen chloride into an ethereal solution of anilin; and by digesting a solution of phenylsulphocarbamide with lead oxide.—Colorless, long needles, arranged concentrically. Fusing point, 36–37°. Difficultly soluble in water, easily soluble in alcohol and ether. Without basic properties. Is spontaneously transformed, even at the ordinary temperature, into the polymeric compound, *triphenylmelamine* $C^3H^3(C^6H^5)^3N^6$, which crystallizes in prisms, fusing at 162–163°.

Diphenylguanidine, $C^{13}H^{13}N^3 = CN^3H^3(C^6H^5)^2$. Is produced from diphenylsulphocarbamide when its solution in alcoholic ammonia is treated with lead oxide.—Long, flattened needles. Fusing point, 147°.

Monatomic base. A base isomeric with this, *β-diphenylguanidine* (melanilin), is produced, in the form of the hydrochlorate, by conducting cyanogen chloride into pure anilin; and by heating an alcoholic solution of phenylcyanamide with anilin hydrochlorate.—Colorless, crystalline laminæ. Fusing point, 131°. Sparingly soluble in water, more easily soluble in alcohol than the α-base.

Triphenylguanidine, $C^{19}H^{17}N^3 = C\left\{\begin{matrix}(NH.C^6H^5)^2\\ N.C^6H^5.\end{matrix}\right.$ Is produced by heating diphenylurea; by heating diphenylsulphocarbamide, either alone or with copper, at 150–160°, or with anilin up to the boiling point of the latter. Is prepared most readily by dissolving 1 molecule diphenylsulphocarbamide and 1 molecule anilin in alcohol, and adding lead oxide, or mercury oxide, or an alcoholic solution of iodine to the boiling liquid. The hydrochlorate is also formed by melting diphenylsulphocarbamide with lead chloride or mercury chloride.—Long, colorless, shiny, rhombic prisms. Fusing point, 143°. Almost insoluble in water even at the boiling temperature, easily soluble in hot alcohol. Monatomic base. Heated with carbon bisulphide at 160–170°, it is converted into phenyl mustard-oil and diphenylsulphocarbamide.

The hydrochlorate of a base isomeric with the preceding, viz.: *β-triphenylguanidine* (Carbotriphenyltriamine) is formed by heating anilin with carbon tetrachloride for a long time at 170–180°. The free base crystallizes in colorless, four-sided plates, that are insoluble in water, difficultly soluble in ether, more easily in alcohol.

Anilin cyanide, $C^{14}H^{14}N^4 = (C^6H^7N)^2(CN)^2$. Is produced by the direct combination of anilin and cyanogen, when an alcoholic solution of anilin is saturated with cyanogen.—Shiny crystalline laminæ; insoluble in water, difficultly soluble in alcohol. Fusing point, 210°. Diatomic base.

Orthodiamidobenzene (Orthophenylendiamine), $C^6H^4(NH^2)^2$, is produced by the reduction of orthonitranilin with iron-filings and acetic acid, or hydriodic acid.—Colorless crystals. Easily soluble in water. Melts at 140°, and boils at 267°.

Paradiamidobenzene (Paraphenylendiamine), $C^6H^4(NH^2)^2$. Is produced by the reduction of dinitrobenzene, or of the paranitranilin obtained from this.—Crystalline mass, which undergoes a change in contact with the air. Easily soluble in water. Fuses at 63°, and boils at 287°. Diatomic base.—The *hydrochlorate*, $C^6H^4(NH^2)^2 2HCl$, crystallizes in fine needles.

Diazobenzene, $C^6H^4N(?)$, is obtained by the decomposition of diazobenzene potassa (see below) with acetic acid.—Thick, yellow, very unstable oil.

Diazobenzene nitrate, $C^6H^4N^2.HNO^3$. A current of nitrous acid is conducted into anilin nitrate, to which is added a quantity of water insufficient for its solution, until caustic potassa no longer causes a precipitate of anilin. The salt is deposited in crystals, which are increased in quantity by the addition of alcohol and ether.—Long, colorless needles, very easily soluble in water, but sparingly in alcohol, insoluble in ether and benzene. Explodes with great violence when heated or struck with a hammer. Is decomposed in a moist atmosphere, and, when boiled with water, yields nitrogen, nitric acid, and phenol.

Diazobenzene sulphate, $C^6H^4N^2.H^2SO^4$, is obtained from anilin sulphate, like the preceding compound; is prepared, however, most readily by treating the latter with dilute sulphuric acid.—Colorless prisms. Detonates at 100°. Conducts itself towards solvents and by boiling with water like the nitrate; by boiling with absolute alcohol it is converted into benzene, nitrogen being evolved, and the alcohol is oxidized to aldehyde; treated with hydriodic acid it yields nitrogen, sulphuric acid, and iodobenzene.

Diazobenzene hydrobromate, $C^6H^4N^2.HBr$, is formed by mixing an ethereal solution of diazoamidobenzene with bromine.—Colorless very unstable laminæ.

When a watery solution of the nitrate is mixed with a solution of bromine in hydrobromic acid, *diazobenzene perbromide*, $C^6H^4N^2.HBr.Br^2$, is produced. Large, yellow lamellæ, insoluble in water and ether, difficultly soluble in alcohol. Heated either alone or with alcohol it yields monobrombenzene.

Diazobenzene potassa, $C^6H^4N^2.KOH$. Is produced by the addition of very concentrated potassa-ley to the nitrate; and can be separated from the saltpetre, that is formed at the same time, by dissolving in alcohol.—Colorless laminæ of a mother-of-pearl lustre, easily soluble in water and alcohol, insoluble in ether. Detonates when heated. A freshly-prepared aqueous solution gives with silver nitrate a grayish-white, very explosive precipitate of *diazobenzene silver-oxide*, $C^6H^4N^2.AgOH$.

Diazo-amidobenzene, $C^{12}H^{11}N^3 = C^6H^4N^2.C^6H^7N$, is produced by mixing an aqueous solution of diazobenzene nitrate with anilin; by conducting nitrous acid into a cooled alcoholic solution of anilin, and by pouring a cooled, slightly alkaline solution of sodium nitrite gradually on anilin hydrochlorate.—Golden-yellow, shiny lamellæ. Fuses at 91°, and detonates when strongly heated. Insoluble in water, easily soluble in ether, benzene, and hot alcohol, less readily in cold alcohol. Nitrous acid, containing nitric acid, converts it into diazobenzene nitrate; concentrated hydrochloric acid into anilin hydrochlorate, phenol, and nitrogen.

Diazobenzenimide, $C^6H^5N^3$, is produced by treating diazobenzene-perbromide with aqueous ammonia. —Slightly yellow-colored oil, volatile with water vapor. Nascent hydrogen converts it into anilin and ammonia.

Diazochlor-, Diazobrom-, Diazoiodo-, and **Diazonitrobenzene-compounds** are formed by the action of nitrous acid on the substitution-products of anilin. They conduct themselves in every respect like the diazo-compounds described.

Azobenzene, $C^{12}H^{10}N^2$, is produced from nitrobenzene by the action of sodium-amalgam, alcoholic potassa or acetic acid, and a great deal of iron; from anilin hydrochlorate by oxidation with potassium hypermanganate.—Large, red crystals, fusing at 66.5°. Distills without decomposition at 293°. Insoluble in water, easily soluble in alcohol and ether.—Combines with bromine, without elimination of hydrogen, forming golden-yellow needles of $C^{12}H^{10}Br^2N^2$, which fuse at 205°. Nitric acid converts azobenzene into nitro-substitution-products.

Amido-azobenzene, $C^{12}H^{11}N^3 = C^{12}H^9.NH^2.N^2$ (Amidodiphenylimide), is produced from the isomeric compound diazoamidobenzene, when the latter is allowed to stand for several days with alcohol and some salt of anilin. It is hence formed together with diazoamidobenzene, and under certain circumstances exclusively by treating anilin with nitrous acid. It is also formed by the oxidation of anilin with sodium stannate, or bromine vapor.—Yellow, rhombic prisms. Almost insoluble in water, easily soluble in alcohol and ether. Fuses at 127.4°. Monatomic base.—Is the principal constituent of the dye known as anilin-yellow.

Azoxybenzene, $C^{12}H^{10}N^2O$, is produced from nitrobenzene, like azobenzene and usually both are formed together.—Long, yellow needles, insoluble in water, easily soluble in alcohol and ether. Fuses at 36°, and yields by distillation, anilin and azobenzene.—Treated with reducing substances, it is converted into azobenzene and hydrazobenzene.

Hydrazobenzene, $C^{12}H^{12}N^2$, is formed by treating azobenzene or azoxybenzene with hydrosulphuric acid, ammonium sulphide, or sodium-amalgam.—Crystallizes in plates, that fuse at 131°. Almost insoluble in water, easily soluble in alcohol and ether. It is resolved by heating into azobenzene and anilin, and, when treated with oxidizing substances, is very readily converted into azobenzene. It does not combine with acids, is converted by them, however, into an isomeric body, benzidine.

Sulphobenzolic acid, $C^6H^5.SO^2.OH + 1\frac{1}{2}H^2O$. Benzene is shaken with weak fuming sulphuric acid, until it is dissolved, the solution is diluted with water, neutralized with barium or lead carbonate, and the metal afterward removed from the solutions of the easily soluble salts by means of sulphuric acid or hydrosulphuric acid. Is also produced from sulphanilic acid by replacement of the NH^2 group by hydrogen.—Small, colorless, four-sided plates, easily soluble in water and alcohol, deliquescent.

Barium sulphobenzolate, $(C^6H^5.SO^3)^2Ba + H^2O$. Plates of a mother-of-pearl lustre, easily soluble in water.

Ethyl sulphobenzolate, $C^6H^5.SO^2.O.C^2H^5$, crystallizes in fine, colorless needles.

Sulphobenzolchloride, $C^6H^5.SO^2Cl$, is thrown down, when sodium sulphobenzolate is intimately mixed with phosphorus pentachloride, the mass gently heated, and then thrown into water.—Colorless oil of specific gravity 1.371; boils at 246–247°, at the same time undergoing partial decomposition. Crystallizes below 0°, in large rhombic crystals. Boiling water decomposes it slowly, forming sulphobenzolic acid and hydrochloric acid. Treated with ammonia or ammonium carbonate, it yields *sulphobenzolamide*, $C^6H^5.SO^2.NH^2$, colorless laminæ, fusing at 149°.

Chlor-, Brom-, Iodo-, Nitro-, and **Amidosulphobenzolic acid,** are produced by dissolving the monosubstitution-products of benzene in weak fuming sulphuric acid. They all belong to the para-series.

Benzenesulphurous acid, $C^6H^5.SO^2H$. The sodium salt is produced by treating an ethereal solution of sulphobenzolchloride with sodium-amalgam. Hydrochloric acid separates the free acid from this.—Large, colorless prisms of a high lustre. Difficultly soluble in cold water, easily soluble in hot water, alcohol, and ether. Fuses at 68–69°, and is decomposed at a higher temperature. With chlorine or bromine it yields sulphobenzolchoride or bromide; and, in contact with the air, is transformed slowly into sulphobenzolic acid, rapidly by means of oxidizing agents. Monobasic acid.

Paradisulphobenzolic acid, $C^6H^4(SO^2.OH)^2$, is formed by heating sulphobenzolic acid or benzonitrile with fuming sulphuric acid. The *barium salt*, $C^6H^4.S^2O^6Ba+1\frac{1}{2}H^2O$, forms easily soluble microscopic crystals.

Paradisulphobenzolchloride, $C^6H^4(SO^2Cl)^2$. Is produced by the action of phosphorus pentachloride on sodium paradisulphobenzolate.—Large, colorless crystals. Fusing point, 62°.

Diphenyl, $C^{12}H^{10}$. Is formed when sodium is allowed to act upon a solution of monobrombenzene in ether or benzene. Is also produced when benzene-vapor is passed through an ignited tube; by heating potassium benzoate with phenol potassium; and in small quantity, together with benzene, by heating benzoic acid with lime.—Large, colorless, crystalline laminæ, insoluble in water, easily soluble in hot alcohol. Fuses at 70.5°, and boils at 239–240°.

Dibromdiphenyl, $C^{12}H^8Br^2$, is produced by the action of bromine on diphenyl under water.—Large, colorless prisms, that fuse at 164°, and can be distilled without decomposition. Insoluble in cold alcohol, difficultly soluble in boiling alcohol, easily in benzene.

Dinitrodiphenyl, $C^{12}H^8(NO^2)^2$, is formed by pouring cold fuming nitric acid on diphenyl.—Fine, colorless needles, difficultly soluble in alcohol. Fuses at 213°.—A compound of the same composition, *isodinitrodiphenyl*, is formed at the same time with dinitrodiphenyl; it is more easily soluble in alcohol, and forms large colorless crystals that fuse at 93.5°.

Diamidodiphenyl (Benzidine), $C^{12}H^8(NH^2)^2$. Is obtained by the reduction of dinitrodiphenyl with ammonium sulphide or tin and hydrochloric acid. Is further formed by the treatment of the isomeric hydrazobenzene (p. 269) with acids; by heating azobenzene with concentrated hydrochloric acid at 115°; by treating monobromanilin with sodium; and, together with anilin, by conducting sulphuretted hydrogen into an alcoholic solution of nitrobenzene in the presence of copper or lead.—Colorless laminæ of a silvery lustre, fusing at 118°, but slightly soluble in cold water, more readily in hot water, and easily in alcohol. Sublimable, but undergoing partial decomposition.—*Benzidine sulphate*, $C^{12}H^{12}N^2.H^2SO^4$, is almost insoluble in water and alcohol.

Carbazol, $C^{12}H^9N$ (probably Imidodiphenyl $\left.\begin{matrix} C^6H^4 \\ \dot{C}^6H^4 \end{matrix}\right\} NH$). Is obtained as a secondary product in the process for the purification of crude anthracene on the large scale. Can be artificially prepared by conducting the vapor of anilin or diphenylamine through red-hot tubes.—Crystals, that resemble those of anthracene; fusing point, 238°; boiling point, 338°.—By the action of hydriodic acid on carbazol, there is formed a base *carbazolin*, $C^{12}H^{15}N$, that crystallizes in large, white needles, fuses at 96°, and boils at 286°.

Disulphodiphenylic acid, $C^{12}H^{8}(SO^{2}.OH)^{2}$, is formed by dissolving diphenyl in concentrated sulphuric acid. Long, colorless prisms, that fuse at 72.5°. Very easily soluble in water. The *potassium salt*, $C^{12}H^{8}.S^{2}O^{6}K^{2} + 2\frac{1}{2}H^{2}O$, crystallizes in large, colorless prisms, moderately difficultly soluble in cold water. The *barium* and *lead salts* are insoluble in water.

Diphenylbenzene, $C^{18}H^{14} = C^{6}H^{4}\left\{\begin{matrix}C^{6}H^{5}\\ C^{6}H^{5}.\end{matrix}\right.$ Is produced like diphenyl by the action of sodium on a mixture of mono- and dibrombenzene.—Colorless, crystalline mass. Fusing point, 205°; boiling point, 400°.

Mercuryphenyl, $(C^{6}H^{5})^{2}Hg$. Is produced, when, to a solution of monobrombenzene in benzene, sodium-amalgam is added, and the whole then heated for a few hours in connection with an inverted condenser; the formation takes place particularly easily in the presence of a little acetic ether. (See Mercury ethyl, p. 62).—Colorless, rhombic prisms, that become yellow in contact with the air. Fusing point, 120°. Insoluble in water, easily soluble in chloroform, carbon bisulphide, and benzene, more difficultly in ether and boiling alcohol. When carefully heated, it can be partially sublimed without decomposition; it is, however, partially resolved into mercury, benzene, diphenyl, and carbon. Treated with two molecules chlorine, bromine, or iodine, it is resolved into monochlor-, brom-, or iodobenzene, and mercury chloride, bromide, or iodide; treated with only one molecule of the halogenes, or heated with mercury chloride, bromide, or iodide and alcohol at 110°, it is converted into *mercurymonophenyl chloride*, $C^{6}H^{5}.Hg.Cl$ (rhombic plates, fusing point, 250°), *bromide*, $C^{6}H^{5}.Hg.Br$ (rhombic plates, fusing point, 275–276°), and *iodide*, $C^{6}H^{5}.Hg.I$ (rhombic plates, fusing point, 265–266°). Hydrogen, sodium, and alkaline sulphides regenerate mercury phenyl from these compounds.—When the chloride is heated with moist silver oxide *mercurymonophenyl hydroxide*, $C^6H^5.Hg.OH$, is produced. This crystallizes in small, white, rhombic prisms, and is a stronger base than ammonia.

Tin triethylphenyl, $C^6H^5(C^2H^5)^3Sn$. Is obtained by treating a solution of monobrombenzene and tintriethyl iodide in ether with sodium.—Colorless liquid, of a not unpleasant odor; boiling point, 254°; easily soluble in ether and absolute alcohol, difficultly in dilute alcohol, insoluble in water. It possesses a strong refracting power; specific gravity at 0°, 1.2639; burns with a luminous flame, leaving a residue of metallic tin. Is reduced by silver nitrate to diphenyl. Hydrochloric acid decomposes it, yielding tintriethyl chloride and benzene.

2. *Toluene (Methylbenzene, Toluol).*
$$C^7H^8 = C^6H^5.CH^3.$$

Preparation. From light oil by partial distillation. By distilling a mixture of toluic acid with an excess of lime. By treating a mixture of monobrombenzene and ethyl iodide with an excess of sodium, the mixture being diluted with ether and kept well cooled.

It is also produced by the dry distillation of tolu-balsam and many resins.

Properties. Colorless liquid of an odor resembling that of benzene; specific gravity, 0.88; boiling point, 111°.—Oxidized with dilute nitric acid or chromic acid, it is converted into benzoic acid.

Substitution-products of Toluene. According as the subtituting chlorine, bromine, and iodine take the place of hydrogen in the benzene residue, or in the methyl, compounds of the same composition but of entirely different properties are formed. Chlorine, etc., that has entered the benzene residue, is held as tenaciously in combination as in chlorbenzene; that which has entered the methyl-group, on the other hand, can be replaced by other monovalent elements or atomic groups with the greatest ease.—When chlorine or bromine is allowed to act on toluene that is kept well cooled or to which is added iodine, substitution takes place only in the benzene residue; at boiling temperature, and in the absence of iodine, on the contrary, the hydrogen of the methyl is replaced. Of the substitution-products of the first class there are, further, certain isomeric modifications possible, the difference of which depends upon the different relative positions of the substituting atoms with reference to each other, and with reference to the methyl-group already present in the molecule (See p. 252). The entrance of one atom of chlorine into toluene can accordingly give rise to the formation of four different compounds, $C^6H^5.CH^2Cl$; and three modifications of $C^6H^4Cl.CH^3$. The direct action of chlorine, bromine, or nitric acid causes chiefly the formation of compounds belonging to the para-series (1 : 4); but, together with these, small quantities of ortho- or meta-compounds are also formed. Both the latter are obtained in a pure condition by treating the substitution-products of the amido-derivatives (toluidins) with nitrous acid, thus converting them into diazo-compounds, and then decomposing the sulphates of the diazo-compounds by boiling with absolute alcohol (See diazobenzenesulphate, p. 266).—The conduct of the monosubtitution-products by oxidation with potassium bichromate and dilute sulphuric acid is very characteristic. The compounds, in which the substitution has taken place in the methyl, are by this means, like toluene itself, converted into benzoic acid: of the other compounds, those belonging to the meta- and para-series are oxidized directly to meta- and para-substitution-products of benzoic acid (by simple oxidation of the group, CH^3 to $CO.OH$); the ortho-compounds on the contrary are completely burnt up without yielding an aromatic acid.

Ortho-, Meta-, and **Parachlortoluene,** $C^6H^4Cl.CH^3$, are very stable liquids, boiling at 156–158°.—*Benzyl chloride* $C^6H^5.CH^2Cl$, a liquid boiling at 176°.

Dichlortoluene, $C^6H^3Cl^2.CH^3$, liquid; boiling point, 196°.—*Chlorbenzyl chloride*, $C^6H^4Cl.CH^2Cl$, liquid; boiling point, 213–214°.—*Benzal chloride* (Chlorobenzol), $C^6H^5.CHCl^2$, is also formed by the action of phosphorus pentachloride on oil of bitter almonds. Liquid, boiling at 206°.

Trichlortoluene, $C^6H^2Cl^3.CH^3$. Colorless crystals; fusing point, 76°; boiling point, 235°.—*Dichlorbenzyl chloride*, $C^6H^3Cl^2.CH^2Cl$. Liquid, boiling at 241°.—*Chlorbenzal chloride*, $C^6H^4Cl.CHCl^2$. Liquid; boiling point, 234°.—*Benzotrichloride*, $C^6H^5.CCl^3$, is also formed by heating benzoyl chloride with phosphorus pentachloride. Liquid; boiling point, 213–214°.

Tetrachlortoluene, $C^6HCl^4.CH^3$; fusing point, 91–92°; boiling point, 271°.—*Trichlorbenzyl chloride*, $C^6H^2Cl^3.CH^2Cl$. Liquid; boiling point, 273°.—*Dichlorbenzal chloride*, $C^6H^3Cl^2.CHCl^2$. Liquid; boiling point, 257°.—*Chlorbenzotrichloride*, $C^6H^4Cl.CCl^3$. Liquid; boiling point, 245°.

Pentachlortoluene, $C^6Cl^5.CH^3$; fusing point, 218°; boiling point, 301°.—*Tetrachlorbenzyl chloride*, $C^6HCl^4.CH^2Cl$. Liquid; boiling point, 296°.—*Trichlorbenzal chloride* $C^6H^2Cl^3.CHCl^2$ Liquid; solidifies below 0°; boiling point, 280–281°.—*Dichlorbenzotrichloride*, $C^6H^3Cl^2.CCl^3$. Liquid; boiling point, 273°.

Pentachlorbenzyl chloride, $C^6Cl^5.CH^2Cl$. Fusing point, 103; boiling point, 325–327°.—*Tetrachlorbenzal chloride*, $C^6HCl^4.CHCl^2$. Liquid; boiling point, 305–306°.—*Trichlorbenzotrichloride*, $C^6H^2Cl^3.CCl^3$. Fusing point, 82°; boiling point, 307–308°.

Pentachlorbenzal chloride, $C^6Cl^5.CHCl^2$. Fusing point, 109°; boiling point, 334°.—*Tetrachlorbenzotrichloride*, $C^6HCl^4.CCl^3$. Fusing point, 104°; boiling point, 316°.

When the attempt is made to replace the last hydrogen-atom in toluene, the molecule breaks up, and perchlorbenzene is formed.

Bromine substitution-products. Parabromtoluene, $C^6H^4Br.CH^3$ (colorless crystals; fusing point, 28.5°; boiling point, 181°), and **Dibromtoluene** (colorless needles; fusing point, 107–108°; boiling point, 245°) are produced by the action of bromine on toluene without the aid of heat.—*Orthobromtoluene*, $C^6H^4Br.CH^3$. From diazoorthobromtoluene sulphate with absolute alcohol. Liquid, boiling at 182–183°.—*Metabromtoluene* $C^6H^4Br.CH^3$, from diazometabromtoluene sulphate with absolute alcohol. Liquid, boiling at 182°. Yields, with bromine, a liquid, dibromtoluene, boiling at 238–239°, that does not congeal at —20°.—*Benzyl bromide*, $C^6H^5.CH^2Br$, is obtained by the action of bromine on boiling toluene; and by the decomposition of benzyl alcohol by means of hydrobromic acid. Colorless liquid; gives off fumes in contact with the air and excites to tears; boiling point, 198–199°.—*Benzal bromide*, $C^6H^5.CHBr^2$, is produced by the action of phosphorus pentabromide on oil of bitter almonds. A liquid that cannot be distilled without suffering partial decomposition.

Paraiodotoluene, $C^6H^4I.CH^3$ (laminæ; fusing point, 35°; boiling point, 211.5°), and **Orthoiodotoluene** (a liquid, boiling at 201°) are produced by the action of hydriodic acid on the diazotoluene sulphates, prepared from the corresponding toluidins. *Benzyl iodide*, $C^6H^5.CH^2I$ (colorless crystals, fusing at 24°, not volatile without decomposition), is produced by the action of hydriodic acid on benzyl chloride at the ordinary temperature.

Benzyl cyanide, $C^6H^5.CH^2.CN$, is obtained by boiling benzyl chloride with alcohol and potassium cyanide; and by distilling potassium benzylsulphate, with potassium cyanide.—Colorless liquid, boiling at 229°.

Paranitrotoluene, $C^6H^4(NO^2).CH^3$, and **Orthonitrotoluene** are produced by treating toluene with fuming nitric acid. The former forms almost colorless prisms (fusing point, 54°; boiling point, 236°); the latter, a liquid boiling at 222–223°.—*Metanitrotoluene* is produced by boiling diazonitrotoluene sulphate (from metanitro-paratoluidin) with absolute alcohol. Crystalline; fusing point, 16°; boiling point, 230–231°.

Dinitrotoluene, $C^6H^3(NO^2)^2.CH^3$. Is produced from toluene, para- and orthonitrotoluene by treating with nitric-sulphuric acid.—Long, almost colorless needles; fusing point, 71°.—An isomeric dinitrotoluene (needles, fusing point, 60°) is produced from metanitrotoluene by the same treatment.

Trinitrotoluene, $C^6H^2(NO^2)^3.CH^3$. Almost colorless needles, but sparingly soluble in cold alcohol. Fusing point, 82°.

Amidotoluene (Toluidins), $C^6H^4(NH^2).CH^3$. The three modifications are prepared from the three isomeric nitrotoluenes, like anilin from nitrobenzene. The commercial crude toluidin is a mixture of ortho- and paratoluidin.

Orthotoluidin (Pseudotoluidin). Colorless liquid, of specific gravity 1.00. Boiling point, 197°. But slightly soluble in water. Does not congeal at —20°. —Gives, with acetyl chloride, an *acettoluide*, $C^6H^4(NH.C^2H^3O).CH^3$, that crystallizes in needles, and fuses at 107°.

Metatoluidin. Colorless liquid, of specific gravity 0.998. Boiling point, 197°. Does not congeal at —13°. With acetyl chloride it gives an acettoluide, that crystallizes in fascicles and fuses at 65.5°.

Paratoluidin. Large, colorless crystals; fusing point, 45°; boiling point, 198°. With acetyl chloride, it gives an acettoluide that crystallizes well, and fuses at 145°.

Benzylamine, $C^6H^5.CH^2.NH^2$. Is produced by the action of nascent hydrogen (zinc and sulphuric acid) on benzonitrile (p. 256); and in small quantity, together with di- and tribenzylamine, by heating benzyl chloride with alcoholic ammonia at 100°.—Clear liquid; boiling point, 183°. Miscible with water, alcohol, and ether in all proportions. From its aqueous solution it is separated by caustic potassa. Attracts carbonic anhydride from the air.—*Dibenzylamine* $(C^6H^5.CH^2)^2NH$. Colorless, thick oil, insoluble in water, easily soluble in ether and alcohol.—*Tribenzylamine* $(C^6H^5.CH^2)^3N$. Colorless laminæ or needles. Fusing point, 91°. Insoluble in water, difficultly soluble in cold alcohol, easily in hot alcohol and in ether. The hydrochlorate is decomposed when heated in a current of dry hydrochloric acid gas, yielding benzyl chloride and dibenzylamine hydrochlorate.

Benzylphosphine, $C^6H^5.CH^2.PH^2$. Two molecules benzylchloride, two molecules phosphonium iodide, and one molecule zinc oxide are heated together for six hours at 160°, and the product then distilled with water vapor. Benzylphosphine and dibenzylphosphine pass over. By means of distillation the benzylphosphine can be prepared from the mixture in a pure condition.—Clear liquid, insoluble in water, easily soluble in ether and alcohol; boiling point, 180°; becomes oxidized in contact with the air, and gives off fumes.—*Dibenzylphosphine* $(C^6H^5.CH^2)^2PH$. Crystallized from alcohol, it forms stellate or fascicular needles, of a high lustre; insoluble in water, difficultly in cold alcohol, more easily in boiling alcohol; fusing point, 205°.

Diamidotoluene (Toluylenediamine), $C^6H^3(NH^2)^2CH^3$, is produced by the reduction of dinitrotoluene.—Long needles; fusing point, 99°; boiling point, 280°. Difficultly soluble in cold water, easily in hot water, in alcohol, and ether.

Anilin-dyes. Rosanilin, $C^{20}H^{19}N^3 = C^6H^4(C^7H^6)^2N^3H^3$. The salts of rosanilin are produced by heating a mixture of anilin and toluidin (commercial anilin) with different oxidizing substances (tin chloride, mercury chloride, mercury nitrate, arsenic acid, etc.). The free base is obtained most readily by adding an excess of ammonia to a hot saturated solution of the acetate. It separates, partially, immediately in the form of a reddish crystalline precipitate. The hot solution, filtered off from this, deposits on cooling another portion of the base in the form of colorless needles or plates, containing one molecule of water. These turn red rapidly in contact with the air without changing their weight. But slightly soluble in water, somewhat more easily in alcohol, insoluble in ether. Not volatile without decomposition. Triatomic base.

Tri-acid rosanilin hydrochlorate, $C^{20}H^{19}N^{3}.3HCl$, is obtained by dissolving the monacid salt in hot concentrated hydrochloric acid. Brown needles. By treatment with water or by heating, it is very easily converted into the monacid salt.—*Monacid rosanilin hydrochlorate*, $C^{20}H^{19}N^{3}.HCl$ (Fuchsine), is produced by heating anilin with metallic chlorides. Rhombic plates of a beautiful metallic green color and bright lustre. Sparingly soluble in water, still less in saline solutions, easily and with an intensely red color in alcohol.

Monacid rosanilin nitrate (Azaleine), $C^{20}H^{19}N^{3}.HNO^{3}$, is obtained by heating anilin with mercury nitrate or other nitrates. It resembles the monacid hydrochlorate.

Rosanilin acetate, $C^{20}H^{19}N^{3}.C^{2}H^{4}O^{2}$, crystallizes in large, very beautifully developed crystals of a metallic green color. More easily soluble in water than the hydrochlorate and nitrate.

Triethylrosanilin, $C^{20}H^{16}(C^{2}H^{5})^{3}N^{3}$. The salts of this base are obtained by heating rosanilin or salts of rosanilin with ethyl iodide and alcohol. They dissolve readily, imparting to the solutions a beautiful violet color, and are (especially the hydrochlorate, which forms a semi-crystalline mass of a golden-yellow lustre), very highly valued dyes (anilin-violet, Hofmann's violet).

Triphenylrosanilin, $C^{20}H^{16}(C^6H^5)^3N^3$. The salts are produced by heating rosanilin salts with an excess of anilin at 180°. The free base is a whitish, almost amorphous mass, that turns blue rapidly in contact with the air.—*Triphenylrosanilin hydrochlorate* (anilin-blue), $C^{20}H^{16}(C^6H^5)^3N^3.HCl$, is a bluish-brown, indistinctly crystalline powder; insoluble in water and ether; difficultly soluble in alcohol. The alcoholic solution has a splendid deep blue color. Excellent dye.—Subjected to destructive distillation, it yields diphenylamine (p. 262).

If, in the preparation of anilin-blue, less anilin is employed, or the heating is not long enough continued, there are produced reddish-violet and bluish-violet dyes, which consist of the salts of mono- or diphenylrosanilin.

Leucanilin, $C^{20}H^{21}N^3$, is produced by the action of zinc and hydrochloric acid, or ammonium sulphide on salts of rosanilin.—White powder, difficultly soluble in water; turns a pale red color in contact with the air. Triatomic base. Yields colorless salts, and is very easily reconverted into rosanilin by oxidizing agents.

Chrysanilin, $C^{20}H^{17}N^3$, is formed as a secondary product in the preparation of rosanilin hydrochlorate. Amorphous powder, but slightly soluble in water, easily soluble in alcohol; looks like freshly precipitated lead chromate. Dyes silk and wool golden yellow. —*Chrysanilin nitrate*, $C^{20}H^{17}N^3.HNO^3$, crystallizes in ruby-red needles, that are exceedingly difficultly soluble in water. Cold concentrated nitric acid converts it into the salt, $C^{20}H^{17}N^3.2(HNO^3)$, which forms crystals resembling potassium ferricyanide, easily decomposable by water.

Anilin-green (Iodine-green), $C^{20}H^{16}(CH^3)^3N.(CH^3)^2I^2 + H^2O$, is prepared by heating 1 part of rosanilin acetate, 2 parts of methyl iodide, and 2 parts of methyl alcohol in closed vessels at 100°. The mass is distilled for the purpose of removing volatile products, and the residue exhausted with water, with an addition of common salt, by which means the green dye is dissolved, and a violet dye remains undissolved. Pure iodine-green crystallizes in prisms with a green metallic lustre, resembling that of cantharides.—The anilin-green of commerce consists chiefly of the picrate, prepared by adding picric acid directly to the solution obtained.

Mauveine, $C^{27}H^{24}N^4$. The sulphate (anilin-purple, anileine, indisine, violine), $(C^{27}H^{24}N^4)^2.H^2SO^4$, is produced by mixing a dilute solution of anilin sulphate (containing toluidin) with a dilute solution of potassium bichromate. The base, separated from this by means of potassa, is a crystalline, almost black, glistening powder, that dissolves in alcohol, forming a violet solution, which turns purple on the addition of acids. Very stable monatomic base. Decomposes ammonium salts. The salts crystallize and have a green metallic lustre.

Anilin-brown is obtained by heating anilin-violet or anilin-blue, with anilin hydrochlorate at 240°. *Aldehyde-green* is prepared by heating rosanilin sulphate, sulphuric acid, and aldehyde together, and treating the resulting blue dye with sodium hyposulphite.

Sulphotoluenic acid, $C^6H^4\left\{\begin{matrix}CH^3\\SO^2.OH,\end{matrix}\right.$ is formed in two isomeric modifications (para- and ortho-), when toluene is dissolved in weak fuming sulphuric acid. There is a very large number of varieties of the substitution-products of these acids known.

Sulphobenzylic acid, $C^6H^5.CH^2.SO^2.OH$. The potassium salt of this acid is produced by heating benzyl chloride with a concentrated solution of potassium sulphite.

Benzylbenzene, $C^{13}H^{12} = C^6H^5.CH^2.C^6H^5$. Is produced by the action of zinc-dust on a mixture of benzyl chloride and benzene.—Colorless crystalline mass, consisting of long prismatic needles; fusing point, 26–27°; boiling point, 261–262°; easily soluble in alcohol, ether, and chloroform. Oxidized by means of potassium bichromate and sulphuric acid, it yields benzophenone of fusing point, 26–26.5° (which see).

Benzyltoluene, $C^{14}H^{14} = C^6H^5.CH^2.C^6H^4.CH^3$. Is formed from benzyl chloride and toluene by the same method as the preceding compound.—Colorless liquid, of a pleasant odor, easily soluble in alcohol, ether, chloroform, and acetic acid; boils at 279–280°; specific gravity, 0.955 at 17.5°.

Ditolyl, $C^{14}H^{14} = \begin{matrix} C^6H^4.CH^3 \\ | \\ C^6H^4.CH^3 \end{matrix}$, is obtained, together with an isomeric liquid compound, by the decomposition of parabromtoluene with sodium.—Colorless, monoclinate crystals. Fusing point, 121°.

Dibenzyl, $C^{14}H^{14} = \begin{matrix} C^6H^5.CH^2 \\ | \\ C^6H^5.CH^2 \end{matrix}$, is produced by the action of sodium on benzyl chloride.—Large, colorless prisms. Easily soluble in hot alcohol, but slightly in cold. Fuses at 52°, and boils at 284°.

Stilbene (Toluylene), $C^{14}H^{12} = \begin{matrix} C^6H^5.CH \\ \| \\ C^6H^5.CH \end{matrix}$, is formed by the distillation of benzyl sulphide, benzyl disulphide, di- and tribenzylamine; and by the action of sodium on oil of bitter almonds.—Large, colorless, thin laminæ, Fusing point, 120°. Easily soluble in hot alcohol, less in cold. Combines directly with bromine, forming a crystalline substance, *stilbene bromide*, $C^{14}H^{12}Br^2$, which is also produced when bromine is allowed to act on dibenzyl, no care being taken to keep the substances cool; treated with alcoholic potassa, it yields *monobromstilbene*, $C^{14}H^{11}Br$ (colorless crystals, fusing point 25°), and tolan.—When treated with hydriodic acid, stilbene is converted into dibenzyl.

Tolan, $C^{14}H^{10}$. Is produced by heating stilbene bromide with alcoholic potassa.—Large, transparent, colorless crystals. Very easily soluble in ether and hot alcohol; melts at 60°. Combines with bromine, forming a crystalline substance, *tolan dibromide* $C^{14}H^{12}Br^2$.—The *tetrachloride*, $C^{14}H^{10}Cl^4$, is produced by heating chlorobenzyl with phosphorus pentachloride. Sodium-amalgam reduces it to tolan.

3. *Hydrocarbons*, C^8H^{10}.

a. Dimethylbenzenes (Xylenes, Xylols). $C^6H^4(CH^3)^2$.

The three modifications, the possibility of the existence of which is indicated by the theory, are all known. That portion of light oil that boils between 136–139° consists essentially of a mixture of meta- and paraxylene, which cannot be separated from each other. Metaxylene forms the largest portion of this mixture.

1. Orthoxylene. Is obtained by distilling a mixture of paraxylylic acid with lime.—Colorless liquid, boiling at 140–141°. Oxidized with nitric acid, it yields orthotoluic acid. Chromic acid burns it up completely.

---

2. Metaxylene (Isoxylene). Is obtained in a pure state by distilling a mixture of xylylic acid, or mesitylic acid with lime.—Liquid boiling at 137°. Dilute nitric acid does not act upon it, chromic acid oxidizes it to isophtalic acid.

Monobrommetaxylene, $C^6H^3Br(CH^3)^2$. Liquid boiling at 204–205°.—*Dibrommetaxylene,* $C^6H^2Br^2(CH^3)^2$. Colorless, shining, crystalline laminæ; fusing point, 72°; boiling point, 256°.—*Tetrabrommetaxylene,* $C^6Br^4(CH^3)^2$. Long, fine needles, difficultly soluble in alcohol. Fusing point, 241°.

Nitrometaxylene, $C^8H^9(NO^2)$. Pale yellow liquid, boiling at 237–239°. Congeals at a low temperature, and melts again at +2°.—*Dinitrometaxylene,* $C^8H^8(NO^2)^2$, is easily produced by heating metaxylene with concentrated nitric acid. Colorless, needly crystals; easily soluble in hot alcohol; fusing point, 93°.—*Trinitrometaxylene,* $C^8H^7(NO^2)^3$, is obtained by pouring metaxylene into a mixture of concentrated sulphuric acid and concentrated nitric acid. Colorless needless, very difficultly soluble in boiling alcohol; fusing point, 176°.

Amidometaxylene (Metaxylidin), $C^8H^9(NH^2)$. Colorless liquid, boiling at 216°. Yields salts that crystallize well.—*Amidonitrometaxylene,* $C^8H^8(NO^2)NH^2$. Reddish-yellow, monoclinate crystals, difficultly soluble in hot water, easily soluble in alcohol; fusing point, 123°. Weak, monatomic base.—*Dinitroamidometaxylene,* $C^8H^7(NO^2)^2NH^2$. Yellow crystals, very sparingly soluble in water, easily soluble in alcohol. Fusing point, 192°. Hardly possesses basic properties.—*Diamidometaxylene,* $C^8H^8(NH^2)^2$. Fine, colorless needles; easily soluble in hot water and in alcohol; fusing point, 152°. Changes its color rapidly in contact with the air. Strong, diatomic base.—*Nitrodiamidometaxylene,* $C^8H^7(NO^2)(NH^2)$. Large, red, shiny prisms; almost insoluble in cold water, more easily soluble in hot water and in alcohol; fusing point, 213°. Weak base.

---

3. Paraxylene. Prepared, like toluene, by the decomposition of a mixture of parabromtoluene or paradibrombenzene, and methyl iodide with metallic sodium.—Colorless liquid, boiling at 136°. At a low temperature, solid and crystalline. Fusing point, 15°. Dilute nitric acid oxidizes it to paratoluic acid; chromic acid to terephtalic acid.

Dibromparaxylene, $C^6H^2Br^2(CH^3)^2$, resembles dibrommetaxylene in all its properties, and melts like this at 72°.—*Tollylenebromide*, $C^6H^4(CH^2Br)^2$, is formed by the action of bromine on boiling paraxylene.—Colorless, lamellar crystals; fusing point, 145–147°.

Dinitroparaxylene, $C^8H^8(NO^2)^2$. Is formed by the action of fuming nitric acid on paraxylene. Two isomeric modifications are produced at the same time, of which the one forms long, thin needles, more difficultly soluble in alcohol, fusing at 123.5°; the other large, monoclinate crystals, more easily soluble in alcohol, fusing at 93°.—*Trinitroparaxylene*, $C^8H^7(NO^2)^3$. Long, colorless needles. Fusing point, 137°. Moderately easily soluble in hot alcohol, but sparingly in cold.

b. Ethylbenzene.

$C^6H^5.CH^2.CH^3$.

Is obtained by the action of sodium on a mixture of brombenzene and ethyl bromide, which is diluted with ether.—Colorless liquid, boiling at 134°; specific gravity, 0.866.—Oxidized either with dilute nitric acid or chromic acid, it yields benzoic acid.

Bromethylbenzene, $C^6H^4Br.C^2H^5$. Colorless liquid, boiling at 199°.—*Benzene-ethyl bromide*, $C^6H^5.CH^2.CH^2Br$, and *chloride*, $C^6H^5.CH^2.CH^2Cl$, are produced by the action of bromine or chlorine on ethylbenzene with the aid of heat. Liquids that cannot be distilled without undergoing decomposition. The chloride is converted into *benzene-ethyl cyanide*, $C^6H^5.CH^2.CH^2.CN$, by boiling with potassium cyanide and alcohol.

Para- and **Orthonitrethylbenzene,** $C^6H^4(NO^2).C^2H^5$, are formed simultaneously when ethylbenzene is treated with fuming nitric acid. Both are liquid; the former boils at 245–246°, the latter at 227–228°. With tin and hydrochloric acid they yield liquid bases.

4. *Hydrocarbons*, C^9H^{12}.

a. Trimethylbenzenes.
$C^6H^3(CH^3)^3$.

That portion of light coal-oil that boils at 163–168° contains, together with other unknown hydrocarbons, two isomeric trimethylbenzenes, pseudocumene, and mesitylene. They cannot be separated from the mixture.

1. Mesitylene (1 : 3 : 5). Is produced, together with other bodies, by the distillation of a mixture of acetone and sulphuric acid, and can be separated from the oily distillate by means of partial distillation.—Colorless liquid, boiling at 163°.—Yields mesitylic and uvitic acids, when oxidized by dilute nitric acid. When heated with phosphonium iodide at 250–300°, it is converted into a hydrocarbon C^9H^{18} (boiling point, 136°), which, under the influence of oxidizing agents, yields the same products as mesitylene.

Monochlormesitylene, $C^6H^2Cl(CH^3)^3$. Colorless liquid; does not congeal at —20°; boiling point, 204–206°.—*Dichlormesitylene*, $C^6HCl^2(CH^3)^3$. Prisms; fusing point, 59°; boiling point, 243–244°.—*Trichlormesitylene*, $C^6Cl^3(CH^3)^3$. Long, fine needles. Fusing point, 204–205°.

Monobrommesitylene, $C^6H^2Br(CH^3)^3$. Colorless liquid, boiling at 225°, congeals below 0°.—*Dibrommesitylene*, $C^6HBr^2(CH^3)^3$, and *tribrommesitylene*, $C^6Br^3(CH^3)^3$, are crystalline. The former fuses at 60°, the latter at 224°.

Nitromesitylene, $C^9H^{11}(NO^2)$. Almost colorless prisms; fusing point, 41°; distillable without decomposition; easily soluble in alcohol.—*Dinitromesitylene*, $C^9H^{10}(NO^2)^2$. Fine, colorless needles, of a bright lustre; fusing point, 86°—*Trinitromesitylene*, $C^9H^9(NO^2)^3$. Needles that fuse at 232°, and are very difficultly soluble in alcohol.

Amidomesitylene, $C^9H^{11}.NH^2$. Colorless liquid; does not congeal at 0°.—*Nitroamidomesitylene*, $C^9H^{10}(NO^2).NH^2$. Long, yellow needles; fusing point, 100°.—*Dinitroamidomesitylene*, $C^9H^9(NO^2).NH^2$. Short, yellow prisms; fusing point, 193–194°.—*Diamidomesitylene*, $C^9H^{10}(NH^2)^2$. Long, colorless needles; fusing point, 90°.—*Nitrodiamidomesitylene*, $C^9H^9.NO^2(NH)^2$. Large, red, monoclinate crystals; fusing point, 184°.

---

2. Pseudocumene (1 : 3 : 4). Is produced by the decomposition of a mixture of brompara- or brommetaxylene and methyl iodide with sodium.—Colorless liquid; boiling point, 166°.—When oxidized with nitric acid, it is converted into xylylic and xylidinic acids.

Monobrompseudocumene, $C^9H^{11}Br = C^6H^2Br(CH^3)^3$. Colorless laminæ; easily soluble in hot alcohol, but slightly soluble in cold alcohol; fusing point, 73°.—*Tribrompseudocumene*, $C^6Br^3(CH^3)^3$. Fine, colorless needles, very difficultly soluble in alcohol; fusing point, 224°.

Nitropseudocumene, $C^9H^{11}(NO^2)$. Long needles; easily soluble in hot alcohol; fusing point, 71°; boiling point, 265°.—*Trinitropseudocumene*, $C^9H^9(NO^2)^3$. Colorless, quadratic prisms; fusing point, 185°.

Amidopseudocumene, $C^9H^{11}.NH^2$. Colorless needles, of a silky lustre; sparingly soluble in water, easily soluble in alcohol; fusing point, 62°.—*Nitro-*

amidopseudocumene, $C^9H^{10}(NO^2).NH^2$. Golden-yellow, shiny needles; fusing point, 137°.

---

b. Parethylmethylbenzene (Ethyltoluene).

$$C^6H^4\left\{\begin{array}{l}CH^3\\CH^2.CH^3.\end{array}\right.$$

Is obtained, like ethylbenzene, from a mixture of parabromtoluene and ethyl iodide.—Boiling point, 159°. Yields the same products as paraxylene when oxidized.

c. Propylbenzene (Cumene).

$$C^6H^5.C^3H^7.$$

Is obtained by distilling cuminic acid with lime.—Colorless liquid; boiling at 151°. Under the influence of oxidizing agents it is converted into benzoic acid.

By the decomposition of a mixture of monobrombenzene and normal propyl bromide with sodium, a hydrocarbon is obtained that is very similar to cumene, but has the boiling point 157°. Cumene is perhaps isopropylbenzene.

5. *Hydrocarbons*, $C^{10}H^{14}$.

a. Tetramethylbenzene (Durene).

$$C^6H^2(CH^3)^4.$$

Is produced by decomposing a mixture of monobrompseudocumene and methyl iodide with sodium.—Colorless crystals, easily soluble in alcohol; fusing point, 79–80°; boiling point, 189–191°. When oxidized with dilute nitric acid, it yields cumylic acid and cumidinic acid.

b. Ethyldimethylbenzene (Ethylxylene).

$$C^6H^3\left\{\begin{array}{l}(CH^3)^2\\C^2H^5.\end{array}\right.$$

Is obtained, like the preceding compound, from bromxylene and ethyl bromide.—Colorless liquid; boiling point, 183–184°.

c. Paradiethylbenzene.

$$C^6H^4(C^2H^5)^2.$$

Is obtained by the decomposition of a mixture of bromethylbenzene and ethyl bromide with sodium.—Colorless liquid; boiling point, 178–179°.—Subjected to oxidation, it yields ethylbenzoic acid and terephtalic acid.

d. Cymene (Parapropylmethylbenzene).

$$C^6H^4 \left\{ \begin{array}{l} CH^3 \\ C^3H^7. \end{array} \right.$$

Is contained in the oil of Roman cumin (the volatile oil of the seed of *Cuminum cyminum*), and several other volatile vegetable oils. Is produced together with toluene, xylene, mesitylene, and other hydrocarbons by the distillation of camphor over zinc chloride or phosphoric anhydride; terpine (which see), heated with bromine, loses water and hydrogen, and is converted into cymene. Can be most readily obtained in a pure condition by gently heating camphor with phosphorus pentasulphide.—Liquid, that boils at 178°.—When oxidized, it yields toluic and terephtalic acids.

e. Isobutylbenzene.

$$C^6H^5.C^4H^9.$$

Is obtained in the same way as ethylbenzene.—Colorless liquid, boiling at 159–161°. Yields benzoic acid by oxidation.

6. *Hydrocarbons containing a larger number of Carbon-atoms.*

Amylbenzene.

$$C^{11}H^{16} = C^6H^5(C^5H^{11}) = C^6H^5.CH^2CH^2.CH(CH^3)^2.$$

Is prepared like ethylbenzene.—Liquid, that boils at 193°.—When oxidized, it yields benzoic acid.

An isomeric amylbenzene (*diethylized toluene*), $C^6H^5.CH(C^2H^5)^2$, is produced by the action of zinc ethyl on benzal chloride, (p. 275.)—Liquid, that boils at 175–180°.

Amylmethylbenzene, $C^{12}H^{18} = C^6H^4 \begin{cases} CH^3 \\ C^5H^{11} \end{cases}$, and *amyl-dimethylbenzene*, $C^{13}H^{20} = C^6H^3 \begin{cases} (CH^3)^2 \\ C^5H^{11} \end{cases}$, are prepared like amylbenzene. The former boils at 213°, the latter at 232–233°.

B. PHENOLS.

The phenols are the hydroxyl-derivatives of the benzene-hydrocarbons. They bear the same relation to the latter as the alcohols to the hydrocarbons of the marsh-gas series. They differ from these alcohols in their conduct towards aqueous solutions of the alkalies, the hydrogen of the hydroxyl groups contained in them being readily replaced by metals. From these bodies so formed, however, even carbonic acid sets the phenol free. The entrance of chlorine, bromine, iodine, or hyponitric acid into the composition of the phenols causes their conversion into stronger acids.

a. Monatomic Phenols.

1. *Phenol (Phenyl alcohol, Carbolic acid).*
$C^6H^6O = C^6H^5.OH.$

Occurrence and formation. Is contained in castoreum; and sometimes in the urine of graminivorous animals; in human urine after taking benzene. Is formed by the dry distillation of coal, bones, wood, and a number of resins; by heating salicylic acid and the acids isomeric with it; by heating potassium sulphobenzolate with caustic potassa; by boiling diazobenzene nitrate with water.

Preparation. Most practicably from the "dead oil." This is shaken with potassa-ley, the insoluble oil removed, and from the alkaline solution, the phenol, mixed with cresol and other bodies, reprecipitated. It is purified by means of partial distillation and by cooling that portion of the distillate that passes over between 180–190° down to —10°; it is thus deposited in crystals, from which the mother liquor is poured and pressed off.

Properties. Large, colorless prisms of a peculiar odor and burning taste. Very difficultly soluble in water, easily in alcohol. Fuses at 37.5°, and boils at 182–183°. Poisonous.—By the action of phosphorus chloride or bromide, it yields substitution-products of benzene.

Phenol-potassium, $C^6H^5.OK$, is produced by dissolving potassium in phenol and by mixing phenol with concentrated potassa-ley.—Fine, white needles, easily soluble in water, alcohol, and ether.

Phenolether, $C^6H^5.O.C^6H^5$. Is formed when diazobenzene sulphate (p. 266) is mixed with an excess of phenol.—Long, colorless needles. Fusing point, 28°; boiling point, 246°. Insoluble in water, easily soluble in alcohol and ether.

Phenol-methylether (Anisol), $C^6H^5.O.CH^3$, is produced by heating phenol-potassium with methyl iodide or potassium methylsulphate at 100–120°; and by the distillation of anisic acid or methylsalicylic acid with baryta.—Colorless liquid of a pleasant odor, boiling at 152°. By the action of bromine or hyponitric acid there are produced substitution-products; heated with hydriodic acid to 130°, it yields phenol and methyl iodide.

Phenol-ethylether (Phenetol), $C^6H^5.O.C^2H^5$, and *phenol amylether* (phenamylol), $C^6H^5.O.C^5H^{11}$, are produced by the action of ethyl or amyl iodide on phenol-potassium. Both compounds are liquid; the former boils at 172°, the latter at 225°.

Phenol-ethylenether, $(C^6H^5.O)^2C^2H^4$, is produced in the same way from phenol-potassium and ethylene bromide.—Colorless crystals; fusing point, 95°.

Phenol-acetate, $C^6H^5.O.C^2H^3O$. Is obtained by heating phenol with acetyl chloride.—Colorless liquid, boiling at 190°.

Phenol-succinate, $(C^6H^5.O)^2C^4H^4O^2$, is obtained by the action of succinyl chloride on phenol.—Laminæ of a mother-of-pearl lustre; fusing point, 118°; boiling point, 330°.

Phenol-carbonate, $(C^6H^5.O)^2CO$. By heating phenol with carbonyl chloride at 140–150°.—Colorless, shiny needles; fusing point, 78°.

Phenol-cyanate, $CO.N.C^6H^5$, is obtained by distilling ethyl phenylcarbamate (p. 263) with phosphoric anhydride.—Colorless liquid; boiling point, 163°; yields diphenylurea when brought together with water.

Chlorine substitution-products of phenol. *Orthochlorphenol*, $C^6H^4Cl.OH$, is produced by the action of chlorine or sulphuryl chloride (SO^2Cl^2) on phenol.—Colorless crystals, insoluble in water, easily soluble in alcohol; fusing point, 41°; boiling point, 218°.—*Dichlorphenol*, $C^6H^3Cl^2.OH$. Colorless, six-sided needles; fusing point, 42–43°; boiling point, 209°.—*Trichlorphenol*, $C^6H^2Cl^3.OH$. Long, colorless needles, fusing point, 67–68°; boiling point, 244°. Moderately strong acid.—*Pentachlorphenol*, $C^6Cl^5.OH$. Shiny, white needles; fusing point, 185°.

Bromine substitution-products. *Orthobromphenol*, $C^6H^4Br.OH$. Colorless liquid; cannot be distilled without decomposition.—*Dibromphenol*, $C^6H^3Br^2.OH$. Colorless crystalline mass; fusing point, 40°.—*Tribromphenol*, $C^6H^2Br^3.OH$. Long, fine, colorless needles; fusing point, 95°.—*Tetrabromphenol*, $C^6HBr^4.OH$, and *pentabromphenol*, $C^6Br^5.OH$, are produced by heating tribromphenol with bromine at 180–220°. Both compounds are crystalline; the former fuses at 120°, the latter at 225°.

Iodine substitution-products. *Monoiodophenol*, $C^6H^4I.OH$. When a mixture of iodine, iodic acid, and phenol is dissolved in an excess of dilute caustic potassa, there are produced two isomeric compounds, *orthoiodophenol* and *metaiodophenol*, of which only the orthoiodophenol is known in a pure condition. It is also produced by boiling diazoiodobenzene sulphate with water. Flat, shiny needles.—A third isomeric modification, *paraiodophenol*, separates in the form of fine, colorless, very stable needles when paradiazoiodobenzene, sulphate is boiled with water.—*Diiodophenol*, $C^6H^3I^2.OH$. Is most easily obtained by adding iodine and mercury oxide to an alcoholic solution of phenol.—Colorless needles, that sublime at 150°.—*Triiodophenol*, $C^6H^2I^3.OH$. Colorless needles; fusing point, 156°; not sublimable.

Nitrosubstitution-products. *Mononitrophenol*, $C^6H^4(NO^2).OH$. When phenol is added to dilute nitric acid two isomeric compounds, *nitrophenol* and *isonitrophenol* (orthonitrophenol), are formed; of these only the former is volatile with water vapor. Nitrophenol forms large prisms of a sulphur-yellow color; sparingly soluble in water, easily soluble in alcohol; fusing point, 45°; boiling point, 214°. Isonitrophenol crystallizes in long, colorless needles that fuse at 110°.

Dinitrophenol, $C^6H^3(NO^2)^2.OH$. Is produced from phenol by treatment with concentrated nitric acid, and also by boiling dinitrochlor- or dinitrobrombenzene with caustic potassa or sodium carbonate.—Almost colorless laminæ or plates. Fusing point, 114°.

Trinitrophenol (Picric acid), $C^6H^2(NO^2)^3.OH$. Is produced by the action of an excess of concentrated nitric acid on phenol and numerous other bodies: indigo, anilin, salicylic acid, several resins, etc.; and by heating trinitrochlorbenzene with water, or more quickly, with a solution of sodium carbonate.—Yellow, shiny prisms or laminæ, of an exceedingly bitter taste. Fuses at 122.5° ; when carefully heated it is sublimable; detonates when rapidly heated. Difficultly soluble in cold water, more easily in hot water, and still more easily in alcohol. Dyes wool and silk yellow.—Strong acid. With bases it yields yellow salts that crystallize well. The salts explode violently when heated, and some of them by percussion.

Potassium picrate, $C^6H^2(NO^2)^3.OK$, crystallizes in long needles, very difficultly soluble in warm water. The *sodium*, *ammonium*, and *barium salts* are easily soluble in water.

Picrocyamic acid (Isopurpuric acid), $C^8H^5N^5O^6$. The free acid cannot be prepared. The *potassium salt* $C^8H^4N^5O^6.K$ is produced by dropping a hot solution of picric acid (1 part in 9 parts of water) into a warm (60°) solution of potassium cyanate (2 parts of potassium cyanate in 4 parts of water). Brownish-red scales with a green metallic lustre. Sparingly soluble in cold water, soluble in hot water and in alcohol, forming a deep-red solution (test for hydrocyanic acid and cyanides). Detonates with a loud report when heated.

By the action of nitric acid on chlor-, brom-, and iodophenols there are produced *nitrochlorine*, *nitrobromine*, and *nitroiodine substitution-products*, of which a very large number is known.

Amido compounds. *Amidophenol*, $C^6H^4(NH^2).OH$, and the isomeric compound, *isoamidophenol* (orthoamidophenol), are produced by the reduction of the corresponding nitro-compounds, most readily by means of tin and hydrochloric acid. Amidophenol crystallizes in colorless, rhombic scales; isoamidophenol, which is also produced by heating amidosalicylic acid, forms colorless needles, that turn brown easily. Both compounds are difficultly soluble in cold water, more readily in alcohol; fuse at 170°, and yield with acids salts that crystallize well.—*Dinitroamidophenol* (Picramic acid) $C^6H^2(NO^2)^2(NH^2)OH$. The ammonium salt is produced by conducting sulphuretted hydrogen into an alcoholic solution of ammonium picrate; by decomposing this with acetic acid the free acid is obtained. —Red needles; fusing point, 165°; slightly soluble in water, more readily in alcohol and mineral acids.—*Diamidonitrophenol* $C^6H^2(NO^2)(NH^2)^2.OH$. Is obtained, like the preceding compound, when aqueous solutions are employed.—Dark yellow needles or narrow laminæ. Yields salts both with bases and acids.

Sulphophenolic acid, $C^6H^4\begin{cases} OH \\ SO^2.OH. \end{cases}$ Phenol dissolves readily in concentrated sulphuric acid, two isomeric acids, *parasulphophenolic* and *metasulphophenolic acids*, being formed. At the ordinary temperature the meta-acid is formed almost exclusively, but with the aid of heat this is readily converted into the para-acid. The acids can be best separated by the preparation and partial crystallization of their potassium salts. Potassium parasulphophenolate crystallizes first in long, hexagonal plates that contain no water. From the mother-liquor, potassium metasulphophenolate is deposited in long, colorless, spicular crystals that contain two molecules of water of crystallization. The other salts of parasulphophenolic acid are also, as a rule, more difficultly soluble in water than those of metasulphophenolic acid. The para-acid is also obtained by decomposing diazobenzenesulphuric acid.—The free acids are not known in a free state.

Disulphophenolic acid, $C^6H^3\begin{cases} OH \\ (SO^2.OH)^2. \end{cases}$ Is formed by heating phenol or the sulphophenolic acids with an excess of concentrated sulphuric acid; and by the action of concentrated sulphuric acid on diazobenzene sulphate (p. 266).—The acid, separated from the barium or lead salt, crystallizes in very deliquescent, concentrically-arranged needles of a silken lustre. The solutions of the free acids, as well as those of its salts, are colored ruby-red on the addition of iron chloride.

Barium disulphophenolate, $C^6H^4S^2O^7Ba + 4H^2O$. Colorless, shiny, monoclinate prisms. Easily soluble in hot water, less soluble in cold.

Phenyl sulphydrate (Benzene sulphydrate), $C^2H^6S = C^6H^5.SH$. Is produced by the action of hydrogen (tin and hydrochloric acid, zinc and dilute sulphuric acid) on benzene sulphochloride (p. 269); by the distillation of phenol over phosphorus pentasulphide; and by the distillation of sodium sulphobenzolate.—Colorless liquid, of an unpleasant odor; boiling point, 166–168°; specific gravity, 1.08. Insoluble in water, easily soluble in alcohol and ether. Dissolves sodium easily; and, when treated with mercury oxide, gives a compound $(C^6H^5S)^2Hg$, that crystallizes from alcohol in white, shiny needles.

Parabromphenyl sulphydrate, $C^6H^4Br.SH$. Is formed in the same way from parabrombenzene sulphochloride.—Colorless, lamellar crystals; fusing point, 93.5°.

Phenyl sulphide (Benzene sulphide), $(C^6H^5)^2S$. Is formed, together with benzene and phenyl sulphydrate, in the destructive distillation of sodium sulphobenzolate; and in the distillation of phenol over phosphorus sulphide. Is further produced by heating several of the metallic compounds of phenyl sulphydrate.—Colorless liquid, of an unpleasant odor; boiling point, 292°; specific gravity, 1.12. Insoluble in water, easily soluble in hot alcohol and ether.

Phenyl disulphide, $(C^6H^5)^2S^2$. Is produced in small quantity in the preparation of phenyl sulphydrate from benzene sulphochloride; and can be readily obtained from the sulphydrate by oxidation with dilute nitric acid. Is also formed, when a solution of the sulphydrate, in alcoholic ammonia, is allowed to evaporate spontaneously in the air. It is further formed when iodine is added to an aqueous solution of the sodium compound of the sulphydrate; by the action of potassium cyanide on an alcoholic solution of benzene sulphochloride; and, together with other products, by treating the sulphydrate with phosphorus chloride.—Colorless, shiny needles, that fuse at 60°, and are distillable without decomposition. Insoluble in water, easily soluble in alcohol and ether. Nascent hydrogen reconverts it into phenyl sulphydrate; when further oxidized it yields sulphobenzolic acid.

Phenyl oxysulphide (Sulphobenzide), $(C^6H^5)SO^2$. Is formed, together with sulphobenzolic acid, by treating benzene with sulphuric anhydride or fuming sulphuric acid; by the oxidation of phenyl sulphide with chromic acid; and in small quantity by the distillation of sulphobenzolic acid.—Crystallizes from alcohol in rhombic plates, from water in fine prisms. Fuses at 128°, and distils without decomposition. Very difficultly soluble in water, difficultly soluble in cold alcohol, easily soluble in hot alcohol and in ether.—Concentrated sulphuric acid dissolves it; and converts it, with the aid of heat, into sulphobenzolic acid. Heated with phosphorus chloride it yields benzene sulphochloride and monochlorbenzene. The same products are formed by the action of chlorine on heated sulphobenzide.

Oxysulphobenzide, $(C^6H^4.OH)^2SO^2$. Is produced when a mixture of equal parts of phenol and concentrated sulphuric acid is heated, from five to six hours, at 190°, and the cooled, tenacious mass poured into from two to three times its volume of water.—Stellate, colorless needles; almost insoluble in cold water, easily in boiling water, and in alcohol and ether.—It gives compounds with bases, in which only one of the two hydrogen-atoms of the hydroxyl groups is replaced; on the other hand it yields others in which both the hydrogen-atoms are replaced by alcohol radicles.

Phenyl oxydisulphide, $(C^6H^5)^2S^2O^2$. Is produced together with sulphobenzolic acid by heating benzenesulphurous acid (p. 270) with water at 130°.—Long, shiny, four-sided needles. Insoluble in water and alkalies, easily soluble in ether and hot alcohol. Fuses at 36°.

2. *Cresols.*

$$C^7H^8O = C^6H^4 \begin{cases} CH^3 \\ OH. \end{cases}$$

a. **Orthocresol.** Is obtained by melting potassium orthosulphotoluenate with caustic potassa, dissolving the mass in acids, and exhausting with ether.—Limpid liquid; boiling point, 188–190°; does not congeal at a low temperature. Heated for a long time with caustic potassa, it is converted into salicylic acid.

b. **Metacresol.** Is produced together with propylene gas by heating thymol (p. 300) with phosphoric anhydride. The product, that consists chiefly of cresol-phosphate, is decomposed by means of potassium hydroxide, the mass acidified and exhausted with ether.—Colorless liquid of an odor like that of phenol. Boiling point, 195–200°. Does not congeal even in a mixture of solid carbonic anhydride and ether. Fused with caustic potassa it is converted into oxybenzoic acid.

Metacresol-ethylether, $C^7H^7.O.C^2H^5$. Is prepared like phenol-ethylether.—Liquid, of boiling point 188–191°.

c. **Paracresol.** Is prepared from parasulphotoluenic acid like orthocresol. Is also produced by boiling with water the diazotoluene sulphate obtained from paratoluidin (p. 277.)—Colorless prisms of a phenol odor, reminding of decayed urine; fusing point, 36°; boiling point, 198°; very difficultly soluble in water. The aqueous solution gives a blue color with iron chloride. Fusing caustic potassa converts it into paraoxybenzoic acid.

Paracresol-methylether, $C^7H^7.O.CH^3$. Colorless liquid boiling at 174°. Is oxidized to anisic acid by chromic acid.

Paracresol-ethylether, $C^7H^7.O.C^2H^5$. Colorless liquid; boiling point, 188°.

Paracresol-acetate, $C^7H^7.O.C^2H^3O$. Liquid; boiling point, 208–211°.

Dinitro-paracresol, $C^6H^2(NO^2)^2 \left\{ \begin{matrix} CH^3 \\ OH. \end{matrix} \right.$ Is produced by the action of nitrous acid on paratoluidin. Yellow crystals, that fuse at 84°.

The cresol contained in coal-tar and wood-tar together with phenol is liquid, and does not congeal. It is either ortho- or meta-cresol, or more probably a mixture of two or all three of the cresols.

3. *Phenols*, $C^8H^{10}O$.

a. Dimethyl-phenols (*Xylenols*).

$$C^6H^3 \left\{ \begin{matrix} (CH^3)^2 \\ OH. \end{matrix} \right.$$

Of the many modifications possible according to the theory, only three are as yet known—

1. Solid Xylenol (Metaxylene-phenol). Is produced together with the following compound, when the mixture of meta- and paraxylene, that is obtained from coal-oil, is converted into sulpho-acids by dissolving in sulphuric acid, and the potassium salts of these acids melted with caustic potassa. It is also produced by heating oxymesitylic acid with caustic potassa.—Colorless crystals; fusing points, 75°; boiling point, 216°.

2. Liquid Xylenol. Is produced together with the preceding compound.—Colorless liquid, boiling point, 211.5°.

3. Phlorol. Is formed in the destructive distillation of barium phloretate; and is perhaps contained in the creosote of coal-tar and beech-wood tar.—Colorless liquid; boils at 220°; specific gravity, 1.037.

---

b. Ethyl-phenol.

$$C^6H^4 \left\{ \begin{array}{l} C^2H^5 \\ OH. \end{array} \right.$$

From potassium sulphethylbenzolate by fusing with caustic potassa.—Large, colorless, prismatic crystals of an odor resembling that of phenol; fusing point, 47–48°; boiling point, 211°. In contact with water it becomes instantaneously liquid. But slightly soluble in water; in alcohol and ether in all proportions. Yields with bromine tetrabromethyl-phenol $C^6Br^4 \left\{ \begin{array}{l} C^2H^5 \\ OH, \end{array} \right.$ which crystallizes in shiny prisms, that fuse at 105–106°.

4. *Phenols*, $C^{10}H^{14}O$.

Two phenols of this composition are known, both of which are methyl-propyl phenols, $C^6H^3 \left\{ \begin{array}{l} CH^3 \\ C^3H^7 \\ OH. \end{array} \right.$

a. **Thymol.** Occurs in thyme-oil (from *Thymus serpyllum*), in the oil of *Monarda punctata* and of *Ptychotis ajowan*, together with the hydrocarbons cymene ($C^{10}H^{14}$) and thymene ($C^{10}H^{16}$). It is extracted from these oils by means of concentrated soda-ley and the aqueous solution of the sodium compound decomposed with hydrochloric acid.—Large, colorless crystals of a pleasant odor, like that of thyme; fusing point, 44°; boiling point, 230°. Sparingly soluble in water, easily soluble in alcohol. Is decomposed when heated with phosphoric anhydride, yielding propylene and meta-cresol-phosphate (p. 298.)

b. **Cymophenol.** From potassium sulphocymolate with fusing potassa.—Yellowish, thick oil; boiling point, 230°.

Benzyl-phenol, $C^{13}H^{12}O = C^6H^4 \begin{cases} C^7H^7 \\ OH. \end{cases}$ Is formed from benzyl chloride and phenol, like benzylbenzene (p. 282) from benzyl chloride and benzene.—White needles of a silky lustre; fusing point, 84°; distils, undergoing partial decomposition.

b. Quinones.

The quinones are derived from the hydrocarbons by the replacement of the hydrogen-atoms of two neighboring carbon-atoms by means of two united oxygen-atoms. Nascent hydrogen and other reducing agents, even sulphurous anhydride, convert them into phenols belonging to the ortho-series. The latter treated with oxidizing substances are readily reconverted into quinones.

1. *Quinone.*

$$C^6H^4O^2 = C^6H^4 {O \atop O}>$$

Formation and preparation. By oxidation of hydroquinone, quinic acid, anilin, orthodiamidobenzene, benzidine and orthoamidophenol. Is further produced by the distillation of a number of vegetable extracts. Is prepared most readily by heating quinic acid (1 part) with manganese peroxide (4 parts), and sulphuric acid (1 part diluted with ½ part of water).

Properties. Golden-yellow prisms; fusing point, 116°. Very easily sublimable; volatilizes even at the ordinary temperature; has a penetrating odor; and excites to tears. Moderately difficultly soluble in water, more readily in alcohol.

Chlorine substitution-products of quinone are formed by the action of chlorine on quinone; and by the distillation of quinic acid with a chlorine-mixture. *Monochlorquinone* $C^6H^3ClO^2$. Long, yellow needles.—*Dichlorquinone* $C^6H^2Cl^2O^2$. Is also produced by the action of chlorous anhydride on benzene; and by treating trichlorphenol with nitric acid. Large, yellow prisms; fusing point, 120°.—*Trichlorquinone* $C^6HCl^3O^2$. Large, yellow laminæ, almost insoluble in water; fusing point, 165–166°.—*Tetrachlorquinone* (chloranile), $C^6Cl^4O^2$, is produced, together with trichlorquinone, also from a number of other organic compounds (phenol, anilin, salicylic acid, isatine, etc.) by treatment with chlorine, or hydrochloric acid and potassium chlorate. Yellow, lamellar crystals, sublimable without decomposition, insoluble in water, but slightly soluble in cold alcohol, more readily in hot.—Heated with phosphorus pentachloride, it yields perchlorbenzene, (C^6Cl^6) (p. 254). Dissolves in dilute caustic potassa, thus causing the formation of potassium chloride and the difficultly soluble purplish-red potassium salt of *chloranilic acid*, $C^6Cl^2(OK)^2O^2 + H^2O$, from which by means of sulphuric acid the free acid, $C^6Cl^2(OH)^2O^2 + H^2O$, may be obtained in the form of reddish-white, shiny scales, resembling mica. The same acid is also obtained by treating trichlorquinone in the same way.

Quinhydrone (Green hydroquinone), $C^{12}H^{10}O^4$, is formed by the action of an insufficient quantity of sulphurous acid on a solution of quinone; or by mixing solutions of quinone and hydroquinone; and may hence be considered as a compound of equal molecules of quinone and hydroquinone: $C^6H^4\left\{\begin{matrix} O \quad . \quad O \\ OH\ HO \end{matrix}\right\}C^6H^4$. In general terms, it is always produced when hydrogen is eliminated from hydroquinone, as, for instance, by means of chlorine water, iron chloride, nitric acid, etc.—Long, thin prisms of a beautiful green metallic lustre, of an odor somewhat resembling that of quinone. It fuses easily, sublimes partially, is slightly soluble in water, easily in alcohol, forming a yellow solution. Further treatment with oxidizing substances converts it readily into quinone; with reducing agents into hydroquinone.

2. *Toluquinone*, $C^7H^6O^2$, is not known. Substitution-products of it—*trichlortoluquinone* and *tetrachlortoluquinone*—are produced by the action of hydrochloric acid and potassium chlorate on creosol (p. 309) and the cresol (p. 299) contained in coal-tar.

3. *Phlorone*, $C^8H^8O^2$. Is obtained from the phenols $C^8H^{10}O$, contained in coal-tar and beech-wood tar, by distilling them with manganese peroxide and sulphuric acid.—Yellow, needly crystals. Easily sublimable. Slightly soluble in cold water, more readily in hot water, easily soluble in alcohol and ether. Its vapor attacks the eyes and mucous membranes violently.

4. *Thymoquinone*, $C^{10}H^{12}O^2$. Is obtained by distilling a solution of thymol, diluted with water, with manganese peroxide.—Yellow, prismatic plates; fusing point, 45.5°; boiling point, 200°; volatile with water vapor. Has a peculiar penetrating odor. Yields, with bromine, *mono-* and *dibromthymoquinone*. The former crystallizes in long, yellow needles, which, when heated with potassa-ley, are converted into *oxythymoquinone*, $C^{10}H^{11}(OH)O^2$; the latter forms bright-yellow laminæ that fuse at 73.5°.

c. *Diatomic Phenols.*

1. *Dioxybenzenes.*

$$C^6H^6O^2 = C^6H^4(OH)^2.$$

a. **Hydroquinone** (Ortho-dioxybenzene). Is produced from quinic acid by destructive distillation; or by the addition of lead superoxide to its aqueous solution; by treating arbutine (see Glucosides) with dilute sulphuric acid; by heating orthoiodophenol (p. 293) with caustic potassa at 180°; and is prepared most readily by treating quinone with sulphurous or hydriodic acid.

Colorless prisms. Easily soluble in water, alcohol, and ether; fusing point, 177.5°. When carefully heated it is sublimable.—Combines with sulphuretted hydrogen and sulphurous anhydride, forming crystalline compounds that are easily decomposed by water. Oxidizing substances convert it into quinone.

Chlorine substitution-products. These cannot be prepared directly from hydroquinone. They are produced by treating the corresponding substitution-products of quinone with sulphurous acid.—*Monochlorhydroquinone* $C^6H^3Cl(OH)^2$ is also produced by evaporating a solution of quinone in concentrated hydrochloric acid.—*Dichlorhydroquinone* $C^6H^2Cl^2(OH)^2$. Stellate groups of colorless needles, fusing at 157–158°.—*Trichlorhydroquinone* $C^6HCl^3(OH)^2$. Colorless prisms, easily soluble in boiling water. Fusing point, 134°.—*Tetrachlorhydroquinone* $C^6HCl^4(OH)^2$. Laminæ, insoluble in water; fusing point, above 200°.

Dinitrohydroquinone, $C^6H^2(NO^2)^2(OH)^2$, is produced by boiling dinitroarbutine (see Glucosides) with dilute sulphuric acid.—Golden-yellow, shiny laminæ; but slightly soluble in cold water, easily soluble in boiling water and in alcohol. The aqueous solution turns deep blue on the addition of alkalies or ammonia.

Disulphohydroquinonic acid, $C^6H^6S^2O^8 = C^6H^2\left\{\begin{array}{l}(OH)^2\\(SO^2.OH)^2\end{array}\right.$. Is produced by treating quinic acid with fuming sulphuric acid.—Non-crystallizing syrup, very easily soluble in water and alcohol. Bibasic acid. Yields salts that crystallize well. Its aqueous solution is colored deep blue by iron chloride.

Dichlordisulphohydroquinonic acid, $C^6Cl^2\left\{\begin{array}{l}(OH)^2\\(SO^2.OH)^2\end{array}\right.$. The potassium salt of this acid $C^6Cl^2(OH)^2.(SO^2.OK)^2 + 2H^2O$ (shiny, difficultly soluble scales) is produced when chloranile is added to a warm dilute solution of potassium bisulphite. Its solution, as well as that of its salts, is colored indigo-blue by iron chloride. The potassium salt, together with potassa-ley, in contact with the air, is rapidly converted into yellow *potassium euthiochronate*, $C^6(O^2)\left\{\begin{matrix}(OK)^2\\(SO^2.OK)^2\end{matrix}\right. +$ $2H^2O$.

Thiochronic acid, $C^6\left\{\begin{matrix}OH\\O.(SO^2.OH)\\(SO^2.OH)^4.\end{matrix}\right.$ The yellow potassium salt, $C^6(OH)O(SO^2.OK)^5 + 4H^2O$, is formed, together with potassium dichlordisulphohydroquinonate, on adding chloranile to a warm solution of potassium bisulphite or sulphite. Boiled with hydrochloric acid, and heated with water at 130–140°, it is resolved into potassium bisulphate and *potassium β-disulphohydroquinonate* $C^6H^2(OH)^2(SO^2.OK)^2 + 4H^2O$. The free acid, isolated from the latter salt, crystallizes in deliquescent, thick plates. It is isomeric with the disulphohydroquinonic acid described above.

---

b. **Pyrocatechin** (Meta-dioxybenzene, Oxyphenic acid). Is contained in the leaves of *Ampelopsis hederacea*. Is produced by the destructive distillation of morintannic, catechuic, protocatechuic, and oxysalicylic acids, and a number of vegetable extracts (catechu, kino, etc.). Is furthermore formed from metaiodophenol (p. 293) and metasulphophenolic acid (p. 295) by fusing with caustic potassa; and by heating cellulose and other hydrocarbons for a long time with water at 200°.—Crystallizes in quadratic prisms, that are easily soluble in water, alcohol, and ether. Fuses at 112°; sublimes in colorless, shiny laminæ; and boils without decomposition at 240–245°. The aqueous solution is colored dark green by iron chloride, and then turns purple on the addition of sodium bicarbonate or tartaric acid, or ammonia.

Guaiacol (Pyrocatechin-monomethylether), $C^7H^8O^2 = C^6H^4\left\{\begin{matrix}O.CH^3\\OH.\end{matrix}\right.$ Is produced by heating equal molecules of pyrocatechin, potassium hydroxide, and potassium methylsulphate at 170–180°; subjecting guaiacum to destructive distillation; and is contained in beech-wood tar (creosote).—Colorless liquid, boiling at 200°. Slightly soluble in water, easily soluble in acetic acid and alkalies. Forms, like phenol, crystallizing, easily soluble, and easily decomposable salts with the alkalies and with ammonia. When heated with hydriodic acid (or iodine and phosphorus) it yields methyl iodide and pyrocatechin. The latter substance is also produced when guaiacol is added to fusing potassium hydroxide.

Pyrocatechin-dimethylether, $C^6H^4(O.CH^3)^2$. Is obtained by heating guaiacol-potassium with methyl iodide.—Liquid, boiling at 205–206°.

Diacetylpyrocatechin, $C^6H^4(O.C^2H^3O)^2$, is produced by the action of acetyl chloride on pyrocatechin.—Needles, easily fusible, insoluble in water, soluble in alcohol.

Tetrabrompyrocatechin, $C^6Br^4(OH)^2$, is produced when pyrocatechin is mixed with an excess of bromine.—Reddish-brown, rhombic needles, insoluble in water, soluble in alcohol.

---

c. **Resorcin** (Para-dioxybenzene). Is formed by adding a number of resins (galbanum, assafœtida, sagapenum, acaroid) to fusing caustic potassa; is extracted from the fused mass by acidifying with sulphuric acid and shaking with ether, and purified by distillation. Is further produced from parachlor- and parabromsulphobenzolic acids, paradisulphobenzolic acid, paraiodophenol, and parasulphophenolic acid by fusing with caustic potassa.—Plates or columns, easily soluble in water, alcohol, and ether. Fuses at 104°, and boils at 271°, evaporates at a lower temperature. The aqueous solution is colored dark-purple by iron chloride.

Diacetylresorcin, $C^6H^4(O.C^2H^3O)^2$, is produced by the action of acetyl chloride on resorcin.—Colorless liquid, insoluble in water.

Trinitroresorcin (Oxypicric acid, Styphnic acid), $C^6H(NO^2)^3(OH)^2$. Is produced by the action of nitric acid on morintannic acid, a number of gum-resins, (galbanum, sagapenum, ammonia-gum), and a number of vegetable extracts (of sapon-wood, Brazil-wood, etc.) Is obtained from orcin by the action of nitric acid at a low temperature.—Pale yellow prisms or lamellæ; sublimable when carefully heated; difficultly soluble in water; fusing point, 175.5°.—Strong, bibasic acid; yields salts that crystallize well and explode violently when heated.

Thioresorcin, $C^6H^4(SH)^2$. Is produced when para-disulphobenzolchloride (p. 270) is heated gently with tin and hydrochloric acid.—Crystalline mass, easily volatile with water-vapor; fusing point, 27°; boiling point, 243°.

Umbelliferone, $C^6H^4O^2$ (or $C^9H^6O^3$). Isomeric with quinone. Is produced in the destructive distillation of a number of resins, chiefly of umbelliferous plants, as galbanum.—Colorless, rhombic prisms, sparingly soluble in cold water, easily soluble in alcohol, and ether. The aqueous solution exhibits, by reflected light, a splendid blue color. Melts at 240°; sublimes without decomposition. Yields resorcin when fused with caustic potassa.

2. *Orcin.*

$$C^7H^8O^2 = C^6H^3\left\{\begin{array}{l}CH^3\\(OH)^2.\end{array}\right.$$

It appears to be ready formed in a number of lichens. Is formed from orsellic acid and other acids (lecanoric, evernic, erythric acids) that occur in various lichens, and bear a close relation to orsellic acid, either by heating them alone, or by boiling them with strong bases. It is further produced when aloes is melted with caustic potassa.—In order to prepare it in large quantity a lichen, belonging to the species *roccella* or *lecanora*, is boiled with milk of lime, filtered, and the filtrate evaporated to about one-fourth. The lime is now precipitated by means of carbonic anhydride, and the solution evaporated nearly to dryness over the water-bath. The residue is boiled several times with benzene, the orcin extracted from its solution in benzene by shaking with water, and the aqueous solution evaporated. Crystallizes in large, colorless, six-sided prisms with 1 molecule of water of crystallization. It has a repulsive, sweet taste. Easily soluble in water, alcohol, and ether. With its water of crystallization it fuses at 58°, anhydrous at 86°; it boils at 290° without undergoing decomposition. In contact with the air it turns red. Its aqueous solution is colored deep violet by iron chloride.

Orcin combines with dry ammonia, forming a crystalline compound. Exposed to the simultaneous influence of moist air and ammonia, it is converted into a dark brown substance *orcein*, $C^7H^7NO^3$, which dissolves in alkalies, forming solutions of a beautiful red color; from these solutions acetic acid precipitates the dissolved orcin. Upon this conduct depends the employment of a number of lichens in the preparation of the beautiful red dyes, known as *archil*, *cudbear*, *persio*. These dyes are obtained by mixing the finely-ground lichens with decaying urine and lime, or with ammonia-water, and allowing the mixture to stand for a long time in contact with the air. *Litmus* is prepared in the same way, particularly from *Lecanora tartarea*.

Orcin-monethylether, $C^7H^6\left\{\begin{matrix}O.C^2H^5\\OH,\end{matrix}\right.$ and **-diethylether,** $C^7H^6(O.C^2H^5)^2$, are produced by the action of caustic potassa and ethyl iodide on orcin. Both compounds are syrupy liquids. The diethylether boils without decomposition at 240–250°.

Diacetylorcin, $C^7H^6(O.C^2H^3O)^2$, is formed, even at the ordinary temperature, by pouring acetyl chloride on orcin.—Colorless needles; fusing at 25°; sublimes almost without decomposition. Scarcely soluble in water, easily soluble in alcohol and ether.

Trinitro-orcin, $C^6(NO^2)^3 \left\{ \begin{matrix} CH^3 \\ (OH)^2. \end{matrix} \right.$ Is produced by dissolving orcin in well-cooled nitric acid, and pouring the solution into concentrated sulphuric acid at —10°; when this mixture is poured into a large quantity of water the nitro-compound separates.—Long, yellow needles. Easily soluble in hot water, but slightly in cold. Fuses at 162°, and at a slightly higher temperature it decomposes with a weak explosion. Strong, bibasic acid. Yields salts that crystallize well, and are for the greater part easily soluble.

Creosol, $C^8H^{10}O^2 = C^7H^6 \left\{ \begin{matrix} O.CH^3 \\ OH, \end{matrix} \right.$ is formed, together with its homologue, guaiacol (p. 305), by the distillation of beech-wood and guaiacum; and can be separated from it by partial distillation.—Colorless liquid, very similar to guaiacol; boiling point, 219°. Treated with hydriodic acid, it yields methyl iodide and a non-crystallizing body, isomeric with orcin (homopyrocatechin).

3. *Phenols*, $C^8H^{10}O^2 = C^8H^8(OH)^2$.

a. **Hydrophloron.** Is obtained by the action of sulphurous acid on phlorone (p. 303) that is suspended in water.—Colorless laminæ, of a mother-of-pearl lustre. Fusible and sublimable. Easily soluble in water, alcohol, and ether. Oxidizing substances convert it readily into phlorone.

b. **Betaorcin** is formed from beta-usnic acid and a few other acids, occurring in lichens, in the same manner as orcin.—Quadratic prisms, sublimable, easily soluble in alcohol and ether. Turns red in contact with the air.

c. **Veratrol** is produced by heating veratric acid with an excess of baryta.—Colorless oil, of an aromatic odor; boils at 202–205°, and congeals in crystalline form at +15°.

---

4. *Thymohydroquinone*, $C^{10}H^{14}O^2 = C^{10}H^{12}(OH)^2$. Is obtained from thymoquinone by treating with sulphurous acid.—Clear, four-sided prisms, of a vitreous lustre. Fusing point, 139.5° Sublimes without decomposition. Difficultly soluble in cold water, easily in boiling water. Oxidizing substances convert it easily into thymoquinone.

d. Triatomic Phenols.

Pyrogallol (Pyrogallic Acid).
$C^6H^6O^3 = C^6H^3(OH)^3$.

Formation. By heating gallic acid alone, most advantageously in an atmosphere of carbonic anhydride, at 210–220°, or with two to three times its weight of water, in a closed vessel, at 200–210°. In smaller quantity by heating gallotannic acid.

Properties. Shiny, colorless laminæ or needles of a bitter taste. Poisonous. Sublimable without decomposition when the air is not allowed to have access. Easily soluble in water. In the presence of alkalies it takes up oxygen rapidly from the air, and decomposes, yielding carbonic anhydride, acetic acid, and brown, amorphous substances. It gives a blackish-blue color with iron sulphate, a red color with iron chloride. It reduces the metals rapidly from gold, silver, and mercury salts.

Triacetylpyrogallol, $C^6H^3(O.C^2H^3O)^3$, is produced by dissolving pyrogallol in an excess of acetyl chloride, and remains behind on evaporation in small, sublimable crystals, insoluble in water.

Tribrompyrogallol, $C^6Br^3(OH)^3$, is produced by mixing pyrogallol with bromine.—Shiny, flat, rhombic needles, of a bright leather-color. Very difficultly soluble in cold water, more easily soluble in hot water.

The following substance is isomeric with pyrogallol:

Phloroglucin, $C^6H^6O^3 = C^6H^3(OH)^3$. Is produced by heating phloretin, quercetin (see Glucosides), dragon's blood, gamboge, kino, etc. with caustic potassa.—Rhombic crystals, with two molecules of water of crystallization, of sweet taste. They effloresce in dry air, give up their water at 100°, fuse at 220°, and sublime almost without decomposition. Easily soluble in water, alcohol, and ether. The aqueous solution turns a deep violet color on the addition of iron chloride. Combines with the alkalies, forming deliquescent salts.

Triacetylphloroglucin, $C^6H^3(O.C^2H^3O)^3$. Small, colorless prisms, but slightly soluble in water.

Phloramine, $C^6H^7NO^2 = C^6H^3 \left\{ \begin{matrix} NH^2 \\ (OH)^2 \end{matrix} \right.$. Is formed by dissolving phloroglucin in heated aqueous ammonia, and by conducting dry ammonia gas over heated phloroglucin.—Thin, shiny laminæ, resembling mica. But slightly soluble in cold water, easily in alcohol. The solution turns rapidly brown in contact with the air. Well characterized base; combines with acids, forming crystallizing salts.

e. Tetratomic Phenols.

These are as yet unknown, though a few substitution-products of tetroxybenzene, $C^6H^2(OH)^4$, have been discovered.

Dichlortetroxybenzene (Hydrochloranilic acid), $C^6Cl^2(OH)^4$. Is produced by the action of nascent hydrogen (sodium-amalgam and hydrochloric acid, tin and hydrochloric acid) on chloranilic acid (p. 302); can be prepared most readily by heating chloranilic acid with a concentrated solution of sulphurous acid at 100°.—Colorless needles. But slightly soluble in cold water, easily soluble in alcohol, and ether. In a moist condition it is reconverted into chloranilic acid in contact with the air. With acetyl chloride, it yields an ether, $C^6Cl^2(O.C^2H^3O)^4$, that crystallizes well, fuses at 235°, and is very stable.

Disulphotetroxybenzolic acid, $C^6 \begin{cases} (OH)^4 \\ (SO^2.OH)^2 \end{cases}$. The alkaline salts of this acid are produced by boiling the salts of euthiochronic acid (p. 305) with tin and hydrochloric acid. The *potassium salt*, $C^6(OH)^4(SO^2.OK)^2 + 2H^2O$, crystallizes in colorless columns, which, when dry, are stable in the air, but when moist or in solution are oxidized, and turn red in contact with the air. The free acid is not known.

C. Alcohols.

The aromatic alcohols are isomeric with the phenols. They differ from the phenols, in that the hydroxyl groups do not replace hydrogen-atoms of the benzene nucleus, but of the substituting methyl, ethyl groups, etc. They conduct themselves in every way analogously to the alcohols of the marsh-gas series.

1. *Benzyl Alcohol.*

$C^7H^8O = C^6H^5.CH^2.OH.$

Occurrence. In the form of benzyl benzoate and cinnamate in Peru- and Tolu-balsams.*

Formation and preparation. From oil of bitter almonds by means of nascent hydrogen (sodium-amalgam and water); or by mixing with an alcoholic solution of potassium hydroxide, it being thus resolved into benzyl alcohol and potassium benzoate, an evolution of heat accompanying the action. After distilling

* Peru- and Tolu-balsams are tenacious yellow or reddish-brown liquids, which are obtained in Mexico and Peru from the branches and bark of *Myroxylon peruiferum* and *Myroxylon toluiferum* by means of soaking or boiling with water, or, less frequently, from incisions, from which they flow spontaneously.

off the alcohol and adding water, the benzyl alcohol is extracted by means of ether. Benzyl chloride (p. 274), when heated with an alcoholic solution of potassium acetate, yields benzyl acetate, which is transformed into potassium acetate and benzyl alcohol, by boiling with an alcoholic solution of potassium hydroxide.

Properties. Colorless liquid of a weak, pleasant odor; specific gravity, 1.06; boiling point, 207°. It is liquid at —18°.

Oxidizing substances convert it into oil of bitter almonds and benzoic acid; hydrochloric and hydrobromic acids into benzyl chloride or bromide (pp. 274 and 276). When distilled with a concentrated solution of potassa, it is resolved into benzoic acid and toluene. Sulphuric acid and other dehydrating agents convert in into a resin.

Benzylic ether, $(C^7H^7)^2O$, is produced by heating benzyl alcohol with anhydrous boracic acid; and by heating benzyl chloride with water at 190°.—Colorless oil, boiling above 300°.

Benzyl acetate, $C^7H^7.O.C^2H^3O$, is formed by mixing benzyl alcohol with acetic and sulphuric acids; and by heating benzyl chloride with potassium acetate.—Colorless liquid of a pleasant odor, boiling at 210°. Heavier than water.

Parachlorbenzyl alcohol, $C^6H^4Cl.CH^2.OH$. The liquid ether (boiling point, 240°) of this alcohol is produced by heating chlorbenzyl chloride (p. 275) with silver acetate. This, heated to 100° with ammonia, yields the alcohol.—Long, colorless, spicular crystals. Insoluble in cold water, difficultly soluble in boiling water. Fuses at 66°, and boils without undergoing decomposition.

Paradichlorbenzyl alcohol, $C^6H^3Cl^2.CH^2.OH$, is prepared from dichlorbenzyl chloride (p. 275) like the preceding compound.—Colorless needles, but slightly soluble in water; fusing point, 77°.

27

Metanitrobenzyl alcohol, $C^6H^4(NO^2).CH^2.OH$, is formed together with potassium nitrobenzoate by heating nitrobenzylic aldehyde with alcoholic potassa.—Thick oil, that cannot be distilled without decomposition.

Paranitrobenzyl alcohol, $C^6H^4(NO^2).CH^2.OH$. The acetic ether of this alcohol (long, pale yellow needles, fusing at 78°) is produced by adding benzyl acetate to cold concentrated nitric acid. By heating with aqueous ammonia to 100°, the alcohol is obtained from this.—Colorless, fine needles; fusing point, 93°. Easily soluble in hot water and ammonia, but slightly in cold water. Dissolved in very concentrated nitric acid, it is converted into *dinitrobenzyl alcohol*, $C^6H^3(NO^2)^2.CH^2.OH$. (Colorless needles, fusing at 71°.)

Benzyl sulphydrate (Benzylmercaptan), $C^6H^5.CH^2.SH$. Is producing by mixing an alcoholic solution of potassium sulphydrate with benzyl chloride, a spontaneous evolution of heat accompanying the action. It is thrown down on the addition of water.—Colorless, highly refracting liquid of an unpleasant leeky odor; boiling point, 194–195°. Yields with mercury oxide a mercaptide, that crystallizes well.

Benzyl sulphide, $(C^6H^5.CH^2)^2S$, is formed when an alcoholic solution of potassium sulphide is mixed with benzyl chloride, a strong evolution of heat accompany-the action.—Long, colorless needles or laminæ. Insoluble in water, easily soluble in alcohol and ether. Fuses at 49°. Not volatile without decomposition.—Nitric acid converts it into *benzyl sulphoxide*, $(C^6H^5.CH^2)^2SO$, a substance that crystallizes in colorless laminæ, fusing at 130°.

Benzyl disulphide, $(C^6H^5.CH^2)^2S^2$, is formed from benzyl sulphide by oxidation in contact with the air, particularly when a solution of the latter containing ammonia is evaporated in the air.—Colorless, shiny laminæ. Insoluble in water, difficultly soluble in cold alcohol, easily in hot. Fuses at 66–67°. Nascent hydrogen converts it into benzyl sulphydrate. When heated it is resolved into toluene, stilbene (p. 282), and other products. The same substances are formed by heating benzyl sulphide.

Saligenin (Ortho-oxybenzyl alcohol), $C^7H^8O^2 = C^6H^4 \begin{cases} OH \\ CH^2.OH. \end{cases}$ Is produced from salicin (see Glucosides) by means of treating with emulsin or saliva and by the action of nascent hydrogen on salicylous acid (p. 322).—Tables, having a pearly lustre; easily soluble in hot water, in alcohol, and ether. Fuses at 82°, and sublimes at 100°. Its solution is colored deep blue by iron chloride. Dilute acids convert it into a resin, *saliretin*, $C^{14}H^{14}O^3$. Oxidizing agents convert it into salicylous and salicylic acids.

Anise alcohol (Methylparaoxybenzyl alcohol), $C^8H^{10}O^2 = C^6H^4 \begin{cases} O.CH^3 \\ CH^2.OH. \end{cases}$ Is prepared from anisic aldehyde (p. 324) in the same manner as benzyl alcohol from the oil of bitter almonds.—Colorless, shiny prisms, that fuse at 20°, and distill without decomposition at 250°. Of a faint odor and burning taste. Oxidizing substances convert it into anisic aldehyde and anisic acid; hydrochloric acid into a liquid chloride, $C^6H^4 \begin{cases} O.CH^3 \\ CH^2Cl. \end{cases}$

2. *Tolyl Alcohol* (*Paramethylbenzyl Alcohol*).

$$C^8H^{10}O = C^6H^4 \begin{cases} CH^3 \\ CH^2.OH. \end{cases}$$

Is prepared from paratolylic aldehyde like benzyl alcohol.—Colorless needles, but slightly soluble in water, easily soluble in alcohol. Fuses at 59°, and boils at 217°. With hydrochloric acid, it yields liquid *tolyl chloride*, $C^6H^4 \begin{cases} CH^3 \\ CH^2Cl, \end{cases}$ which is converted into *tolyl cyanide* by boiling with an alcoholic solution of potassium cyanide.

The following substances are isomeric with tolyl alcohol:—

Styryl alcohol (primary phenylethyl alcohol), $C^8H^{10}O = C^6H^5.CH^2.CH^2.OH$. Is prepared from benzene-ethyl bromide (p. 285) in the same manner as benzyl alcohol from benzyl chloride.—Liquid, boiling at 225°.

Secondary phenylethyl alcohol, $C^6H^5.CH(OH).CH^3$. Is produced by the action of sodium-amalgam on a solution of acetophenone in water and alcohol.—Long, colorless spiculæ; fusing point, 120°; distils almost without decomposition.

---

3. *Cumine alcohol*, $C^{10}H^{14}O = C^6H^4 \begin{cases} C^3H^7 \\ CH^2.OH. \end{cases}$ Is produced from the cuminic aldehyde (contained in the oil of Roman cumin), by heating with alcoholic potassa.—Colorless liquid of a pleasant odor; boiling at 243°. Insoluble in water; mixes with alcohol in all proportions.

4. *Sycoceryl alcohol*, $C^{18}H^{30}O$. That portion of the resin of *Ficus rubiginosa* which is insoluble in cold alcohol consists of *sycoceryl acetate*, $C^{18}H^{29}.O.C^2H^3O$. This crystallizes in flat prisms or scales, fuses at 118–120°, and yields sycoceryl alcohol, when boiled with alcoholic potassa.—Colorless, fine crystals, insoluble in water and alkalies, easily soluble in ether and alcohol. Fuses at 90°. Not distillable without partial decomposition.

Benzhydrol, $C^{13}H^{12}O = C^6H^5.C(OH).C^6H^5$. Is obtained by the action of sodium-amalgam on a solution of benzophenone in dilute alcohol.—Needles of a silky lustre. Fusing point, 67.5°; boils at 297–298°, at the same time being partially decomposed into water and benzhydrolic ether $(C^{13}H^{11})^2O$. But slightly soluble in water, easily soluble in alcohol and ether. Oxidizing substances reconvert it into benzophenone.

Benzhydrol acetate, $C^{13}H^{11}.O.C^2H^3O$. Colorless liquid, boiling at 301–302°; does not congeal at —15°.

Tollylene alcohol, $C^8H^{10}O^2 = C^6H^4 \left\{ \begin{array}{l} CH^2.OH \\ CH^2.OH. \end{array} \right.$ Is obtained by heating tollylenebromide (p. 285) with water at 170–180°.—Colorless needles. Fusing point, 112–113°. Easily soluble in water. Diatomic alcohol. Oxidizing substances convert it into terephtalic acid.

Tollylene acetate, $C^6H^4(CH^2.O.C^2H^3O)^2$. Hard, shiny laminæ. Fusing point, 47°.

D. Aldehydes.

1. *Benzylic Aldehyde (Oil of Bitter Almonds).* $C^7H^6O = C^6H^5.CHO$.

Formation and preparation. Together with hydrocyanic acid and sugar by the action of dilute acids or emulsin (an albuminous substance contained in almonds) on amygdalin (see Glucosides). By the distillation of a mixture of calcium benzoate and formate. By the oxidation of benzyl alcohol with nitric acid; by heating benzal chloride (p. 275) with water at 130–140°, with alcoholic potassa or with mercury oxide; by dissolving benzal chloride in concentrated sulphuric acid at 50°, and afterward adding water; by boiling benzyl chloride with dilute nitric acid, or, better, with a dilute solution of lead nitrate; by conducting the vapor of benzoic or phtalic acids over heated powdered zinc.—In order to prepare it, bitter almonds or other vegetable substances, containing amygdalin, freed of fixed oil by pressing, are stirred up with water, allowed to stand a day, and the mass then distilled. The oil passes over with the water, together with hydrocyanic acid, and remains partially dissolved in the water (*aqua amygdalarum amararum, aqua laurocerasi*); the greater part collects below the water. In order to separate it from hydrocyanic acid, it is shaken with a concentrated solution of sodium bisulphite, with which it (like the other aldehydes) combines, forming a difficultly soluble, crystalline compound, $C^7H^5.SO^3Na + 1\frac{1}{2}H^2O$. This is purified by pressing, and washing with alcohol, and then decomposed with sodium carbonate.

Properties. Colorless, highly refracting, thin oil, of a peculiar pleasant odor. Specific gravity, 1.063. Boiling point, 180°. Soluble in 30 parts of water. The pure oil, free of hydrocyanic acid, is not poisonous.—It combines, like acetic aldehyde, with acetic anhydride, forming a crystalline compound, $C^6H^5.CH(O.C^2H^3O)^2$, fusing at 45–46°; the same compound is also formed by the action of silver acetate on benzal chloride. It combines with ammonia and amides with elimination of water.—Oxidizing agents convert it into benzoic acid.—When distilled with phosphorus chloride or phosphorus bromide, it yields benzal chloride or benzal bromide (p. 274 and 276).—Nascent hydrogen (from sodium-amalgam and water) converts it into benzyl alcohol, hydrobenzoïn and isohydrobenzoïn (p. 320).—When boiled with an alcoholic solution of potassa, it yields benzyl alcohol and benzoic acid.

Orthochlorbenzylic aldehyde, $C^6H^4Cl.CHO$. Is produced by heating orthochlorbenzal chloride (see Salicylic aldehyde, p. 322) with water at 170°.—Liquid, boiling at 210°.

Parachlorbenzylic aldehyde, $C^6H^4Cl.CHO$. Is produced by conducting chlorine into oil of bitter almonds, containing iodine; by boiling chlorbenzyl chloride (p. 275) with a solution of lead nitrate, and by heating chlorbenzal chloride (p. 275) with water.—Colorless liquid, distillable without decomposition.

Dichlorbenzylic aldehyde, $C^6H^3Cl^2.CHO$, and *trichlorbenzylic aldehyde*, $C^6H^2Cl^3.CHO$, are obtained by heating di- or trichlorbenzal chloride (p. 275) with water at 200–260°.—Both crystallize in colorless needles and are volatile with water-vapor. The former fuses at 68°, the latter at 110–111°.

Metanitrobenzylic aldehyde, $C^6H^4(NO^2).CHO$. Is produced by dropping oil of bitter almonds into cold, very concentrated nitric acid, or a mixture of nitric and sulphuric acids.—Colorless, shiny needles, that fuse at about 50°. But slightly soluble in cold water, more easily soluble in hot water.

Sulphobenzylic aldehyde (Sulphobenzene), $C^7H^6S = C^6H^5.CHS$, is produced by heating benzal chloride with an alcoholic solution of potassium sulphydrate.—Crystallizes from alcohol in colorless laminæ; from ether in transparent four-sided prisms. Fusing point, 68–70°. Is decomposed at a high temperature, yielding stilbene (p. 282) and other products.

Hydrobenzamide, $C^{21}H^{18}N^2 = (C^6H^5.CH)^3N^2$, is produced by continued action of concentrated aqueous ammonia on oil of bitter almonds or benzal chloride.—Colorless, inodorous and tasteless octahedral crystals; insoluble in water, soluble in alcohol; fusing point, 110°. When boiled with water or alcohol it is decomposed, yielding ammonia and oil of bitter almonds.

Amarin, $C^{21}H^{18}N^2$, a base isomeric with hydrobenzamide, is produced by conducting ammonia into an alcoholic solution of benzylic aldehyde; further by heating hydrobenzamide for several hours at 130°; and by boiling it with potassa-ley.—Crystallizes from alcohol in colorless, lustrous prisms, that fuse at 100°. Insoluble in water. Poisonous. Forms very difficultly soluble salts with acids.

Lophin, $C^{21}H^{16}N^2$, is produced when amarin or hydrobenzamide are distilled, and by heating di- and tribenzylamine (p. 278).—Long, colorless needles, that fuse at 270°, are insoluble in water and difficultly soluble in alcohol, particularly in cold. Combines with acids, forming salts which are very difficultly soluble in water and more readily soluble in alcohol.

Hydrobenzoin, $C^{14}H^{14}O^2$, is produced from oil of bitter almonds by the action of nascent hydrogen (sodium-amalgam, zinc and hydrochloric acid).—Large rhombic plates, which fuse at 132.5°, and are volatile without decomposition. Hydrobenzoïn conducts itself like a diatomic alcohol, $C^{14}H^{12}(OH)^2$.—*Diacetylhydrobenzoïn*, $C^{14}H^{12}(O.C^2H^3O)^2$, is produced by the action of acetyl chloride on hydrobenzoïn and by heating stilbene bromide (p. 283) with silver acetate. Needly crystals, insoluble in water, easily soluble in alcohol.

Isohydrobenzoin, $C^{14}H^{14}O^2$. Is formed together with the preceding compound by the action of sodium-amalgam on a solution of oil of bitter almonds in dilute alcohol.—Long, colorless needles. Fusing point, 119.5°. More easily soluble in alcohol than hydrobenzoïn.—Yields an acetic ether, $C^{14}H^{12}(O.C^2H^3O)^2$, with acetyl chloride, that crystallizes in laminæ and fuses at 117–118°.

Benzoin, $C^{14}H^{12}O^2$, is produced by gently heating hydrobenzoïn with concentrated nitric acid; from oil of bitter almonds, which contains prussic acid, by treating it with a concentrated alcoholic solution of caustic potassa, or from that which is free of prussic acid, by mixing it with an alcoholic solution of potassium cyanide, the benzoïn separating in crystalline form.—Colorless, inodorous prisms. Fusing point, 133–134°. Insoluble in water, difficultly soluble in cold alcohol and ether.—When treated with alcoholic potassa it yields hydrobenzoïn and potassium benzilate. It dissolves in acetyl chloride, forming hydrochloric acid and the compound, *acetylbenzoïn*, $C^{14}H^{11}(C^2H^3O)O^2$, which crystallizes well and fuses at 75°.

Desoxybenzoin (Toluylen oxide), $C^{14}H^{12}O$, is formed by the action of zinc and hydrochloric acid on benzoïn and chlorbenzil.—Thin laminæ. Slightly soluble in water, easily in alcohol and ether. Fuses at about 55°. Distils without decomposition. With phosphorus chloride it yields *monochlorstilbene*, $C^{14}H^{11}Cl$; heated with hydriodic acid, dibenzyl (p. 282).

Toluylenhydrate, $C^{14}H^{14}O$. Is produced by the action of sodium-amalgam on desoxybenzoïn; and by heating desoxybenzoïn or hydrobenzoïn with alcoholic potassa.—Long, fine, brittle needles, of a vitreous lustre. Fusing point, 62°. Insoluble in water, easily soluble in alcohol and ether. Nitric acid oxidizes it readily, forming desoxybenzoïn. When boiled with dilute sulphuric acid it is resolved into stilbene and water. With acetyl chloride it yields a thick liquid ether, $C^{14}H^{13}.O.C^2H^3O$.

Benzil, $C^{14}H^{10}O^2$. Is produced by the oxidation of benzoïn with nitric acid or chlorine; and, together with stilbene, by heating stilbene bromide with water, alcohol, or silver oxide.—Large, six-sided columns, tasteless and inodorous, insoluble in water, soluble in alcohol and ether; fusible at 90°. Hydrogen (iron filings and acetic acid, or zinc and hydrochloric acid) reconverts it into benzoïn.

Chlorbenzil, $C^{14}H^{10}Cl^2O$. Is produced by gently heating benzil with phosphorus chloride.—Rhombic prisms or plates. Fusing point, 71°. Insoluble in water, difficultly soluble in alcohol. When heated with concentrated nitric acid, or with water or alcohol to 180°, it yields benzil; when heated with phosphorus chloride to 200°, tolan tetrachloride; with zinc and hydrochloric acid, desoxybenzoïn.

Benzilic acid, $C^{14}H^{12}O^3$, is produced from benzil, when this is heated to boiling with a concentrated alcoholic solution of potassa. After saturating the solution with hydrochloric acid, benzilic acid separates from the hot filtered solution in long, lustrous needles, which fuse at 150°. But slightly soluble in water, easily soluble in alcohol and ether, soluble in concentrated sulphuric acid, forming a deep red solution. Monobasic acid. Its barium salt yields benzhydrol (p. 316) when subjected to destructive distillation.

Benzoylbenzoic acid, $C^{14}H^{10}O^3 = C^6H^5.CO.C^6H^4.COOH$. Is formed by the oxidation of benzyltoluene with potassium bichromate and sulphuric acid.—Beautiful, lustrous laminæ; easily soluble in ether and alcohol; very difficultly soluble in cold water, somewhat more easily in hot water; fuses at 194°; sublimes.

The **barium salt,** $(C^{14}H^9O^3)^2Ba + 2H^2O$. Crystallizes in fascicular needles or in laminæ. Difficultly soluble in cold water, more easily in hot water.

Benzhydrylbenzoic acid, $C^{14}H^{12}O^3 = C^6H^5.CH(OH).C^6H^4.CO.OH$. Is formed by the action of nascent hydrogen (zinc and hydrochloric acid) on benzylbenzoic acid.—Much more easily soluble in water than the preceding acid; easily soluble in alcohol and ether; fuses at 164–165°.

Benzylbenzoic acid, $C^{14}H^{12}O^2 = C^6H^5.CH^2.C^6H^4.CO.OH$. Is formed by heating benzhydrylbenzoic acid with hydriodic acid in sealed tubes, for several hours, at 160°; also by direct oxidation of benzyltoluene with dilute nitric acid.—Crystallizes from alcohol in laminæ or needles of satin lustre; difficultly soluble in cold water, more easily in hot water, easily soluble in ether, alcohol, and chloroform; fusing point, 154–155°; sublimable. Its salts do not crystallize.

Salicylic aldehyde (Ortho-oxybenzylic aldehyde, Salicylous acid) $C^7H^6O^2 = C^6H^4\left\{\begin{matrix} OH \\ CHO. \end{matrix}\right.$ Occurs in all parts of the herbaceous spiræas, and in the larvæ of *chrysomena populi*. Is produced by the oxidation of saligenin, salicin, and populin (see Glucosides).—Colorless oil, of a strong aromatic odor and burning taste; congeals at —20°. Boils at 196°. Difficultly soluble in water; mixes with alcohol in all proportions. The aqueous solution is colored deep violet by iron chloride.—Like oil of bitter almonds, it combines with alkaline bisulphites and with ammonia, forming crystalline compounds; and is converted by oxidizing agents into salicylic acid. With phosphorus chloride, at the ordinary temperature, it yields *orthooxybenzal chloride* $C^6H^4(OH).CHCl^2$ (prisms, fusing at 82°); when heated with an excess of phosphorus chloride, *orthochlorbenzal chloride* $C^6H^4Cl.CHCl^2$ (liquid, boiling at 227–230°, isomeric with the chlorbenzal chloride of the para-series, obtained from toluene, p. 275).—Salicylic aldehyde is dissolved by the alkalies, crystallizing compounds being formed. The *potassium compound* (potassium salicylite), $C^6H^4\left\{\begin{matrix}OK\\CHO\end{matrix}\right.$, crystallizes in quadratic plates, which are easily soluble in alcohol and water, and, when moist, are decomposed rapidly in contact with the air.

Methylsalicylic aldehyde (Methylorthooxybenzylic aldehyde), $C^6H^4\left\{\begin{matrix}O.CH^3\\CHO\end{matrix}\right.$, is obtained by allowing methyl iodide to act upon potassium salicylite.—Liquid, boiling at 238°.

Acetylsalicylic aldehyde, $C^6H^4\left\{\begin{matrix}O.C^2H^3O\\CHO.\end{matrix}\right.$ Is obtained by the action of acetic anhydride on sodium salicylite at the ordinary temperature.—Fine needles. Fusing point, 37°. Boils at 253°, at the same time undergoing partial decomposition.

Chlorosalicylic aldehyde, $C^6H^3Cl(OH).CHO$. By the action of chlorine on salicylic aldehyde.—Yellowish-white lamellæ.

Anisic aldehyde (Methylparaoxybenzylicaldehyde), $C^8H^8O^2 = C^6H^4 \begin{cases} O.CH^3 \\ CHO. \end{cases}$ Is produced by heating oil of anise, oil of fenchel, oil of sternanis, or oil of esdragon (the volatile oils from the seeds of *Pimpinella anisum*, *Anethum fœniculum*, *Illicium anisatum*, and the green portions of *Artemisia dracunculus*) with dilute nitric acid, or potassium bichromate and dilute sulphuric acid, a substance called anethol, $C^{10}H^{12}O$, which is contained in these oils, being oxidized in this process. The aldehyde separates as an oil, and is purified by shaking with alkaline bisulphites, and decomposing the crystalline compound thus formed by sodium carbonate.—Colorless oil, of a spicy odor, boiling at 248°; of specific gravity 1.12.

Dioxybenzylic aldehyde (Protocatechuic aldehyde), $C^7H^6O^3 = C^6H^3 \begin{cases} (OH)^2 \\ CHO. \end{cases}$ Is obtained by boiling dichlorpiperonal (see Piperonal) with water; and together with carbon, by heating piperonal with dilute hydrochloric acid at 200°.—Flat, lustrous crystals. Fusing point, 150°. Easily soluble in water. The aqueous solution is colored deep green by iron chloride.

Methylene-dioxybenzylic aldehyde (Piperonal), $C^8H^6O^3 = C^6H^3 \begin{cases} \begin{matrix} O \\ O \end{matrix} > CH^2 \\ CHO. \end{cases}$ Is produced by distilling a dilute solution of one part potassium piperate with two parts potassium hypermanganate.—Long, lustrous, colorless crystals, of a very pleasant odor; fusing point, 37°; boiling point, 263°; difficultly soluble in cold water, more easily in hot water, very easily soluble in alcohol. Combines with alkaline bisulphites. Nascent hydrogen converts it into *piperonyl alcohol*, $C^8H^8O^3$, and two isomeric compounds, corresponding to hydrobenzoin (p. 320). When heated with three molecules phosphorus chloride it yields a liquid body, *dichlorpiperonal chloride*, $C^8H^4Cl^4O^2$, which, with cold water, yields *dichlorpiperonal*, $C^8H^4Cl^2O^3$, and hydrochloric acid, and when boiled with water, is resolved into carbonic anhydride and protocatechuic aldehyde.

---

2. *Paratolylic aldehyde*, $C^8H^8O = C^6H^4 \left\{ \begin{array}{l} CH^3 \\ CHO. \end{array} \right.$ Is obtained by distilling a mixture of calcium paratoluate and formate.—Colorless liquid, boiling at 204°. Yields paratoluic acid by oxidation.

3. *Cuminic aldehyde* (Cuminol), $C^{10}H^{12}O =$ $C^6H^4 \left\{ \begin{array}{l} C^3H^7 \\ CHO. \end{array} \right.$ Occurs, together with cymene, in oil of Roman cumin and in the oil from the seeds of *Cicuta virosa*. Is obtained from these oils by shaking with alkaline bisulphites, and decomposing the crystalline compounds with sodium carbonate.—Colorless oil, of a pleasant odor, boiling at 237°. When added to fusing potassic hydrate, or boiled with alcoholic potassa, it yields cuminic acid: in the latter case cuminic alcohol is also formed. Yields by oxidation terephtalic acid.

E. Acids.

a. Monobasic, Monatomic Acids.

1. *Benzoic Acid.*

$$C^7H^6O^2 = C^6H^5.CO.OH.$$

Occurrence. In a number of resins, particularly in gum-benzoin; occasionally in the urine of herbivorous animals.

Formation. From monobrombenzene by the simultaneous action of sodium and carbonic acid; the ethyl ether, by the decomposition of a mixture of monobrombenzene and ethyl chlorocarbonate with sodium. By the oxidation of all hydrocarbons, alcohols, aldehydes, and acids in which only one hydrogen-atom of the benzene is replaced by a monovalent carbon-group (for instance, toluene, ethyl benzene, benzyl chloride, benzyl alcohol, oil of bitter almonds, alphatoluic acid, hydrocinnamic acid, cinnamic acid) by means of dilute nitric acid or chromic acid; by heating a mixture of equal parts, by weight, of potassium sulphobenzolate and sodium formate to fusion; by heating benzotrichloride (p. 275) with water to 150°; by heating a mixture of equal molecules of calcium phtalate and calcium hydroxide to 330–350°; by treating hippuric acid and populin with acids or bases; by the action of acids on cocain; by the oxidation of albuminoid substances.

Preparation. By fusing gum-benzoin. The best way is to heat the gum in a shallow basin, over which is placed a paper cone, made of blotting paper: the acid condenses in this cone in the form of needly crystals. More readily by boiling the powdered gum with calcium hydroxide, filtering, and concentrating the resulting solution of calcium benzoate, and decomposing the latter with hydrochloric acid; the benzoic acid thus separating in crystalline form. It can be purified by recrystallization or sublimation. Most advantageously from hippuric acid. (See Preparation of Glycocol, p. 84.)

Properties. Lustrous, white, long, very thin, somewhat flexible needles and laminæ.—Fuses at 120°, and boils at 250°. Difficultly soluble in cold water, easily soluble in hot water and in alcohol. Easily sublimable. Passes over with the vapor of water on heating its aqueous solution. Its vapor and its boiling solution possess a peculiar odor, that excites coughing.

Most of its salts are soluble in water. Their solutions give a reddish precipitate with iron chloride, consisting of iron benzoate.

Calcium benzoate, $(C^7H^5O^2)^2Ca + 3H^2O$, crystallizes in lustrous, colorless, radiating prisms. Easily soluble in water.

Silver benzoate, $C^7H^5O^2.Ag$, is very difficultly soluble in cold water; crystallizes from hot water.

Ethyl benzoate, $C^7H^5O.O.C^2H^5$. Colorless, viscid, fragrant liquid; specific gravity, 1.054; boiling point, 213°.

Benzoyl chloride, $C^6H^5.COCl$. Is produced by the action of phosphorus chloride on benzoic acid; and of chlorine on oil of bitter almonds.—Colorless oil, boiling at 199°, of an exceedingly pungent odor. Is decomposed by water and by contact with moist air, yielding benzoic and hydrochloric acids. Distilled with bromides, iodides, or cyanides, it yields benzoyl bromide, iodide, and cyanide, all of which are crystallizing compounds. Heated with an excess of phosphorus pentachloride, it is converted into benzotrichloride, $C^6H^5.CCl^3$ (p. 275).

Benzamide, $C^6H^5.CO.NH^2$. Is formed by continued action of ammonia on ethyl benzoate or benzoic anhydride; and by bringing benzoyl chloride together with concentrated aqueous ammonia or dry ammonium carbonate.—Colorless, lustrous crystals; fuses at 125°; and sublimes without decomposition. But slightly soluble in cold water, easily soluble in hot water and in alcohol.

Benzhydroxamic acid, $C^6H^5.CO.NOH.H$. Is obtained by the action of benzoyl chloride on an aqueous solution of hydroxylamine hydrochlorate, which is saturated with sodium carbonate.—Colorless rhombic crystals. Comparatively difficultly soluble in cold water (44½ parts at 6°), much more readily in warm water, very easily in alcohol. Has an acid reaction; fuses at 124–125°, and decomposes at a higher temperature suddenly and violently. By heating with dilute hydrochloric or sulphuric acid, it is decomposed into benzoic acid and hydroxylamine salt.—Monobasic acid; yields crystallizing salts.

Dibenzhydroxamic acid, $(C^6H^5.CO)^2NOH$. Is formed together with the preceding compound in the described reaction.—Lustrous, rhombic crystals. Almost insoluble in water, difficultly soluble in cold alcohol, more readily in hot, very slightly in ether. Has an acid reaction, fuses at 145–146°, and decomposes with violence at a higher temperature. Monobasic acid; yields crystallizing salts.

Tribenzhydroxylamine, $(C^6H^5.CO)^2.N.O(C^6H^5.CO)$. Is formed by the action of benzoyl chloride on dry hydroxylamine hydrochlorate, which is dissolved in a hydrocarbon boiling at 110°; also when potassium dibenzhydroxamate is heated with benzoyl chloride.—Lustrous prisms; fusing point, 141–142°; decomposes at 190°; insoluble in water, ether, and benzene; very difficultly soluble in cold alcohol, much more readily in hot alcohol.

Benzoic anhydride, $(C^7H^5O)^2O$. Is produced by the action of benzoyl chloride on sodium benzoate; and by heating 6 parts dry sodium benzoate with 1 part phosphorus oxichloride to 150°. The salts (sodium metaphosphate and sodium chloride), that are formed are extracted with water.—Oblique prisms, insoluble in cold water, soluble in alcohol, forming a neutral solution. Fuses at 42°, and distils at 310°. When boiled with water, it is gradually converted into benzoic acid; and when heated in hydrochloric acid gas, is decomposed, yielding benzoic acid and benzoyl chloride.

Substitution-products of benzoic acid. Those substitution-products which are formed by the direct action of chlorine, bromine, etc., on benzoic acid, belong to the meta-series; the isomeric ortho-compounds are obtained from salicylic acid; the para-compounds by oxidation of the para-substitution-products of toluene. By the latter method the meta-compounds can also be obtained, but not the ortho-compounds (cf. p. 274).

Orthochlorbenzoic acid (Chlorsalylic acid), $C^7H^5ClO^2 = C^6H^4Cl.CO.OH$. The chloride (chlorsalyl chloride), $C^6H^4Cl.COCl$ (a colorless oil, boiling at 240°), is produced by the action of phosphorus chloride on salicylic acid. This yields the acid when treated with water.—Needles, that fuse at 137°; more readily soluble in water than the isomeric compounds. Fuses under boiling water.

Metachlorbenzoic acid. Is produced from benzoic acid by heating with hydrochloric acid and potassium chlorate or antimony chloride or calcium hypochlorite; by the decomposition of chlorhippuric with hydrochloric acid; by boiling cinnamic acid with a solution of bleaching lime; and by oxidation of meta-chlortoluene.—Colorless needles, that fuse at 152°, and sublime without decomposition. Very difficultly soluble in cold water.

Parachlorbenzoic acid (Chlordracylic acid), formed by the oxidation of parachlortoluene.—Sublimes in colorless scales, that fuse at 236–237°.

Dichlorbenzoic acid, $C^6H^3Cl^2.CO.OH$. Is produced from meta- and parachlorbenzoic acids by boiling with a solution of bleaching lime, or by treating with antimony chloride; by oxidation of dichlortoluene, dichlorbenzyl chloride, and dichlorbenzal chloride (p. 275) with chromic acid; and by heating dichlorbenzotrichloride (p. 275) with water.—Colorless needles fusing at 201–202°.

Trichlorbenzoic acid, $C^6H^2Cl^3.CO.OH$, and *Tetrachlorbenzoic acid*, $C^6HCl^4.CO.OH$, are obtained by heating tri- and tetrachlorbenzotrichloride (p. 275) with water to 260–280°. Both crystallize in colorless needles; the former fuses at 163°, the latter at 187°.

Metabrombenzoic acid, $C^7H^5BrO^2$, is formed by heating benzoic acid with bromine and water to 130–160°.—Colorless needles; fuse at 152–153°; but slightly soluble in water.—*Parabrombenzoic acid* (Bromdracylic acid), $C^7H^5BrO^2$, is obtained by the oxidation of parabromtoluene.—Small, colorless needles, almost insoluble in cold water. Fusing point, 251°.

Dibrombenzoic acid, $C^7H^4Br^2O^2$ (fusing point, 223–227°), *Tribrombenzoic acid*, $C^7H^3Br^3O^2$ (fusing point, 234–235°), and *Pentabrombenzoic acid*, $C^7HBr^5O^2$ (fusing point, 234–235°), are formed by heating benzoic acid with bromine to 200° and over.

Paraiodobenzoic acid, $C^7H^5IO^2$. From paraiodotoluene by oxidation.—Colorless scales; fusing point, 250°.

Fluorbenzoic acid, $C^7H^5FlO^2$. Is produced by treating diazoamidobenzoic acid with hydrofluoric acid.—Rhombic prisms; fusing point, 182°.

Orthonitrobenzoic acid, $C^7H^5(NO^2)O^2$. Is obtained by oxidation of nitrocinnamic acid (which see).—Easily soluble in water, fuses at 232°.—*Metanitrobenzoic acid* is formed by treating benzoic acid with hot very concentrated nitric acid, or with a mixture of sulphuric and nitric acids.—Crystallizes in fine needles or laminæ, which fuse at 141–142°.—*Paranitrobenzoic acid* (Nitrodracylic acid) is produced by the oxidation of paranitrotoluene.—Slightly yellowish colored laminæ, that fuse at 240°. Much less easily soluble in water than the two isomeric compounds.

Dinitrobenzoic acid, $C^7H^4(NO^2)^2O^2$. By continued heating of metanitrobenzoic acid with a mixture of nitric and sulphuric acids.—Crystallizes from water in large, very thin quadratic plates; from alcohol in prisms. Fusing point, 204–205°.

By treating chlor- or brombenzoic acids with nitric acid, there are formed *chlornitro-* and *bromnitrobenzoic acids.* From metabrombenzoic acid are formed simultaneously two isomeric modifications *α-bromnitrobenzoic acid* (fusing point, 246–248°, but very slightly soluble in water), and *β-bromnitrobenzoic acid* (fusing point, 140–141°, more easily soluble in water).

Ortho-amidobenzoic acid (Anthranilic acid), $C^6H^4(NH^2).CO.OH$. Is formed, when indigo (1 part) is boiled with soda-ley (10 parts, of 1.38 specific gravity) for several days, finely powdered black oxide of manganese being gradually added, and the evaporated water being replaced, until the color of the mass has become bright yellow. This is then dissolved in water, the solution neutralized with sulphuric acid, filtered, evaporated to dryness, and the sodium anthranilate extracted by means of alcohol. The salt that remains behind after the evaporation of the alcohol is then dissolved in water and decomposed by acetic acid.—It is also formed by the action of sodium-amalgam on the bromamidobenzoic acids (obtained by reduction of the two bromnitrobenzoic acids).—Thin, colorless prisms or laminæ, but slightly soluble in cold water, easily in hot water and in alcohol. Fuses at 144°, and decomposes at a higher temperature, yielding anilin and carbonic anhydride.

Meta-amidobenzoic acid is formed by heating an alcoholic solution of metanitrobenzoic acid with ammonium sulphide, and decomposing the ammonium salt thus obtained with acetic acid.—Is obtained more readily by gently heating metanitrobenzoic acid with tin and concentrated hydrochloric acid. After the action is over the solution is precipitated with an excess of sodium carbonate, and the concentrated solution acidified with acetic acid.—Small, colorless prisms, easily soluble in hot water, slightly in cold. Fuses at 164–165°; and is resolved, by heating with caustic potassa, into carbonic anhydride and anilin. Yields crystallizing salts with bases, as well as with acids.

Para-amidobenzoic acid (Amidodracylic acid) is obtained from paranitrobenzoic acid in the same way as the meta-acid.—Long, fine, lustrous needles. Fusing point, 186–187°; moderately easily soluble in water.

Diamidobenzoic acid, $C^6H^3(NH^2)^2.CO.OH$. Is obtained from dinitrobenzoic acid by reduction with tin and hydrochloric acid.—Almost colorless, long, thin needles; fusing point, 240°; not volatile without decomposition. Difficultly soluble in cold water; combines with bases and acids, forming salts.

Azobenzoic acid, $C^{14}H^{10}N^2O^4 + \frac{1}{2}H^2O$. Is formed by the action of sodium-amalgam on an aqueous solution of sodium metanitrobenzoate, and is precipitated by hydrochloric acid after the completion of the action.—Amorphous, bright yellow powder, very slightly soluble in water, alcohol, and ether; not volatile without decomposition. Very stable, bibasic acid; yields crystallizing, yellow colored salts, and ethers.—*Parazobenzoic acids* (azodracylic acid), $C^{14}H^{10}N^2O^4$. Is obtained from paranitrobenzoic in the same way as azobenzoic acid.—Flesh-colored, amorphous powder very similar to azobenzoic acid.

Hydrazobenzoic acid, $C^{14}H^{12}N^2O^4$. Is formed, when a solution of iron sulphate is added to sodium azobenzoate, dissolved in an excess of soda-ley. The acid is then precipitated from the filtered solution by means of hydrochloric acid.—Yellowish-white, indistinctly crystalline flocks. Insoluble in water, difficultly soluble in boiling alcohol. Weak acid. In aqueous solution, its salts absorb oxygen from the air and are converted into azobenzoates. When heated with concentrated hydrochloric acid, it is resolved into azobenzoic and amidobenzoic acids.—*Parahydrazobenzoic acid*, (hydrazodracylic acid), $C^{14}H^{12}N^2O^4$. Small, lustrous, crystalline needles. Is prepared like hydrazobenzoic acid, and conducts itself like this.

Azoxybenzoic acid, $C^{14}H^{10}N^2O^5$, is produced by boiling an alcoholic solution of metanitrobenzoic acid, to which is added solid caustic potassa.—Microscopical needles or laminæ. Insoluble in water; difficultly soluble in alcohol and ether. Bibasic acid.

Diazobenzoic acid, $C^7H^4N^2O^2$. Is precipitated as a yellow, very unstable mass, when an alkali is added to a solution of nitric-diazobenzoic acid.—*Nitric-diazobenzoic acid*, $C^7H^4N^2O^2.HNO^3$, is thrown down, when a current of nitrous acid is conducted into meta-amidobenzoic acid dissolved in cold nitric acid. Colorless prisms, very easily soluble in cold water. Is decomposed by boiling with water, yielding nitrogen, nitric acid, and meta-oxybenzoic acid. Explodes violently when heated.

Diazobenzoic-Amidobenzoic acid, $C^7H^4N^2O^2 + C^7H^5(NH^2)O^2$. Is produced by mixing aqueous solutions of nitric-diazobenzoic acid and meta-amidobenzoic acid. Can be prepared most readily by conducting nitrous acid into an alcoholic solution of meta-amidobenzoic acid, or by mixing this solution at 30° with ethyl nitrite, the acid in this case being thrown down immediately.—Orange-yellow crystalline granules, or small microscopical prisms. Inodorous and tasteless. Almost insoluble in water, alcohol, and ether. Is decomposed at 180°, the decomposition being accompanied by a detonation. Weak, bibasic acid. The salts are easily decomposed in aqueous solution, nitrogen being evolved. Heated with hydrochloric acid the acid is decomposed below 100°, yielding chlorbenzoic acid and meta-amidobenzoic acid hydrochlorate. Hydrobromic and hydriodic acids cause an analogous decomposition.

Para-amidobenzoic acid conducts itself like meta-amidobenzoic acid when treated with nitrous acid, and yields diazo-compounds, which are isomeric with those just described, and completely analogous to them.

Meta-sulphobenzoic acid, $C^7H^6SO^5 = C^6H^4 \left\{ \begin{matrix} CO.OH \\ SO^2.OH. \end{matrix} \right.$ Is formed, together with a small quantity of the para-acid, by the action of fuming sulphuric acid on benzoic acid, and when the vapor of sulphuric anhydride is conducted upon dry benzoic acid. Separated from the barium salt, it forms a crystalline, colorless, very deliquescent, strongly acid mass. Very stable bibasic acid. *The neutral barium salt*, $C^7H^4SO^5Ba$, is very easily soluble; the *acid salt*, $(C^7H^5SO^5)^2Ba + 3H^2O$, crystallizes in difficultly soluble, oblique rhombic prisms.

A mixture of concentrated nitric and sulphuric acids converts it into *nitrosulphobenzoic acid*, $C^6H^3(NO^2)\left\{\begin{matrix} CO.OH \\ SO^2.OH \end{matrix}\right.$—well developed crystals, easily soluble in water—which, when treated with ammonium hydrosulphide, is transformed into *amidosulphobenzoic acid*, $C^6H^3(NH^2)\left\{\begin{matrix} CO.OH \\ SO^2.OH \end{matrix}\right.$—radiating, colorless needles. When distilled with phosphorus chloride, sulphobenzoic acid yields metachlorbenzoyl chloride.

Parasulphobenzoic acid, $C^7H^6SO^5 = C^6H^4\left\{\begin{matrix} CO.OH \\ SO^2.OH. \end{matrix}\right.$ Is formed in varying quantities, together with the preceding compound, in the preparation of the latter; and by oxidizing parasulphotoluene with potassium bichromate and sulphuric acid. The free acid is very similar to the meta-acid; is not, however, deliquescent. The *acid barium salt*, $(C^7H^5SO^5)^2Ba + 3H^2O$, crystallizes in long, flat needles, which are very difficultly soluble in water.

Disulphobenzoic acid, $C^6H^3\left\{\begin{matrix} CO.OH \\ (SO^2.OH)^2. \end{matrix}\right.$ Is formed by the action of concentrated sulphuric acid and phosphoric anhydride on benzoic acid in sealed tubes.—Crystalline, deliquescent mass. The *neutral barium salt*, $(C^7H^3S^2O^8)^2Ba^3 + 7H^2O$, crystallizes in small, well-formed prisms.

Thiobenzoic acid, $C^6H^5.CO.SH$. Is obtained by the action of benzoyl chloride on an alkaline solution of potassium sulphite and precipitation with hydrochloric acid.—White, radiating, crystalline mass. Fusing point, 24°. But slightly soluble in warm water. Not distillable alone, but very easily with water vapor. In ethereal solution, in contact with the air, it easily becomes oxidized, forming *benzoyl disulphide* $(C^6H^5.CO)^2S^2$.

A *thiobenzoic acid*, $C^6H^5.CS.OH$, isomeric with the foregoing, is formed, together with benzoic acid, by the oxidation of sulphobenzylic aldehyde (p. 319)—colorless needles, united in fascicles, which, under the influence of heat, decompose without melting.

Dithiobenzoic acid, $C^6H^5.CS.SH$. The potassium salt is formed by mixing benzotrichloride (p. 275) with an alcoholic solution of potassium sulphite. The free acid is a heavy, violet, very unstable oil.

ACETONES.

Benzophenone (Diphenylketone), $C^{13}H^{10}O = C^6H^5.CO.C^6H^5$, is formed, together with benzene, by the destructive distillation of calcium benzoate. Is also formed by heating mercury-phenyl (p. 272) with benzoyl chloride.—Colorless, rhombic prisms, insoluble in water, easily soluble in alcohol. Fuses at 48°, and boils at 295.° Hydrogen, in *statu nascendi*, converts it into benzhydrol (p. 316).—Under certain conditions, the nature of which is not understood, a second modification of benzophenone is formed. This fuses at 26–26.5°, and appears to belong to the monoclinic system. It is very easily converted into the rhombic modification. The reverse transformation has not been observed.

Methylbenzophenone, $C^{14}H^{12}O = C^6H^5.CO.C^6H^4.CH^3$. Is formed, together with benzoylbenzoic acid, by the oxidation of benzyltoluene with a mixture of potassium bichromate and dilute sulphuric acid.—Colorless oil, of a weak aromatic odor; insoluble in water, easily soluble in alcohol or ether. It boils at 307–312°.—Yields benzoylbenzoic acid (p. 322) by oxidation.

Acetophenone (Methylphenylketone), $C^6H^5.CO.CH^3$. Is obtained by distilling a mixture of calcium benzoate and acetate; and by the action of benzoyl chloride on zincmethyl.—Colorless, large, crystal plates. Fusing point, 14°; boiling point, 198°. Treated with chlorine at a slightly elevated temperature, it is converted into *chloracetyl benzene* $C^6H^5.CO.CH^2Cl$ (crystalline; fusing point, 41°; boiling point, 246°). Yields nitrosubstitution-products with nitric acid. Hydrogen, in *statu nascendi*, converts it into secondary phenylethyl alcohol (p. 316). Oxidizing agents convert it into benzoic and carbonic acids.

Ethylphenylketone, $C^6H^5.CO.C^2H^5$. Is prepared by the action of benzoyl chloride on zincethyl.—Boiling point, 208–212°. Insoluble in water. Yields by oxidation benzoic and acetic acids.

Hippuric acid (Benzoylglycocol) $C^9H^9NO^2 = CH^2 \left\{ \begin{matrix} NH.C^7H^5O \\ CO.OH. \end{matrix} \right.$ Occurs in small quantity in normal human urine, in large quantity in the normal urine of graminivorous animals.—Toluene, benzoic acid, cinnamic acid, and oil of bitter almonds, taken into the system, are converted into hippuric acid in all animals; quinic acid, in the case of man and graminivorous animals, likewise undergoes the same change.—Obtained artificially, by the action of benzoyl chloride on glycocol-zinc or glycocol-silver (p. 85).—To prepare it, fresh urine of horses or cows is evaporated to about one-fourth its volume, and then acidified with hydrochloric acid, the hippuric acid being thus thrown down as a crystalline magma. The yield varies very much, according to the fodder of the animals, and according as they have lived in stalls or in the open air. The crude acid is washed out with cold water, pressed, digested with a large quantity of chlorine water, and finally dissolved in it at boiling temperature. On cooling it separates in colorless needles. Or the crude acid is dissolved in boiling weak soda-ley, sodium hypochlorite gradually added until the color is removed, and then, when the solution has ceased boiling, hydrochloric acid is added until the whole has an acid reaction. It is completely purified by recrystallizing from water.

Large, colorless, rhombic prisms, of a weak taste; soluble in 600 parts of cold water, much more readily in boiling water, and in alcohol. Fusible without decomposition. Heated above its fusing point, it is decomposed, and yields hydrocyanic and benzoic acids and benzonitrile.

By boiling with acids or alkalies, it is resolved into benzoic acid and glycocol, the elements of water being assimilated. The same decomposition is effected by ferments. By heating with manganese superoxide and dilute sulphuric acid, it yields benzoic acid, carbonic anhydride, and ammonia. Nitrous acid converts it into *benzoylglycolic acid*, $C^9H^8O^4 = CH^2 \begin{cases} O.C^7H^5O \\ CO.OH, \end{cases}$ which crystallizes in thin, colorless prisms, difficultly soluble in cold water, easily in hot water, and alcohol.

Monobasic acid. Most of the hippurates, even the silver and lead salts, are soluble in water, and crystallizable.

Ethyl hippurate, $C^9H^8NO^3.C^2H^5$, is produced by saturating a boiling solution of hippuric acid in alcohol with hydrochloric acid gas. On the addition of water, the ether subsequently separates as an oil, which soon becomes crystalline.—It crystallizes in long, colorless prisms of a silky lustre, but slightly soluble in water, easily in alcohol and ether; fuses at 44°; not volatile without decomposition.

Chlorhippuric acid, $C^9H^8ClNO^3$, and **Dichlorhippuric acid**, $C^9H^7Cl^2NO^3$, are produced, when to hippuric acid, in a vessel containing concentrated hydrochloric acid, potassium chlorate is added and the whole gently heated. The former is oleaginous, viscid, uncrystalline; the latter crystallizes gradually, when left in contact with the air or under water. Boiled with acids or alkalies, chlorhippuric acid yields glycocol and metachlorbenzoic acid;—dichlorhippuric acid yields glycocol and dichlorbenzoic acid. Chlorhippuric acid occurs in the urine after metachlorbenzoic acid is taken into the system.

Nitrohippuric acid, $C^9H^8(NO^2)NO^3$, is formed, when hippuric acid is added to a mixture of equal volumes of concentrated sulphuric and nitric acids; and separates on the addition of water, and partial neutralization of the acid with sodium carbonate.—Fine, white prisms of a silky lustre; fuses between 150° and 160°; difficultly soluble in cold water, easily soluble in hot water and in alcohol. Boiled with hydrochloric acid, it is resolved into glycocol and nitrobenzoic acid; ammonium hydrosulphide reduces it to *amidohippuric acid*, $C^9H^8(NH^2)(NO^3)$, which crystallizes in small, white laminæ, difficultly soluble in water.

2. *Acids*, $C^8H^8O^2$.

a. Toluic Acids.

$$C^6H^4\left\{\begin{matrix}CH^3\\ CO.OH.\end{matrix}\right.$$

1. Ortho-toluic acid. Is obtained by oxidation of ortho-xylene with dilute nitric acid, and is purified in the same way as para-toluic acid (see below). Also by distilling potassium ortho-sulphotoluenate with potassium cyanide, and treating the cyanide thus formed with alcoholic potassa.—Long, very fine needles; fusing point, 102°; difficultly soluble in cold water, easily in hot water. When warmed with chromic acid (potassium bichromate and dilute sulphuric acid), it is burned completely, yielding carbonic anhydride and water.

Calcium ortho-toluate, $(C^8H^7O^2)^2Ca + 2H^2O$, crystallizes in easily soluble needles.

---

2. Meta-toluic acid (Isotoluic acid). Is produced together with para-toluic acid by oxidation of the xylenes (p. 283) contained in coal-tar; it cannot, however, be separated from the para-acid. It is obtained in a pure condition by the action of sodium-amalgam on a solution of brommeta-toluic acid.—Colorless needles; fusing point, 90–93°. Chromic acid oxidizes it, forming isophtalic acid.

Calcium meta-toluate, $(C^8H^7O^2)^2Ca + 2H^2O$. Needles, easily soluble in water.

Brommeta-toluic acid, $C^6H^3Br \begin{cases} CH^3 \\ CO.OH. \end{cases}$ Is formed together with the isomeric compound, brompara-toluic acid, when the mixture of brommeta- and brompara-xylene, obtained by the action of bromine on xylene from coal-tar, is boiled for a long time with potassium bichromate and dilute sulphuric acid. By preparing the barium salt, which is comparatively difficultly soluble in water, it can be readily separated from the para-acid.—Crystalline powder, difficultly soluble even in boiling water. Fusing point, 205–206°.

---

3. Para-toluic acid. Is produced from parabromtoluene by the simultaneous action of sodium and carbonic acid; from para-xylene or cymene by boiling for several days with dilute nitric acid (mixture of 1 volume nitric acid of specific gravity, 1.4 with 2–3 volumes water), in a flask connected with a reversed condensing apparatus. The acid, that separates on cooling, still contains impurities in the form of nitro-substitution-products. In order to free it from these, it is suspended in water and this distilled, the acid passing over with the vapors; or the crude acid is heated for some time with tin and concentrated hydrochloric acid, and the undissolved portion crystallized from boiling water.—Fine, colorless, needly crystals. But slightly soluble in cold water, comparatively easily in boiling water, but less so than benzoic acid, very readily soluble in alcohol. Fuses at 176°, and sublimes easily. Chromic acid oxidizes it, forming terephtalic acid.

Calcium para-toluate, $(C^8H^7O^2)^2Ca + 3H^2O$, crystallizes in lustrous, colorless needles, that are easily soluble in water.

b. Alphatoluic Acid (Phenylacetic Acid).
$C^6H^5.CH^2.CO.OH$.

Is produced by boiling benzyl cyanide (p. 276) with alkalies; by the action of hydriodic acid on mandelic acid; together with methyl alcohol and oxalic acid by boiling vulpic acid with barium hydroxide; by melting atropic acid with caustic potassa; its ethyl ether by heating a mixture of monobrombenzene and ethyl chloracetate with copper to 180–200°.—Crystallizes in broad, lustrous laminæ. Very similar to benzoic acid. Fuses at 76.5°, and boils without decomposition at 261–262°.—Oxidized with chromic acid, it is converted into benzoic acid.

When bromine and nitric acid are allowed to act upon alphatoluic acid without the aid of heat, substitution-products result which consist principally of members of the para-series, and by oxidation yield parabrom- or paranitrobenzoic acids. Together with these, in small quantities, are formed isomeric compounds, probably belonging to the meta-series.

When mandelic acid is heated with concentrated hydrochloric acid to 140–150°, and when bromine acts upon heated alphatoluic acid, another class of substitution-products is formed, in which the hydrogen of the acetic acid residue is replaced (for example: phenylchloracetic acid $C^6H^5.CHCl.CO.OH$).

3. *Acids*, $C^9H^{10}O^2$.

1. Mesitylenic acid, $C^6H^3\left\{\begin{matrix}(CH^3)^2\\ CO.OH\end{matrix}\right.$ (1: 3: 5). Is formed by oxidizing mesitylene with dilute nitric acid. The crude acid is purified like para-toluic acid.—Crystallizes from water in small, colorless needles, from alcohol in large, transparent, monoclinate crystals. Almost insoluble in cold water, very difficultly in hot water, very easily in alcohol. Fuses at 166°, and sublimes without undergoing decomposition.—By further oxidation, it is converted into uvitic and trimesic acids. Distilled with an excess of lime, it yields meta-xylene.

Barium mesitylate, $(C^9H^9O^2)^2Ba$, crystallizes in large, lustrous prisms, easily soluble in water.

2. Xylylic acid, $C^6H^3\left\{\begin{matrix}(CH^3)^2\\ CO.OH\end{matrix}\right.$ (1:2:4),* is produced by the simultaneous action of sodium and carbonic acid on monobrommeta-xylene; and together with para-xylylic acid by oxidation of pseudo-cumene. The mixture of acids is purified by distilling off with water vapor, and heating gently with tin and hydrochloric acid; and the two acids then separated by means of partial crystallization of the calcium salts. Calcium para-xylylate separates first, and afterward the xylylate. The acids are precipitated from the solutions of their salts by hydrochloric acid.—Crystallizes from alcohol in large, transparent, monoclinate prisms, from water in fine needles. Fuses at 126°. Very similar to mesitylenic acid. Distilled with lime, it, like mesitylenic acid, yields meta-xylene, but is converted into xylidinic acid by further oxidation.

Calcium xylylate, $(C^9H^9O^2)^2Ca + 2H^2O$, forms large, hard, transparent, monoclinate prisms.

3. Para-xylylic acid, $C^6H^3\left\{\begin{matrix}(CH^3)^2\\ CO.OH\end{matrix}\right.$ (1:3:4).* In regard to the formation and preparation see Xylylic Acid. Separates from boiling water in indistinctly crystalline flocks, from alcohol in lanceolar prisms, concentrically grouped. Fusing point, 163°. More easily soluble in alcohol than xylylic acid. By further oxidation it is converted, like xylylic acid, into xylidinic acid, but yields ortho-xylene by distillation with lime.

Calcium para-xylylate, $(C^9H^9O^2)^2Ca + 3\frac{1}{2}H^2O$, forms soft, untransparent, fascicular crystals.

4. Ethyl-benzoic acid (Para-), $C^6H^4\left\{\begin{matrix}C^2H^5\\ CO.OH,\end{matrix}\right.$ is obtained by the action of sodium and carbonic acid on

* The position of the group CO.OH is designated by 1.

bromethylbenzene; and by oxidation of diethylbenzene with dilute nitric acid.—Colorless, lustrous laminæ, similar to benzoic acid. But slightly soluble in cold water, more readily in hot water, very easily soluble in alcohol. Fuses at 110°. Further oxidation converts it into terephtalic acid.

5. Alphaxylylic acid, $C^6H^4 \left\{ \begin{array}{l} CH^3 \\ CH^2.CO.OH. \end{array} \right.$ Is produced from tolyl cyanide (p. 315) by boiling with alcoholic potassa.—Colorless, lustrous, broad laminæ. Easily soluble in hot water. Fuses at 42°.

6. Hydrocinnamic acid (Phenylpropionic acid), $C^6H^5.CH^2.CH^2.CO.OH$. Is formed by the action of nascent hydrogen (sodium-amalgam) on cinnamic acid; and by boiling benzene-ethyl cyanide (p. 285) with alcoholic potassa.—Crystallizes from water in long, fine needles. Easily soluble in boiling water and in alcohol, difficultly soluble in cold water, but more readily than benzoic acid. Fuses at 47°, and boils without decomposition at 280°.—Chromic acid oxidizes it, forming benzoic acid.

7. Hydratropic acid, $C^6H^5.CH \left\{ \begin{array}{l} CH^3 \\ CO.OH. \end{array} \right.$ Is produced by the action of nascent hydrogen (sodium-amalgam) on atropic acid.—Colorless liquid, which does not congeal at a low temperature.

4. *Acids*, $C^{10}H^{12}O^2$.

1. Durylic acid (Cumylic acid), $C^6H^2 \left\{ \begin{array}{l} (CH^3)^3 \\ CO.OH. \end{array} \right.$ Is obtained by oxidizing durene with dilute nitric acid.—Crystallizes from alcohol in lustrous, hard prisms; fusing point, 149–150°. Almost insoluble in cold water, easily soluble in alcohol and ether. When further oxidized it is converted into cumidinic acid.

2. Cuminic acid, $C^6H^4 \left\{ \begin{array}{l} C^3H^7 \\ CO.OH. \end{array} \right.$ Is produced from cuminol (p. 325) by boiling with alcoholic potassa or by adding to fusing caustic potassa; probably also by boiling cuminol with dilute nitric acid.—Colorless, tabular or prismatic crystals. Almost insoluble in cold water, very difficultly in hot water, easily soluble in alcohol. Fuses at 113°, and sublimes without decomposition in long needles. Is converted into terephtalic acid when oxidized with nitric or chromic acids; and yields cumene when heated with lime.

5. *Acids*, $C^{11}H^{14}O^2$.

Homocuminic acid, $C^6H^4 \left\{ \begin{array}{l} C^3H^7 \\ CH^2.CO.OH. \end{array} \right.$ Is produced from cumyl cyanide (from cumine alcohol, p. 316) by boiling it with alcoholic potassa.—Small crystals, fusing at 52°.

b. Monobasic, Diatomic Acids.

1. *Oxybenzoic Acids.*

$$C^7H^6O^3 = C^6H^4 \left\{ \begin{array}{l} OH \\ CO.OH. \end{array} \right.$$

The three isomeric oxyacids corresponding to the other substitution-products of benzoic acid are well known.

1. Salicylic acid (Ortho-oxybenzoic acid). Is contained in the blossoms of *Spiræa ulmaria;* and in the form of the methyl ether in wintergreen oil (the volatile oil of *Gaultheria procumbens*).—The sodium salt is produced by the direct combination of phenol and carbonic anhydride in the presence of sodium; the ethyl ether, by bringing a mixture of equal parts by weight of phenol and chlorcarbonic ether (p. 222) together with sodium. It is produced further by treating saligenin and salicylous acid (p. 322) with oxidizing agents; by melting ortho-cresol (p. 298) and salicin (see Glucosides) with caustic potassa; by conducting nitrous acid into a dilute aqueous solution of anthranilic acid (p. 330); in small quantity, by the action of fusing caustic potassa on phenol.—Is prepared most advantageously by warming gaultheria-oil with potassa-ley, by which means it is converted into methyl alcohol and potassium salicylate. From the solution of this salt, the acid is precipitated by means of hydrochloric acid; and by recrystallization from hot water it is purified.

Colorless, inodorous prisms, difficultly soluble in cold water; fusing point, 155–156°. Sublimable, when carefully heated; heated rapidly either alone or with water, it is resolved at 220–230° into carbonic anhydride and phenol; heated with hydriodic acid to 140–150°, the same decomposition takes place. Treated with chromic acid it undergoes rapid and complete combustion, yielding carbonic anhydride and water. Its solution turns deep violet when treated with iron chloride.

It conducts itself towards bases as a monobasic acid. Under certain circumstances however the second hydrogen-atom can be replaced by metals. The salts, which are formed in this way, however, can, only with difficulty, be prepared in pure condition, and are decomposed even by carbonic acid.

Methyl salicylate, $C^6H^4\left\{\begin{array}{l}OH\\CO.O.CH^3\end{array}\right.$ By distilling *Gaultheria procumbens* with water.—Colorless oil of a pleasant odor; specific gravity, 1.197; boiling point, 224°. But slightly soluble in water, easily soluble in alcohol. Combines with bases in the cold, forming instable salts, which are decomposed by heat.

Ethyl salicylate, $C^6H^4\left\{\begin{array}{l}OH\\CO.O.C^2H^5\end{array}\right.$, is formed by distilling salicylic acid with alcohol and sulphuric acid.—Colorless oil, boiling at 221°.

Methylsalicylic acid, $C^6H^4\left\{\begin{array}{l}O.CH^3\\CO.OH\end{array}\right.$ By heating 2 parts of gaultheria-oil with 1 part of potassium hydroxide (dissolved in alcohol), and 3–4 parts of methyl iodide at 100–120°, there is produced the liquid (boiling point, 248°) methyl ether of methylsalicylic acid, $C^6H^4 \begin{cases} O.CH^3 \\ CO.O.CH^3, \end{cases}$ which when boiled with soda-ley yields sodium methylsalicylate; from the solution of this salt hydrochloric acid precipitates the free acid.—Large, colorless plates, difficultly soluble in cold water, easily soluble in hot water and in alcohol; fuses at 98.5°, and above 200° is resolved into anisol (p. 291) and carbonic anhydride. Strong monobasic acid. Its salts are just as instable as the salicylates.

Ethylsalicylic acid, $C^6H^4 \begin{cases} O.C^2H^5 \\ CO.OH. \end{cases}$ Is obtained in the same way as methylsalicylic acid.—Crystalline mass; fusing point, 19.5°; is resolved into carbonic anhydride and phenol-ethylether at 300°.

Acetylsalicylic acid, $C^6H^4 \begin{cases} O.C^2H^3O \\ CO.OH, \end{cases}$ is produced by the action of acetyl chloride on salicylic acid or salicylates.—Colorless, fine prisms.

Salicylamide (Salicylamic acid), $C^7H^7NO^2 = C^6H^4 \begin{cases} OH \\ CO.NH^2 \end{cases}$ (isomeric with the amidobenzoic acids) is produced by continued action of ammonia on gaultheria oil, and by heating ammonium salicylate.—Pale yellow, crystalline laminæ, difficultly soluble in water; fusing point, 142°; sublimable.

When salicylic acid is distilled with phosphorus chloride, orthochlorbenzoyl chloride (p. 328) is produced.—Dry chlorine tranforms it, according as salicylic acid or chlorine is in excess, into *chlorsalicylic acid*, $C^7H^5ClO^3$, or *dichlorsalicylic acid*, $C^7H^4Cl^2O^3$; bromine also forms *brom-* or *dibromsalicylic acid*; iodine in alkaline solution or in aqueous solution in the presence of iodic acid converts it into a mixture of *iodo-*, *diiodo-* and *triiodosalicylic acids*, which are difficult of separation. All these acids crystallize well, and are but slightly soluble in water, more readily in alcohol. When distilled (best when previously mixed with sand and baryta), they are decomposed like salicylic acid, yielding carbonic anhydride and substitution-products of phenols. By the action of vapors of sulphuric anhydride and subsequent treatment with water, it is converted into *sulphosalicylic acid*, $C^6H^3(OH)\left\{\begin{matrix}SO^2.OH\\CO.OH.\end{matrix}\right.$

Nitrosalicylic acid (Anilic acid), $C^7H^5(NO^2)O^3$, is formed by treating salicylic acid, indigo, or salicin with nitric acid.—Needly crystals, very difficultly soluble in cold water, more easily in hot water and in alcohol. When boiled with nitric acid it is converted into picric acid.

Amidosalicylic acid, $C^7H^5(NH)^2O^3 =$ $C^6H^3(NH^2)\left\{\begin{matrix}OH\\CO.OH.\end{matrix}\right.$ Is obtained by the reduction of nitrosalicylic acid with tin and hydrochloric acid.—Needles of the lustre of satin. Insoluble in cold water and alcohol, difficultly soluble in hot water. Combines with bases and acids. Easily decomposable. At a high temperature it is resolved into carbonic anhydride and isoamidophenol (p. 294).

---

2. Oxybenzoic acid (Meta-oxybenzoic acid), $C^6H^4\left\{\begin{matrix}OH\\CO.OH.\end{matrix}\right.$ Is produced by conducting nitrous acid into a dilute aqueous solution of amidobenzoic acid; by boiling nitric-diazobenzoic acid (p. 332) with water; and by melting metachlor-, metaiodo-, metasulphobenzoic acids, and meta-cresol with caustic potassa. —Crystalline powder, consisting of small quadratic plates, or large verrucose crystals, without water of crystallization. But slightly soluble in cold water, more readily in hot water. Fuses at 200°, and is decomposed only at a very high temperature.

Ethyl oxybenzoate, $C^6H^4\left\{\begin{array}{l} O.CH^3 \\ CO.OH. \end{array}\right.$ Colorless plates; fusing point, 72°; boiling point, 282°; almost insoluble in cold water, moderately soluble in boiling water. Treated with cold, concentrated soda-ley, it yields a colorless, crystalline, easily soluble sodium compound $C^6H^4\left\{\begin{array}{l} ONa \\ CO.O.C^2H^5. \end{array}\right.$

Methyloxybenzoic acid, $C^6H^4\left\{\begin{array}{l} O.CH^3 \\ CO.OH. \end{array}\right.$ The potassium salt is obtained by heating one molecule oxybenzoic acid with two molecules potassium hydroxide and two molecules methyl iodide to 140°, and decomposing the ether thus formed by means of potassa-ley. The sodium salt is formed by the simultaneous action of sodium and carbonic anhydride on the methyl ether of monobromphenol.—The acid, precipitated from these salts by means of hydrochloric acid, crystallizes in long, colorless needles. But slightly soluble in cold water, easily soluble in hot water and in alcohol. Fuses at 95°, and sublimes without decomposition.

Ethyloxybenzoic acid, $C^6H^4\left\{\begin{array}{l} O.C^2H^5 \\ CO.OH. \end{array}\right.$ Colorless needles; fusing point, 137°.

Acetyloxybenzoic acid, $C^6H^4\left\{\begin{array}{l} O.C^2H^3O \\ CO.OH. \end{array}\right.$ Colorless crystals; fusing point, 127°.

---

3. Para-oxybenzoic acid, $C^6H^4\left\{\begin{array}{l} OH \\ CO.OH. \end{array}\right.$ Is produced by conducting nitrous acid into a boiling, very dilute, aqueous solution of para-amidobenzoic acid; and by fusing anisic acid, paraiodo-, and parasulphobenzoic acids, para-cresol, phloretic acid, amidohydrocinnamic acid, and a number of resins (gum-benzoin, aloes, dragon's blood, acaroïd) with potassium hydroxide. Is much more easily soluble in cold water than salicylic acid, more easily in hot water and in alcohol. Fuses at 210°, but decomposes partially even at this temperature, forming carbonic anhydride and phenol. Its solution gives a yellow, amorphous precipitate with iron chloride, soluble in an excess of the reagent.—Treated with phosphorus chloride it yields parachlorbenzoyl chloride.

Methyl para-oxybenzoate, $C^6H^4\left\{\begin{array}{l}OH\\CO.O.CH^3\end{array}\right.$, is obtained by heating equal molecules of paraoxybenzoic acid, potassium hydrate, and methyl iodide to 120°.—Crystallizes from ether in large tablets; fuses at 17°, and boils at 283°.

Ethyl para-oxybenzoate, $C^6H^4\left\{\begin{array}{l}OH\\CO.O.C^2H^5\end{array}\right.$, is prepared like the methyl ether.—Colorless, crystalline mass; fuses at 113°, and boils at 297°; but slightly soluble in water, easily soluble in alcohol. With soda-ley it yields a solid, easily soluble sodium compound.

Methylpara-oxybenzoic acid (Anisic acid), $C^8H^8O^3 = C^6H^4\left\{\begin{array}{l}O.CH^3\\CO.OH.\end{array}\right.$ Is obtained from paraoxybenzoic acid in the same way as methyloxybenzoic acid from oxybenzoic acid. Is further produced by the oxidation of anisic aldehyde and anethol (cf. p. 324) with nitric acid or a mixture of potassium bichromate and dilute sulphuric acid; and by the oxidation of paracresol-methylether (p. 299) with potassium bichromate and dilute sulphuric acid.—Large, colorless prisms. Almost insoluble in cold water, easily soluble in alcohol; fuses at 175°; sublimes; its salts are almost all soluble in water, and crystallize well.—Heated with hydriodic or hydrochloric acids it yields paraoxybenzoic acid and methyl iodide or chloride. Fusing potassium hydroxide converts it into paraoxybenzoic acid. Heated with lime or baryta it is resolved into anisol (p. 291) and carbonic anhydride.

Chloranisic acid, $C^8H^7ClO^3$, is produced by conducting chlorine into melted anisic acid.—Small, colorless prisms; fuse at 180°; sublimable; insoluble in water, soluble in alcohol.—*Bromanisic acid* and *iodanisic acid* are very similar to the chlorinated acid. Distilled with baryta, they yield substitution-products of anisol.

Nitroanisic acid, $C^8H^7(NO^2)O^3$, is formed when anisic acid, or the oils which anisic acid yields by oxidation, are boiled with nitric acid until completely dissolved.—Small, lustrous prisms, that fuse at 175–180°, and are volatile only with partial decomposition. But slightly soluble in water and cold alcohol, easily in hot alcohol. Treated in alcoholic solution with ammonium sulphide, it is converted into

Amidoanisic acid, $C^8H^7(NH^2)O^3 =$
$C^6H^3(NH^2)\left\{\begin{array}{l}O.CH^3\\CO.OH.\end{array}\right.$ Crystallizes from alcohol in short, four-sided prisms. Difficultly soluble in water, easily in hot alcohol; fuses at 180°; not volatile without decomposition. Combines with bases and acids, forming salts.

Ethylparaoxybenzoic acid, $C^6H^4\left\{\begin{array}{l}O.C^2H^5\\CO.OH,\end{array}\right.$ is obtained from paraoxybenzoic acid and paracresol-ethylether in the same way as anisic acid.—Colorless needles; very difficultly soluble in boiling water. Fuses at 195°, and sublimes without decomposition.

Chlorparaoxybenzoic acid, $C^7H^5ClO^3$, **Iodo-,** and **Diiodoparaoxybenzoic acids,** are crystallizing acids, which are prepared like the substitution-products of salicylic acid.

Nitroparaoxybenzoic acid, $C^7H^5(NO^2)O^3$, is formed by treating paraoxybenzoic acid with very dilute nitric acid. Small, flesh-colored crystals. Treated with tin and hydrochloric acid it is reduced, forming *amido-paraoxybenzoic acid*, $C^7H^5(NH^2)O^3 = C^6H^3(NH^2)\left\{\begin{array}{l}OH\\CO.OH,\end{array}\right.$

which crystallizes in easily decomposable needles with one and a half molcclues water of crystallization.

Tyrosin, $C^9H^{11}NO^3$ (perhaps ethylamidoparaoxybenzoic acid $= C^6H^3(NHC^2H^5)\left\{\begin{matrix}OH\\CO.OH\end{matrix}\right)$. Is produced, together with leucine (p. 98) and other products, by continued boiling of albuminous substances, horn, etc., with hydrochloric acid or dilute sulphuric acid, and by fusing them with potassium hydroxide. It also occurs in the living organism, particularly in a diseased condition of the organism. It is prepared most advantageously from horn, which, in the form of shavings, is kept boiling, for about sixteen hours, with double its weight of concentrated sulphuric acid, previously diluted with from four to four and a half times its volume of water. During the boiling the evaporated water is replaced, the original volume being retained. At the end of the time mentioned the liquid is neutralized with milk of lime. The filtered solution is evaporated to half its volume, then acidified with sulphuric acid, and, after filtering, mixed with enough white lead to form a thin pasty mass. The solution, which contains the tyrosin, in the form of the lead salt, is treated with sulphuretted hydrogen. On evaporating the filtrate from lead sulphide, the tyrosin crystallizes out, and can be easily obtained in a pure condition by repeated recrystallization. Leucine remains in the mother-liquor.

Colorless, long, fragile, usually radiating prisms; very slightly soluble in alcohol, more easily in hot water, insoluble in ether.—Combines with bases and acids.—When heated alone, it is decomposed and yields phenol and other compounds. When fused with caustic potassa it yields paraoxybenzoic and acetic acids and ammonia.—Dilute nitric acid (4 parts water and 4 parts concentrated nitric acid to 1 part tyrosin) converts it, without the aid of heat, into nitrotyrosin nitrate, a crystallizing substance, from the solution of which *nitrotyrosin* $C^9H^{10}(NO^2)NO^3$ may be precipitated by ammonia. It crystallizes in thin, pale-yellow needles, which are very slightly soluble in cold water. It also unites with bases and acids.—When a mixture of tyrosin with nitric acid is evaporated at a slightly elevated temperature, *dinitrotyrosin*, $C^9H^9(NO^2)^2NO^3$, is formed. This crystallizes in golden-yellow laminæ. Simultaneously with these two compounds, a red coloring matter (erythrosin) is produced by the action of nitric acid on tyrosin.—When heated with concentrated sulphuric acid, tyrosin yields several sulpho-acids, the soluble salts of which are colored a beautiful violet by iron chloride.

2. *Acids*, $C^8H^8O^3$.

a. Oxytoluic Acids (Cresotic Acids).

$$C^6H^3 \begin{cases} OH \\ CH^3 \\ CO.OH. \end{cases}$$

Three acids of this composition are known. They are formed, like salicylic acid, by the simultaneous action of sodium and carbonic anhydride on the three modifications of cresol (p. 298).

α-Cresotic acid. From para-cresol. Long, colorless needles; fusing point, 147–150°.

β-Cresotic acid. From ortho-cresol. Long, colorless needles; fusing point, 114°.

γ-Cresotic acid. From meta-cresol. Needles; fusing point, 168–173°.

The solutions of all three acids are colored violet by iron chloride.

b. Oxymethylphenylformic Acid.

$$C^6H^4 \begin{cases} CH^2.OH \\ CO.OH. \end{cases}$$

When para-toluic acid is treated with bromine with the aid of heat, an acid $C^6H^4 \begin{cases} CH^2Br \\ CO.OH \end{cases}$ is produced, which as yet has not been prepared in a pure condition.

When this acid is boiled with baryta water, it yields barium bromide and the barium salt of oxymethylphenylformic acid. Hydrochloric acid throws down the free acid from the solution thus obtained.—Flat needles; fusing point, 176°. Moderately easily soluble in water, particularly in hot water.

c. Mandelic Acid (Phenylglycolic Acid).

$$C^6H^5.CH \left\{ \begin{matrix} OH \\ CO.OH. \end{matrix} \right.$$

Is formed when a solution of oil of bitter almonds, containing hydrocyanic and dilute hydrochloric acids, is heated for thirty to thirty-six hours in a flask connected with an inverted condensing apparatus, and the solution then evaporated. It is also formed by heating amygdalin with concentrated hydrochloric acid. By dissolving it in ether, it may be separated from the sal ammoniac, which is formed at the same time.—Crystallizes in prisms or plates. Easily soluble in water, alcohol, and ether. Heated alone, it yields oil of bitter almonds and a resin. Oxidizing agents convert it into benzoic acid. Hydriodic acid reduces it, forming alphatoluic acid; with hydrochloric and hydrobromic acids it yields water and *chlor-* and *bromalphatoluic acids* (p. 340).

3. *Acids*, $C^9H^{10}O^3$.

1. Oxymesitylenic acid, $C^6H^2 \left\{ \begin{matrix} OH \\ (CH^3)^2 \\ CO.OH. \end{matrix} \right.$ Is produced by heating potassium sulphomesitylenate with potassium hydroxide to 240–250°.—Colorless, fine needles, of a silky lustre. Fusing point, 176°. Almost insoluble in cold water, difficultly in boiling water, easily soluble in alcohol and ether. The solution of the free acid and its salts is colored deep blue by iron chloride. When heated with potassium hydroxide to a high temperature, it is resolved into carbonic anhydride and solid xylenol (p. 299).

2. Phloretic acid, $C^6H^2\begin{cases}OH\\(CH^3)^2\\CO.OH.\end{cases}$ Is formed, together with phloroglucin (p. 311), by evaporating phloretin with potassa-ley. Potassium phloretate is extracted from the residue by means of alcohol, precipitated from this solution by ether, and after dissolving in water, decomposed by hydrochloric acid.—Long, brittle, colorless prisms, difficultly soluble in cold water, easily soluble in hot water and in alcohol. Fuses at 128–130°, and when heated with baryta, is decomposed into carbonic anhydride and phlorol (p. 300). Its solution is colored green by iron chloride.

3. Alorcic acid, $C^6H^2\begin{cases}OH\\(CH^3)^2\\CO.OH.\end{cases}$ In small quantity, together with orcin and paraoxybenzoic acid, in the preparation of orcin from aloes.—Fine, brittle needles; difficultly soluble in cold water, easily in boiling water, in alcohol and ether. Fusing caustic potassa converts it into orcin and acetic acid.

4. Melilotic acid (Hydrocoumaric acid), $C^6H^4\begin{cases}OH\\CH^2.CH^2.CO.OH.\end{cases}$ Is contained in common melilot (*Melilotus officinalis*), in the leaves of *Faham*, sometimes in combination with coumarin, sometimes free; and is produced by treating an aqueous solution of coumarin with sodium-amalgam.—Large, colorless, lanceolar crystals. Very easily soluble in hot water, alcohol, and ether, moderately easily in cold water (in 20 parts of 18°). Fuses at 82°. Its solution is colored bluish for the moment by iron chloride. Heated alone it is resolved into water and its *anhydride*, $C^9H^8O^2$, a substance that crystallizes in rhombic plates, fuses at 25°, and boils at 272°. Fusing potassium hydroxide decomposes it, yielding acetic and salicylic acids, the action being accompanied by an evolution of hydrogen.—Its salts, when carefully heated, yield the anhydride, when more strongly heated, phenol.

5. Hydroparacoumaric acid,
$C^6H^4 \begin{cases} OH \\ CH^2.CH^2.CO.OH. \end{cases}$ Is formed from paracoumaric acid by treating it with sodium-amalgam; and by the action of nitrous acid on amidohydrocinnamic acid.—Small, well-formed, monoclinate crystals; easily soluble in water, alcohol, and ether; fuses at 125°.

6. Tropic acid (Phenylsarcolactic acid),
$C^6H^5.CH \begin{cases} CH^2.OH \\ CO.OH. \end{cases}$ Is formed by heating atropin (see Alkaloids) for several hours with fuming hydrochloric acid to 120–130°.—Fine, colorless, prismatic crystals. Moderately easily soluble in water (in 49 parts at 14.5°), easily soluble in alcohol and ether; fuses at 117–118°. When heated higher with hydrochloric acid or with barium hydroxide, it is converted into atropic and isatropic acids, at the same time giving up water.

7. Phenyllactic acid, $C^6H^5.CH^2.CH \begin{cases} OH \\ CO.OH. \end{cases}$ Is produced by the action of sodium-amalgam in a cold solution of phenylchlor- or phenylbromlactic acid.—Pointed needles, united in hemispherical groups. Exceedingly easily soluble in hot water; fuses at 93–94°; when heated to 180°, it is resolved into water and cinnamic acid; and when its solution is mixed with concentrated hydrochloric, hydrobromic, or hydriodic acid, it is converted into substitution-products of hydrocinnamic acid (p. 342).

Phenylchlorlactic acid, $C^9H^9ClO^3 =$
$C^6H^5.C^2H^2Cl \begin{cases} OH \\ CO.OH. \end{cases}$ The sodium salt is produced, when chlorine gas is conducted into a solution of equal molecules of cinnamic acid and sodium carbonate, until a portion of the liquid, when tested, bleaches vegetable colors. The solution is acidified with hydrochloric acid, filtered and evaporated, and the acid then extracted by shaking with ether.—Crystallizes from water in fine, hexagonal laminæ with one molecule of

water of crystallization. Soluble in hot water in almost every proportion. Fuses at 70–80° while still containing water; in an anhydrous condition at 104°.

Phenylbromlactic acid, $C^9H^9BrO^3$. Is obtained from cinnamic acid dibromide, by boiling with water. —Very similar to the chlorinated acid. Fuses in an anhydrous condition at 125°.

4. *Acids*, $C^{11}H^{14}O^3$.

Thymotic acid, $C^6H^2\begin{cases}OH\\CH^3\\C^3H^7\\CO.OH.\end{cases}$ Is formed from thymol (p. 300) by the simultaneous action of sodium and carbonic anhydride.—Long, fine needles; very difficultly soluble in water; fuses at 120°, and is sublimable without decomposition. The solutions of the acid and those of its salts, particularly that of the ammonium salt, turn a beautiful blue when warmed with iron chloride. When the potassium salt is heated with phosphorus chloride, or when the free acid is heated with phosphoric anhydride, a substance called *thymotide*, $C^{11}H^{12}O^2$, is produced. This crystallizes well, and fuses at 187°.

---

c. Monobasic, Triatomic Acids.

1. *Dioxybenzoic acids.*

$$C^7H^6O^4 = C^6H^3\begin{cases}(OH)^2\\CO.OH.\end{cases}$$

Three isomeric acids of this composition are positively known.

1. Oxysalicylic acid. Is produced when a solution of monoiodosalicylic acid is boiled with concentrated potassa-ley until on acidifying no precipitate is formed. It may now be extracted from the acidified solution by agitating with ether.—Lustrous needles;

moderately difficultly soluble in cold water (in 58 parts at 21°), easily soluble in hot water, alcohol, and ether; contains no water of crystallization; fuses at 183°, and is decomposed at 210–212° into carbonic anhydride and a mixture of hydroquinone and pyrocatechin. Iron chloride turns its solution deep blue, which becomes blood-red on a subsequent addition of a little ammonia. The salts are very unstable, and are decomposed, when left in aqueous solution in contact with the air.

2. Protocatechuic acid. Is formed by the action of melting caustic potassa on iodoparaoxybenzoic acid, bromanisic acid, para- and ortho-cresolsulphuric acids, piperic acid, catechin and a great many resins (guaiacum, gum-benzoïn, dragonsblood, assafœtida, myrrh, acaroïd, etc); the production from resins is usually accompanied by the formation of paraoxybenzoic acid.—Crystallizes from water in colorless laminæ or needles with one molecule of water of crystallization. Difficultly soluble in cold water, more easily in hot water, in alcohol, and ether. Fuses at 199°, and decomposes at a higher temperature into carbonic anhydride and pyrocatechin. Its solution is turned dark green by iron chloride; this color changes to a beautiful blue on the addition of a small quantity of a dilute solution of sodium carbonate, the addition of more of the latter solution giving rise to a dark red. The solutions of its salts turn violet on the addition of salts of iron suboxide.—When mixed with bromine it is converted into *monobromprotocatechuic acid*, $C^7H^5BrO^4$, which crystallizes in fine rhombic needles.

Dimethyl-protocatechuic acid, $C^6H^3 \left\{ \begin{matrix} (O.CH^3)^2 \\ CO.OH. \end{matrix} \right.$

Is obtained by heating 1 part of protocatechuic acid, 4 parts of methyl iodide, and 1 part of potassium hydroxide with methyl alcohol in sealed tubes at 140° about three hours. The mass thus obtained is boiled with caustic soda; and the acid precipitated by means of sulphuric acid.—Fine lustrous needles; gives no reaction with iron chloride; fusing point, 170–171°.

Diethyl-protocatechuic acid, $C^6H^3 \begin{cases} (O.C^2H^5)^2 \\ CO.OH, \end{cases}$ prepared in the same way as the preceding acid, forms lustrous, white needles, which give no reaction with iron chloride, and fuse at 149°.

Piperonylic acid (Methylen-protocatechuic acid), $C^6H^3 \begin{cases} \begin{matrix} O \\ O \end{matrix} > CH^2 \\ CO.OH. \end{cases}$ Is produced by further oxidation of piperonal (p. 324) by means of potassium hypermanganate; and boiling piperonal with alcoholic potassa. Is prepared artificially by heating protocatechuic acid, methylene iodide, and potassium hydroxide together in sealed tubes; boiling the product with potassa-ley, and acidifying the solution.—Colorless needles; fusing point, 228°; sublimable without decomposition; insoluble in cold water, difficultly soluble in boiling water and cold alcohol, more easily soluble in hot alcohol. Monobasic acid. When heated with dilute hydrochloric acid, it is resolved into carbon and protocatechuic acid.

Ethylene-protocatechuic acid, $C^6H^3 \begin{cases} \begin{matrix} O \\ O \end{matrix} > C^2H^4 \\ CO.OH, \end{cases}$ is prepared by heating protocatechuic acid, ethylene bromide, and potassium hydroxide together, and treating the mass thus obtained as in the previous case. This acid resembles the preceding one.

A substance called *carbohydroquinonic acid*, which is obtained by the action of bromine on an aqueous solution of quinic acid (p. 361); and by fusing quinic acid with caustic potassa, is in all probability identical with protocatechuic acid.

3. Dioxybenzoic acid. Obtained by fusing the potassium salt of disulphobenzoic acid (p. 334) with caustic potassa.—Crystallizes from water with 1½ molecules water of crystallization; fusing point above 220°; gives no reaction with iron chloride.

2. *Orsellic Acid.*

$$C^8H^8O^4 = C^6H^2 \begin{cases} (OH)^2 \\ CH^3 \\ CO.OH. \end{cases}$$

Is formed by boiling erythrin with baryta-water and by heating a neutral solution of lecanoric acid in lime-water.—Colorless prisms, soluble in water, alcohol, and ether; fuses at 176°, undergoing decomposition into carbonic anhydride and orcin (p. 307). Its solution is turned purple by iron chloride.

Erythrin (Erythrite biorsellate), $C^{20}H^{22}O^{10} = C^4H^8(C^8H^7O^3)^2O^4$. Is contained in the lichen *Roccella fuciformis*, which is employed in the manufacture of archil (p. 308). It can be extracted from this by means of cold milk of lime. The solution is decomposed rapidly with carbonic acid; and the erythrin extracted from the precipitate with alcohol.—Crystalline, globular mass with 1½ molecule water of crystallization. Almost insoluble in cold water, difficultly soluble in hot water, easily soluble in alcohol. When boiled for a long time with water or baryta, it is decomposed into orsellic acid and *picroerythrin* (erythrite monorsellate), $C^{12}H^{16}O^7 + H^2O$, which forms colorless, bitter tasting crystals, that are easily soluble in water and alcohol.—By continued boiling of erythrin with baryta there are formed carbonic acid, orcin, and erythrite (p. 180).

Lecanoric acid (Diorsellic acid), $C^{16}H^{14}O^7 + H^2O$. Occurs in several lichens, belonging to the genera *Roccella*, *Lecanora*, and *Variolaria*. It can be extracted from these by means of ether or milk of lime, and is then precipitated by hydrochloric acid.—Crystallizes from alcohol and ether in colorless prisms; almost insoluble in water. Dissolved in lime or baryta water and boiled, it is at first converted into orsellic acid; by continued boiling, into carbonic acid and orcin. When its alcoholic solution is boiled ethyl orsellate, a crystalline body, is produced.

$$3.\ \textit{Acids},\ C^9H^{10}O^4 = C^8H^7 \left\{ \begin{array}{l} (OH)^2 \\ CO.OH. \end{array} \right.$$

1. Veratric acid, $C^9H^{10}O^4$. Is contained in sabadilla-seeds (from *Veratrum sabadilla*). To prepare it, the powdered seeds are exhausted with alcohol and a little sulphuric acid, the extract mixed with lime, filtered and the alcohol distilled off from the filtrate. Veratrin (see Alkaloids) separates, and from the filtered solution, which contains calcium veratrate, the free acid is obtained by precipitating with hydrochloric acid. By recrystallization from alcohol it is purified.

Colorless prisms; difficultly soluble in cold water, more readily in hot water, and in alcohol; fusible, and when carefully heated, sublimable.—Gently warmed with an excess of baryta it is resolved into carbonic acid and veratrol (p. 310).

2. Everninic acid, $C^9H^{10}O^4$. In the lichen *Evernia prunastri* there occurs an acid, *evernic acid*, $C^{17}H^{16}O^7$, that crystallizes in small, colorless prisms and is very similar to lecanoric acid. This can be extracted from the lichen by milk of lime and precipitated from the filtered solution by hydrochloric acid. This acid is resolved into orsellic acid (p. 358, or its decomposition-products, orcin and carbonic acid) and everninic acid when boiled with alkalies or baryta-water.—Fine, colorless crystals, resembling those of benzoic acid, almost insoluble in cold water, easily soluble in hot water, in alcohol, and ether; fuses at 157°. Its aqueous solution is colored violet by iron chloride.

3. Umbellic acid, $C^9H^{10}O^4$. Is produced by heating an alkaline solution of umbelliferone (p. 307) with sodium-amalgam.—Colorless, granular crystals; difficultly soluble in cold water, easily soluble in alcohol and ether; fuses below 125°, but suffers partial decomposition even at this temperature; its solution reduces an alkaline solution of copper and an ammoniacal solution of silver; and gives a green reaction with iron chloride. It is decomposed in alkaline solution in contact with the air.

4. Hydrocaffeic acid, $C^9H^{10}O^4$. Is produced by the action of sodium-amalgam on a hot solution of caffeic acid.—Colorless, rhombic crystals; easily soluble in water; the aqueous solution is colored an intense green by iron chloride, this turning to a cherry-color on the subsequent addition of sodium carbonate. Its salts are amorphous, decompose readily in a moist condition in contact with the air, and reduce solutions of copper and silver.

d. Monobasic, Tetratomic Acids.

Gallic Acid.

$$C^7H^6O^5 = C^6H^2 \begin{cases} (OH)^3 \\ CO.OH. \end{cases}$$

Occurrence. In gallnuts, in mango kernels, in dividivi (fruit of *Cæsalpina coriaria*), in tea, in the bark of the root of the pomegranate tree, and in several other plants.

Formation and preparation. From gallotannic acid by boiling with dilute acids or alkalies, and by keeping the solution in contact with the air; by heating diiodosalicylic acid with an excess of an alkaline carbonate at 140–150°; and also probably by evaporating a solution of bromprotocatechuic acid (p. 356) in an excess of potassa-ley.

Properties. Crystallizes from water in fine prisms of a silky lustre with one molecule of water of crystallization; soluble in 100 parts cold, in 3 parts of boiling water, easily in alcohol; fuses by about 200°, and is resolved into carbonic anhydride and pyrogallol (p. 310) at 210–220°. The aqueous solution reduces solutions of gold and silver, throwing down the metals, and yields a blue-black precipitate with iron chloride. Its salts do not undergo change, when in a dry condition or when in acid solution in contact with the air, but, when contained in alkaline solution, they absorb oxygen rapidly and decompose. Heated with phosphorus oxichloride to 120°, it is converted into an amorphous body, *digallic acid*, $C^{14}H^{10}O^9$, which is reconverted into gallic acid when boiled with concentrated hydrochloric acid.

Mono- and **Dibromgallic acids,** $C^7H^5BrO^5$ and $C^7H^4Br^2O^5$, are formed by the action of bromine on gallic acid at the ordinary temperature. Both compounds consist of colorless crystals, which are but slightly soluble in cold water, and are not sublimable.

Rufigallic acid, $C^7H^4O^4 + H^2O$, is formed by the slow heating of gallic acid (1 part) with concentrated sulphuric acid (4 parts) to 140°; and separates in reddish-brown, granular crystals, when the mass is subsequently diluted with water.—Small, lustrous crystals; loses its water of crystallization at 120°, and sublimes at a higher temperature in the form of cinnabar-red prisms.—Soluble in alkalies, forming a red solution, which is decomposed if air is allowed to have access to it; when treated with baryta-water, it becomes indigo-blue without dissolving. Materials mordanted with alumina salts are colored a beautiful red by it.—Fused with potassium hydroxide it yields carbonic acid and a substance called oxyquinone, $C^6H^4O^3$, which crystallizes in straw-colored needles.

The following acid bears a close relation to the preceding acids:—

Quinic acid, $C^7H^{12}O^6$. Occurs principally in cinchona barks (also in the false *Cinchona nova*); further, in the bilberry plant, in coffee-beans, in *Galium mollugo*, and probably in small quantity in a great many other plants.—Is obtained as a secondary product in the preparation of quinine. The extract, obtained from the broken-up bark, with water or dilute sulphuric acid, is treated with milk of lime in order to precipitate the alkaloids. The filtered solution, on being evaporated, leaves calcium quinate behind, and this may be purified by recrystallization, and then decomposed by oxalic acid.

Transparent, colorless, oblique rhombic prisms. Easily soluble in water, but very slightly in absolute alcohol; fuses at 162°, and, when heated above its fusing point, yields hydroquinone, pyrocatechin, benzoic acid, phenol, and other products. Oxidizing agents (manganese peroxide and sulphuric acid) resolve it into quinone (p. 301), carbonic anhydride, and formic acid. Heated with concentrated hydriodic acid, it is reduced to benzoic acid. When quinic acid is taken into the system of man or graminivorous animals, it is converted into hippuric acid.

Monobasic acid. All its salts are soluble in water.

Calcium quinate, $(C^7H^{11}O^6)^2Ca + 10H^2O$, forms large, easily soluble rhombic crystals, that effloresce in contact with the air. Is contained in cinchona barks.

e. Bibasic Acids.

1. *Benzenedicarbonic Acids.*

$$C^8H^6O^4 = C^6H^4 \begin{cases} CO.OH \\ CO.OH. \end{cases}$$

1. Phtalic acid (Ortho-phtalic acid) is produced by the oxidation of naphthalene and several of its derivatives; also of alizarin and purpurin with nitric acid or black oxide of manganese and sulphuric acid. Is also formed by treating benzene or benzoic acid with black oxide of manganese and sulphuric acid.—Colorless laminæ, or short, thick prisms; difficultly soluble in cold water, easily soluble in hot water, in alcohol and ether; fuses at 182°, and when heated to a higher temperature it is resolved into water and *phtalic anhydride*, $C^8H^4O^3$, a substance that crystallizes in long lustrous needles; fusing point, 127–128°.—Heated with an excess of potassa or lime, it breaks up into benzene and carbonic acid; when 1 molecule of its calcium salt is heated with 1 molecule calcium hydroxide at 330–350°, calcium benzoate is the result. When heated with hydriodic acid, it undergoes the same change. With phosphorus chloride it yields *phtalyl chloride* $C^6H^4(CO.Cl)^2$, a light-yellow liquid, boiling at 270°.

Barium phtalate, $C^8H^4O^4Ba$, forms small laminæ, which are very difficultly soluble in water.

Methyl and **ethyl phtalate** are colorless liquids.

Dichlorphtalic acid, $C^6H^2Cl^2(CO.OH)^2$, is prepared by boiling dichlornaphthalene tetrachloride (which see) with ordinary nitric acid.—Slightly yellowish colored, thick, intertangled prisms; fusing point, 183–185°; easily soluble in ether, alcohol, and hot water.

Tetrachlorphtalic acid, $C^6Cl^4(CO.OH)^2$, is obtained by heating pentachlornaphthalene with dilute nitric acid to 180–200°.—Colorless laminæ, or hard, thick plates; fuses at 250°, at the same time breaking up into water and anhydride.

Monobromphtalic acid, $C^6H^3Br(CO.OH)^2$. By heating phtalic acid, for a long time, with an excess of bromine and water at 180–200°.—White, crystalline powder; fusing point, 136–138°; easily soluble in water, alcohol, and ether.

Nitrophtalic acid, $C^6H^3(NO^2)(CO.OH)^2$. By digesting phtalic acid with nitric-sulphuric acid.—Pale yellow prisms; fusing point, 208–210°; easily soluble in water, alcohol, and ether. When heated with tin and hydrochloric acid it is converted into meta-amidobenzoic acid, carbonic anhydride being given off.

Hydrophtalic acid, $C^8H^8O^4$. Is produced by continued action of sodium-amalgam on a cold solution ot

1 part phtalic acid and 1 part crystallized sodium carbonate.—Hard, tabular crystals; difficultly soluble in cold water and ether, more easily in hot water and in alcohol; fuses above 200°, water being given off and phtalic anhydride formed. Is decomposed when heated with soda-lime, yielding benzene, hydrogen, and carbonic acid; when fused with potassa, it yields benzoic acid, hydrogen, and carbonic acid; when it is gently warmed with phosphorus chloride, it yields benzoyl chloride, carbonic oxide, hydrochloric acid, and phosphorus oxichloride; when dissolved in concentrated sulphuric acid, when bromine is allowed to act on its aqueous solution, and when oxidized with dilute nitric acid, it is converted into benzoic acid; when its alcoholic solution is saturated with hydrochloric acid gas, ethyl benzoate is formed.

Tetrahydrophtalic acid, $C^8H^{10}O^4$. The anhydride of this acid ($C^8H^8O^3$, colorless laminæ, fusing at 68°) is formed in the dry distillation of hydropyromellitic acid. When this anhydride is heated with water, the acid is generated.—Easily soluble laminæ; fuses at 96°, being resolved at this temperature into water and the anhydride. Bibasic acid. Bromine, when added to its aqueous solution, converts it into *brommalophtalic acid*, $C^8H^{10}Br(OH)O^4$, which crystallizes in hard crusts, and, when heated with baryta-water, is converted into *tartrophtalic acid*, $C^8H^{10}(OH)^2O^4$ (large, easily soluble prisms).

Hexahydrophtalic acid, $C^8H^{12}O^4$. Is obtained by heating tetrahydrophtalic acid with concentrated hydriodic acid to 230°; or, better, by heating hydrophtalic acid with concentrated hydriodic acid to 240–250°.—Indistinct, small, hard crystals; fusing point, 203–205°; somewhat difficultly soluble in water; bibasic.

---

2. Isophtalic acid (Meta-phtalic acid), $C^8H^6O^4$. Is obtained by oxidizing meta-xylene (p. 283) and meta-toluic acid (p. 338) with potassium bichromate and dilute sulphuric acid.—Is also formed by melting an intimate mixture of potassium metabrom- or meta-sulphobenzoate with sodium formate; and by heating hydropyromellitic and hydroprehnitic acids.—Long colorless, very fine crystals; almost insoluble in cold water, difficultly soluble in boiling water, more easily soluble in alcohol; fuses above 300°, and can be sublimed without undergoing decomposition.

Barium isophtalate, $C^8H^4O^4Ba + 3H^2O$. Crystallizes in colorless, lustrous prisms; easily soluble in water.

Methyl isophtalate, $C^6H^4(CO.O.CH^3)^2$. Colorless needles, fusing at 64–65°. The *ethyl ether* is a colorless liquid, boiling at 285°, and congealing at 0°.

Nitroisophtalic acid, $C^6H^3(NO^2)(CO.OH)^2$. By heating isophtalic acid with fuming nitric acid.—Large, colorless, lustrous, thin laminæ; fusing point, 248–249°; easily soluble in water and alcohol; is converted into *amido-isophtalic acid*, $C^6H^3(NH^2)(CO.OH)^2$, by tin and hydrochloric acid.

---

3. Terephtalic acid (Para-phtalic acid). Is produced from bodies belonging to the para-series: para-xylene, ethylmethylbenzene, cymene, amylmethylbenzene, cuminol, para-toluic acid, cuminic acid, ethylbenzoic acid, oil of turpentine, etc., by oxidizing them with a mixture of potassium bichromate (2 parts) and sulphuric acid (3 parts concentrated acid diluted with three times its volume of water); by boiling paradicyanbenzene (p. 256) with potassa-ley; and by melting a mixture of sodium parasulphobenzoate with sodium formate.—White powder; when allowed to separate slowly it is crystalline; almost insoluble in water, alcohol, and ether; freshly precipitated from a solution of one of its salts, it is moderately easily soluble in hot alcohol, and separates from this solution in crystalline form on cooling; sublimes undecomposed without previously melting.

Calcium terephtalate, $C^8H^4O^4Ca + 3H^2O$, and **Barium terephtalate,** $C^8H^4O^4Ba + 4H^2O$, are crystalline compounds, very difficultly soluble in water.

Methyl terephtalate, $C^6H^4(CO.O.CH^3)^2$. Long prisms, fusing at 140°, and subliming without decomposition; but slightly soluble in cold alcohol, easily soluble in hot alcohol.—The *ethyl ether* crystallizes in prisms that fuse at 44°.

Nitroterephtalic acid, $C^6H^3(NO^2)(CO.OH)^2$. Is formed by boiling terephtalic acid with very concentrated nitric acid.—Cauliflower-like masses; fusing point, 259°; moderately easily soluble in water.

Sulphoterephtalic acid, $C^6H^3(SO^2.OH)(CO.OH)^2$, is formed by heating terephtalic acid with fuming sulphuric acid in sealed tubes for six hours.—The *barium salt* can be purified by recrystallization.

Hydroterephtalic acid, $C^8H^8O^4$, is formed by the action of nascent hydrogen (sodium-amalgam) on terephtalic acid in a strongly alkaline solution.—White powder, very similar to terephtalic acid.

2. *Acids*, $C^9H^8O^4 = C^6H^3 \begin{cases} CH^3 \\ (CO.OH)^2. \end{cases}$

1. Uvitic acid (1 : 3 : 5). Is produced, together with mesitylenic acid (p. 340) by continued boiling of mesitylene with dilute nitric acid; and by boiling pyroracemic acid (p. 175) with barium hydroxide.—Colorless, fine needles; fusing point, 287°; almost insoluble in cold water, difficultly soluble in hot water, easily soluble in ether and alcohol; not volatile with water vapor; when oxidized with chromic acid, it is converted into trimesic acid, and, when heated with an excess of lime, it is resolved into carbonic acid and toluene.

2. Xylidinic acid (1 : 3 : 4).* Is formed from pseudocumene, xylylic, and paraxylylic acids by boiling them for a long time with dilute nitric acid.—Indistinct, colorless crystals; fusing point, 280–283°; almost insoluble in cold water, very slightly in boiling water, more easily in alcohol.

3. Isuvitic acid. Is formed, together with phloroglucin, pyrotartaric, and acetic acids, by fusing gamboge with caustic potassa.—Short, thick, rhombic, columnar crystals; but slightly soluble in cold water, more easily in hot water; fuses at about 160°.

$$3.\ \textit{Acids},\ C^{10}H^{10}O^4 = C^6H^2 \left\{ \begin{matrix} (CH^3)^2 \\ (CO.OH)^2. \end{matrix} \right.$$

Cumidinic acid. Is produced from durene and durylic acid by continued boiling with dilute nitric acid.—Long, transparent prisms; almost insoluble in water, even at the boiling temperature; easily soluble in hot alcohol; sublimes at a high temperature in plates, without previously fusing.

f. Tribasic Acids.

Benzenetricarbonic Acids.

$$C^9H^6O^6 = C^6H^3(CO.OH)^3.$$

1. Trimesic acid (1 : 3 : 5). Is obtained by oxidizing mesitylenic and uvitic acids with potassium bichromate and dilute sulphuric acid; is also produced, together with carbonic anhydride and benzenetetracarbonic acids, by heating hydro- and isohydromellitic acids with concentrated sulphuric acid.—Short, colorless prisms; rather difficultly soluble in cold water, easily soluble in hot water, in alcohol and ether; fuses

* 1 and 4 indicate the position of the CO.OH groups; 3 that of the group CH^3.

above 300°, and sublimes without decomposition; heated with an excess of lime, it is resolved into carbonic acid and benzene.

Barium trimesate. The *neutral salt*, $(C^9H^3O^6)^2Ba^3 + 3H^2O$, is a crystalline precipitate, almost insoluble in water. The *acid salt*, $(C^9H^5O^6)^2Ba + 4H^2O$, is thrown down when barium chloride is added to a solution of the free acid; fine, colorless needles, but slightly soluble in hot water.

Ethyl trimesate, $C^6H^3(CO.O.C^2H^5)^3$. Long prisms, of a silky lustre, fusing at 129°.

2. Hemimellitic acid (1 : 2 : 3). Is produced, together with phtalic anhydride, by heating hydromellophanic acid (p. 370) with concentrated sulphuric acid.—Colorless needles; rather difficultly soluble in water; from its concentrated aqueous solution it is precipitated by hydrochloric acid; fuses at 185°, and, when heated to a higher temperature, yields phtalic anhydride and benzoic acid.

3. Trimellitic acid (1 : 2 : 4). Is formed, together with isophtalic acid and pyromellitic anhydride, by heating hydropyromellitic acid with concentrated sulphuric acid.—Indistinct, verrucous crystals; fusing point, 216°; moderately easily soluble in water and ether.

Barium trimellitate, $(C^9H^3O^6)^2Ba^3 + 3H^2O$, forms difficultly soluble, verrucous crystals.

g. Tetrabasic Acids.

Benzenetetracarbonic Acids.

$$C^{10}H^6O^8 = C^6H^2(CO.OH)^4.$$

1. Pyromellitic acid. Is formed by careful distillation of mellitic acid; and is obtained most readily by heating sodium mellitate with sulphuric acid.—

Crystallizes from water with two molecules of water of crystallization, in colorless prisms; but slightly soluble in cold water, easily in hot water and in alcohol; fuses at 264°, and when distilled is converted into the anhydride, $C^{10}H^{2}O^{6}$, which forms large crystals and fuses at 286°.

Barium pyromellitate, $C^{10}H^{2}O^{8}Ba^{2}$, and **Calcium pyromellitate,** $C^{10}H^{2}O^{8}Ca^{2}$, are white precipitates, insoluble in water.

Ethyl pyromellitate, $C^{6}H^{2}(CO.O.C^{2}H^{5})^{4}$. Short, flat needles, insoluble in water; fusing point, 53°.

Hydropyromellitic acid, $C^{10}H^{10}O^{8}$. Is slowly formed by the action of sodium-amalgam on an aqueous solution of ammonium pyromellitate.—Colorless syrup, gradually congealing in crystalline form; very easily soluble in water; when heated alone, it is converted into the anhydride of tetrahydrophtalic acid (p. 364); when heated with concentrated sulphuric acid, it yields carbonic anhydride, pyromellitic anhydride, trimellitic and isophtalic acids.

2. Prehnitic acid. Is formed, together with carbonic anhydride, trimesic and mellophanic acids, by heating hydro- and isohydromellitic acids (p. 371) with concentrated sulphuric acid.—Large prisms united in groups; contain two molecules of water of crystallization; easily soluble in water; fuses at 237–250°, the anhydride being formed at the same time.

Hydroprehnitic acid, $C^{10}H^{10}O^{8}$. Is obtained like hydropyromellitic acid.—Syrupy. When heated with sulphuric acid, it yields prehnitic and isophtalic acids and carbonic anhydride.

3. Mellophanic acid. Is formed together with the preceding acid.—Small, indistinct crystals, united in crusts, without water of crystallization; fuses at 215–238°, giving rise to the formation of the anhydride; with sodium-amalgam it yields hydromellophanic acid.

h. Hexabasic Acids.

Mellitic Acid.

$$C^{12}H^6O^{12} = C^6(CO.OH)^6.$$

Occurrence and formation. In the mineral kingdom; in honeystone or mellite (found in lignite), which consists of aluminium mellitate crystallized in yellow, quadratic octahedrons. The ammonium salt, which crystallizes well, is prepared from this by boiling with ammonium carbonate; and from the ammonium salt the insoluble barium or silver salt is prepared by precipitation; the salt thus obtained is decomposed by dilute sulphuric or hydrochloric acid.—It can be prepared artificially by oxidizing pure carbon by means of potassium hypermanganate in an alkaline solution.

Properties. Fine needles of a silky lustre; easily soluble in water and alcohol. When heated it melts; when distilled alone it is resolved into carbonic anhydride, water, and pyromellitic anhydride; when heated with an excess of lime it yields carbonic acid and benzene. Very stable acid; is not decomposed by concentrated sulphuric, nitric, and hydriodic acids, nor bromine even at an elevated temperature.

Ammonium mellitate, $C^{12}O^{12}(NH^4)^6 + 9H^2O$. Crystallizes in large, colorless rhombic prisms.—*Barium mellitate,* $C^{12}O^{12}Ba^3 + 3H^2O$, and *Calcium mellitate* are precipitates, insoluble in water, rapidly becoming crystalline.

Methyl mellitate, $C^6(CO.O.CH^3)^6$, crystallizes in colorless laminæ, that fuse at 140°. The *Ethyl ether,* $C^6(CO.O.C^2H^5)^6$, forms lozenge-shaped crystals, that fuse at 69°.

Paramide (Mellimide), $C^{12}H^3N^3O^6 = C^6\left(\begin{matrix}CO\\CO\end{matrix}\right\}NH\Big)^3$. Ammonium mellitate, when heated to 160°, is resolved into water, ammonia, paramide, and ammonium euchronate. Paramide, which is insoluble in water, can be freed of the euchronate by water.—White, amorphous mass, insoluble in water and alcohol; is converted into acid ammonium mellitate when heated with water to 200°.

Euchronic acid, $C^{12}H^4N^2O^8 =$
$C^6\left(\begin{matrix}CO\\CO\end{matrix}\right\}NH\Big)^2(CO.OH)^2$. Is separated from its ammonium salt (see above, Paramide) by means of hydrochloric acid.—Colorless, short prisms; but slightly soluble in cold water; heated with water to 200°, it is converted into acid ammonium mellitate. Its solution, when brought in contact with zinc or nascent hydrogen from any source, throws down a deep-blue, insoluble body, *euchron*, which, when gently heated in the air, is reconverted into colorless euchronic acid, and dissolves in alkalies, forming beautiful, purple-red solutions, which rapidly become colorless in contact with air.

Hydromellitic acid, $C^{12}H^{12}O^{12}$. Is formed slowly by the action of sodium amalgam on ammonium mellitate.—Colorless, indistinct crystals; easily soluble in water; hexabasic. When kept it is slowly converted into *isohydromellitic acid*, $C^{12}H^{12}O^{12}$; the same change takes place rapidly when it is heated with concentrated hydrochloric acid to 180°. Isohydromellitic acid crystallizes in thick, four-sided prisms, is easily soluble in water, and is precipitated from the aqueous solution by hydrochloric acid. By heating hydromellitic acid with concentrated sulphuric acid, there is formed, under certain conditions, a third isomeric acid, *mesohydromellitic acid*, which forms voluminous needles, very difficultly soluble in cold water.

SECOND GROUP.

Cinnamene (Styrol).

$C^8H^8 = C^6H^5.CH:CH^2.$

Is contained in liquid storax, the expressed viscid juice of the bark of *Liquidambar orientale;* and is obtained from this by distilling with water and sodium carbonate. Is formed by heating benzene-ethyl bromide (p. 285) with water or baryta, and by heating acetylene gas (*cf.* Benzene). Is probably also contained in coal-tar.—Colorless, mobile liquid, of an aromatic odor; refracts light strongly. Boiling point, 146°; specific gravity, 0.924. When kept it is slowly converted into *metacinnamene*, a body polymeric with it; the same change takes place rapidly by heating it to 200°. Metacinnamene is a solid, amorphous, transparent mass, which, when distilled, is reconverted into cinnamene. Cinnamene yields benzoic acid when subjected to the influence of oxidizing agents.

Cinnamene chloride, $C^8H^8Cl^2 = C^6H^5.CHCl.CH^2Cl$, and **Cinnamene bromide,** $C^8H^8Br^2$, are produced by the direct combination of cinnamene with chlorine or bromine. The chloride is liquid; the bromide crystallizes in colorless laminæ or needles, that fuse at 67°. Heated alone, or, better, with caustic lime or alcoholic potassa, these compounds are converted respectively into *α-chlorcinnamene*, $C^6H^5.CH:CHCl$, or *α-bromcinnamene*. Both of these latter compounds are heavy liquids, not distillable without decomposition, the vapor of which excites to tears. The isomeric substitution-products, *β-chlorcinnamene*, $C^6H^5.CCl:CH^2$ (liquid, boiling, without decomposition, at 199°, of a pleasant odor like that of hyacinthes), and *β-bromcinnamene*, $C^6H^5.CBr:CH^2$ (boiling point, 228°), are formed by heating phenylchlor- and phenylbromlactic acids (pp. 354 and 355) with water at 200°.

Cinnamene iodide, $C^8H^8I^2$, separates in crystals when a solution of iodine in potassium iodide is added to cinnamene. When kept it is rapidly converted into metacinnamene, iodine being thrown down.

Nitrocinnamene, $C^8H^7(NO^2)$, crystallizes in large prisms.

Styryl Alcohol.

$$C^9H^{10}O = C^6H^5.CH:CH.CH^2.OH.$$

Is obtained by distilling styryl cinnamate (Styracin, p. 375) with concentrated potassa-ley. Is also produced in small quantity by heating styrylic aldehyde with alcoholic potassa.—Colorless, lustrous needles, of a pleasant odor; fuses at 33°, and boils at 250°. When oxidized slowly it is converted into cinnamic acid; when oxidized rapidly, oil of bitter almonds and benzoic acid are formed; when treated with hydrochloric acid gas or phosphorus iodide, liquid *chlorstyryl* C^9H^9Cl or *iodostyryl* C^9H^9I are formed.

Cinnamic Aldehyde (Styrylic Aldehyde).

$$C^9H^8O = C^6H^5.CH:CH.CHO.$$

Is contained in oil of cinnamon or oil of cassia (the volatile oils of the bark of *Persea cinnamomum* and *Persea cassia*); and can be extracted from them by agitating with alkaline bisulphites, and decomposing the separated crystalline compound with dilute sulphuric acid. Is formed by the distillation of a mixture of calcium cinnamate and formate; and when a mixture of acetic aldehyde and oil of bitter almonds is saturated with hydrochloric acid gas.

Colorless oil, heavier than water, does not mix with it. Not distillable alone, but very readily with water vapor. In contact with air it changes to cinnamic acid. Oxidizing agents convert it into oil of bitter almonds and benzoic acid. Combines with dry ammonia, forming water and a crystalline substance, *hydrocinnamide* $(C^9H^8)^3N^2$.

Cinnamic Acid.

$C^9H^8O^2 = C^6H^5.CH:CH.CO.OH.$

Occurrence and formation. In storax, in Tolu- and Peru-balsams (p. 312), and in a few varieties of gum-benzoin. Is formed from cinnamic aldehyde by oxidation; by boiling styracin with potassa; by the simultaneous action of sodium and carbonic anhydride on α-bromcinnamene; and by heating oil of bitter almonds with acetyl chloride.

Preparation. Most advantageously from storax. This is boiled for a long time with a solution of sodium carbonate or with potassa-ley, the clear solution of sodium or potassium cinnamate filtered off from the undissolved resin, and the cinnamic acid precipitated by hydrochloric acid. By recrystallizing from hot water or by subliming, it is purified.

Properties. Crystallized from hot water, it forms fine needly crystals; from alcohol, large, clear, easily cleavable prisms; inodorous; of a weak taste; fuses at 133°, and is distillable almost completely without decomposition. Monobasic acid. Very similar to benzoic acid; its salts also resemble the benzoates very strongly, but give a yellow precipitate with iron chloride; fusing potassium hydroxide converts it into benzoic and acetic acids. Subjected to the influence of oxidizing agents (dilute nitric acid, potassium hypermanganate, potassium bichromate, and sulphuric acid), it yields oil of bitter almonds and benzoic acid. Nascent hydrogen converts it into hydrocinnamic acid (p. 342). Heated with water to 180–200°, and with lime, it is resolved into carbonic acid and cinnamene.

Ethyl cinnamate, $C^9H^7O^2.C^2H^5$, is produced by conducting hydrochloric acid gas into a solution of cinnamic acid in absolute alcohol.

Benzyl cinnamate, $C^9H^7O^2.C^7H^7$, is contained in Tolu- and Peru-balsams; and is produced by heating sodium cinnamate with benzyl chloride.—Lustrous prisms; fusing point, 39°; distillable without decomposition only in a vacuum.

Styryl cinnamate (Styracin), $C^9H^7O^2.C^9H^9$, is contained in the brown resin, the residue from the preparation of cinnamic acid from storax. Can be most readily prepared by digesting storax with dilute soda-ley, at a temperature not higher than 30°, until the residual styracin has become colorless. After it is washed out and dried, it is recrystallized from alcohol, which contains ether.—Fine, colorless needles, united in nodules, insoluble in water; fuses at 50°.

Nitrocinnamic acid, $C^9H^7(NO^2)O^2$. When nitric acid is allowed to act upon cinnamic acid, two nitro-acids are formed, which can be separated by means of crystallization.—One fuses at 265°, is difficultly soluble in water, and yields paranitrobenzoic acid when oxidized. The second is easily soluble in water, and yields orthonitrobenzoic acid when oxidized.

Cinnamic acid dibromide (Dibromhydrocinnamic acid), $C^9H^8Br^2O^2 = C^6H^5.CHBr.CHBr.CO.OH$. Is formed by direct union, when bromine in the form of vapor is allowed to act on cinnamic acid, either at the ordinary temperature or at 100°.—Colorless, rhombic laminæ; insoluble in cold water, easily soluble in alcohol and ether; not fusible without decomposition; when boiled with water it is decomposed, yielding β-bromcinnamene and phenylbromlactic acid (p. 355). Nascent hydrogen converts it into hydrocinnamic acid.

Monobromcinnamic acid, $C^9H^7BrO^2$. Two isomeric modifications of this acid are formed by the addition of alcoholic potassa to a boiling-hot alcoholic solution of cinnamic acid dibromide. They can be separated by means of partial crystallization. The salt of α-bromcinnamic acid is difficultly soluble in water, that of β-bromcinnamic acid is very easily soluble and even deliquescent in the air.—*α-Bromcinnamic acid* crystallizes in long, lustrous, four-sided needles, which are but slightly soluble in cold water, more easily in boiling water, in alcohol in all proportions; fusing point, 130–131°; distil almost entirely without decomposition.—β-*Bromcinnamic acid* crystallizes from boiling water in large, hexagonal, flat crystals, which are easily soluble in boiling water and in alcohol; fuse at 120°; and are converted in α-bromcinnamic acid when subjected to distillation.

The following acids are isomeric with cinnamic acid:—

Atropic and Isatropic acids, $C^9H^8O^2$. Both acids are produced from tropic acid (p. 354) when this is heated with baryta or concentrated hydrochloric acid; and are hence formed, together with tropic acid, in the decomposition of atropin. Atropic acid is particularly formed when baryta is employed; isatropic acid, on the other hand, when hydrochloric acid is the decomposing agent. Atropic acid crystallizes in monoclinic plates, dissolves in 700–800 parts water of the ordinary temperature, easily in boiling water; fuses at 106.5°; yields benzoic acid when oxidized with chromic acid; alphatoluic acid (p. 340) when melted with potassium hydroxide; and hydratropic acid (p. 342) when treated with nascent hydrogen.—Isatropic acid forms thin, rhombic laminæ; is almost insoluble in cold water, but very slightly in boiling water; and also in alcohol it is less soluble than atropic acid. It melts at 200°, is not acted upon by chromic acid, and does not combine with hydrogen.

Phenylangelic acid, $C^{11}H^{12}O^2 = C^6H^5.C^4H^6.CO.OH$. Is formed, like cinnamic acid, by heating oil of bitter almonds with butyryl chloride at 120–130°.—Long, fine, colorless needles; fusing point, 81°; difficultly soluble in water, easily in alcohol. Chromic acid oxidizes it, forming benzoic acid.

Coumarin, $C^9H^6O^2$. Occurs in Tonka beans (the fruit of *Dipterix odorata*), partially in the form of crystals; in *Melilotus officinalis*, *Asperula odorata*, and *Anthoxanthum odoratum;* and can be extracted by means of alcohol. After distilling off the greater part of the alcohol, the residue is mixed with boiling water and filtered, when the greater part of the coumarin separates. It is also formed by the action of acetic anhydride on the sodium compound of salicylic aldehyde (p. 323).—Colorless, lustrous prisms, of a pleasant aromatic odor; but slightly soluble in cold water, more readily in hot water; fuses at 67°, and boils without decomposition at 291°. When its aqueous solution is treated with sodium-amalgam, it is converted into coumaric and melilotic acids (p. 353); if on the other hand, sodium-amalgam is allowed to act on its alcoholic solution, the principal product formed is the sodium salt of hydrocoumaric acid, $C^{18}H^{18}O^6$, which crystallizes in fine needles, difficultly soluble in cold water. This acid, when heated, is resolved into water and its anhydride, *hydrocoumarin*, $C^{18}H^{14}O^4$ (needles; fusing point, 222°).

Coumarin dichloride, $C^9H^6Cl^2O^2$ (syrupy mass), and **Coumarin dibromide,** $C^9H^6Br^2O^2$ (colorless prisms), are formed by the action of chlorine or bromine on solutions of coumarin in chloroform or carbon bisulphide. When treated with alcoholic potassa, they yield substitution-products of coumarin. The latter are also formed by the action of phosphorus chloride or of bromine on coumarin at a slightly elevated temperature. When the sodium compound of chlor- or bromsalicylic aldehyde is treated with acetic anhydride, substitution-products of coumarin are obtained that are isomeric with those formed by direct action of the reagents.

Coumarin is probably the anhydride of coumaric

acid, and constituted according to the formula
$C^6H^4\left\{\begin{matrix} O - CO \\ CH:CH \end{matrix}\right.$

When sodium salicylic aldehyde is heated with butyric and valeric anhydrides, compounds homologous with coumarin are formed: *butyric coumarin*, $C^{11}H^{10}O^2$ (needles; fusing point, 70–71°; boiling point, 296–297°) and *valeric coumarin*, $C^{12}H^{12}O^2$ (long prisms; fusing point, 54°; boiling point, 301°).

Coumaric acid, $C^9H^8O^3 = C^6H^4\left\{\begin{matrix} OH \\ CH:CH.COOH \end{matrix}\right.$ Is contained in *Melilotus officinalis*, and in the leaves of *Faham*. Is formed from coumarin by boiling with very concentrated potash-ley. Colorless, lustrous prisms, easily soluble in hot water and in alcohol; fuses at 195°; not volatile without decomposition. Fused with potassa, it yields potassium salicylate and acetate. The solutions of its alkaline salts are markedly fluorescent.

Paracoumaric acid, $C^9H^8O^3 =$ $C^6H^4\left\{\begin{matrix} OH \\ CH:CH.COOH \end{matrix}\right.$ Is produced by boiling an aqueous solution of aloes, to which has been added sulphuric acid, and, after filtering, it can be extracted from the solution by means of ether. Colorless, lustrous, brittle needles. But slightly soluble in cold water, easily soluble in hot water and in alcohol; fuses at 179–180°. Fused with potassium hydroxide, it is converted into paraoxybenzoic acid. It combines with nascent hydrogen, forming hydroparacoumaric acid (p. 3[illegible]). When boiled with fuming nitric acid, it yields picric acid.

Caffeic acid, $C^9H^8O^4 = C^6H^3\left\{\begin{matrix} (OH)^2 \\ COOH \end{matrix}\right.$ Is formed, together with sugar, by boiling caffetannic acid (or extract of coffee) with potash-ley, and separates from the solution on the subsequent addition of sulphuric acid. Yellowish, lustrous prisms or laminæ, but slightly soluble in cold water, more easily in hot water, very easily in alcohol. The aqueous solution reduces a silver solution when warm, but does not change an alkaline solution of copper; becomes intensely green on the addition of iron chloride, and the color changes to dark red if sodium carbonate now be added. When subjected to dry distillation, it yields pyrocatechin (p. 306); when fused with potassium hydroxide, protocatechuic (p. 546) and acetic acids. It combines with nascent hydrogen, forming hydrocaffeic acid (p. 560).

THIRD GROUP.

Acetenylbenzene or Phenylacetylene,
$C_8H_6 = C_6H_5.C≡CH$,

is produced by heating cinnamene bromide, α-bromcinnamene and acetophenone chloride ($C_6H_5.CCl_2.CH_3$, from acetophenone (p. 557) with phosphorus chloride) with alcoholic potash to 120°, by heating phenyl propiolic acid with water to 120° and by subjecting barium phenylpropiolate to dry distillation. Colorless liquid, boiling point 139-140°. Its dilute alcoholic solution yields a bright yellow flocculent precipitate, $(C_8H_5)_2Cu_2$, with an ammoniacal solution of copper; a white precipitate $(C_8H_5Ag)_2Ag_2O$, with an ammoniacal solution of silver. Hydrochloric acid separates the acetenylbenzene from both compounds. When sodium is added to its ethereal solution, it yields a white sodium compound, C_8H_5Na, which in contact with air takes fire and gradually ... .

Diacetenylphenyl, $C_{16}H_{10}$, is formed when the copper compound of acetenylbenzene is gradually shaken with alcoholic ammonia, the air being allowed to have free access. Large brilliant, colorless and transparent needles, fusing point 97°, insoluble in water, [illegible], soluble in alcohol and ether.

Phenylpropiolic Acid.

$C^9H^6O^2 = C^6H^5.C{:}C.CO.OH.$

Is formed by the simultaneous action of sodium and carbonic anhydride on β-bromcinnamene; by bringing the sodium compound of acetenylbenzene together with carbonic anhydride; and by heating α-bromcinnamic acid with alcoholic potassa.—Long, colorless needles; fusing point, 136–137°; easily sublimable; but slightly soluble in cold water, easily soluble in boiling water, in alcohol and ether. It combines directly with four atoms of bromine, forming an acid that crystallizes only with great difficulty. Nascent hydrogen (sodium-amalgam) converts it into hydrocinnamic acid. Chromic acid oxidizes it, forming benzoic acid. Heated with water to 120°, it is resolved into carbonic anhydride and acetenylbenzene.

In connection with this group a few aromatic compounds will be here described, that have not been so well investigated. They also, for the greater part, differ from the compounds of the first group by containing a smaller number of hydrogen-atoms, the carbon-atoms being combined more closely instead.

Anethol, $C^{10}H^{12}O \left(= C^6H^4 \begin{cases} O.CH^3 \\ C^3H^5 \end{cases} ?\right)$. Is contained in oils of anise, fenchel and tarragon, and separates on cooling these oils.—Colorless, lustrous laminæ; fusing point, 21°; boiling point, 232°. When heated with hydriodic acid, it yields methyl iodide and a resin; heated with potassium hydroxide for a long time at 200–230°, it yields paraoxybenzoic acid and a substance, having the character of phenols, *anol*, $C^9H^{10}O \left(= C^6H^4 \begin{cases} OH \\ C^3H^5 \end{cases} ?\right)$, which forms white, lustrous laminæ; fusing point, 92.5°; not distillable without decomposition. When anethol is oxidized with nitric or chromic acids, it is converted into anisic aldehyde

(p. 324) and anisic acid. Under the influence of concentrated phosphoric acid, concentrated sulphuric acid, antimony chloride, iodine in potassium iodide, etc., it is transformed into isomeric or polymeric compounds. An isomeric liquid compound also appears to be contained in the oils mentioned, together with the solid variety.

Eugenol (Eugenic acid), $C^{10}H^{12}O^{2}$ ($= C^{6}H^{3}\left\{\begin{matrix} OH \\ O.CH^{3} \\ C^{3}H^{5} \end{matrix}\right. ?\Big)$. Is contained in oil of cloves (from the bud blossoms of *Caryophyllus aromaticus*), in oil of pimenta (from the fruit of *Myrtus pimenta*), in the volatile oils of *Persea caryophyllata*, and in the bark of *Canella alba*. Can be obtained therefrom by dissolving in caustic potassa, removing the undissolved portions of the oils by filtering and heating, and reprecipitating with carbonic anhydride.—Colorless liquid, boiling at 253°; of an aromatic odor and sharp taste; insoluble in water; becomes brown when kept. Conducts itself chemically like a phenol. When distilled with hydriodic acid, it yields methyl iodide and a red resin, $C^{9}H^{10}O^{2}$; fused with potassium hydroxide, protocatechuic acid (p. 356) and acetic acid.

Eugetic acid, $C^{11}H^{12}O^{4}$ ($= C^{9}H^{7}\left\{\begin{matrix} O.CH^{3} \\ OH \\ CO.OH \end{matrix}\right. ?\Big)$. Is formed by the simultaneous action of sodium and carbonic anhydride on eugenol.—Thin, colorless prisms; difficultly soluble in cold water, easily in alcohol and ether; its aqueous solution is turned deep blue by iron chloride; fuses at 124°, and is resolved by stronger heat into carbonic anhydride and eugenol.

Sinapic acid, $C^{11}H^{12}O^{5}$. Is formed by boiling the salts of sinapin (see Alkaloids) with barium hydroxide. —Small, colorless prisms, not volatile without decomposition. But slightly soluble in cold water and cold alcohol, more easily soluble in the hot liquids. Bibasic acid.

Ferulic acid, $C^{10}H^{10}O^4$. Is contained in assafœtida. The alcoholic solution of the latter is precipitated with an alcoholic solution of lead acetate and the precipitate, after being repeatedly pressed and washed with alcohol, decomposed with dilute sulphuric acid.—Colorless, long, four-sided needles; almost insoluble in cold water, easily soluble in hot water and in alcohol; fusing point, 153–154°; not sublimable. The aqueous solution gives a yellowish-brown precipitate with iron chloride; monobasic, diatomic acid; yields protocatechuic and acetic acids when fused with potassa.

Hemipinic acid, $C^{10}H^{10}O^6 \left(C^6H^2 \left\{ \begin{matrix} (O.CH^3)^2 \\ (CO.OH)^2 \end{matrix} \right. ?\right)$. Is produced, together with cotarnin, meconin, and opianic acid, by the oxidation of narcotin (see Alkaloids) with dilute nitric acid or black oxide of manganese and sulphuric acid.—Large, four-sided prisms with varying quantities of water of crystallization; difficultly soluble in cold water, more easily in alcohol; fuses at 180°, and sublimes without decomposition. When heated with hydriodic or hydrochloric acids, it yields methyl iodide or chloride, carbonic anhydride, and two isomeric acids, *opinic acid* and *isopinic acid*, $C^{14}H^{10}O^8 + 3H^2O$.

Opianic acid, $C^{10}H^{10}O^5$. Is formed from narcotin together with the preceding compound.—Colorless, fine prisms, but slightly soluble in cold water; fuses at 140°; conducts itself in most reactions like an aldehyde; yields meconin and hemipinic acid when heated with potassa-ley; and when oxidized is completely converted into hemipinic acid.

Meconin, $C^{10}H^{10}O^4$. Is contained in opium; is formed together with cotarnin (see Alkaloids) from narcotin by heating with water at 100°; and from opianic acid by the action of nascent hydrogen or by treatment with potassa-ley.—Lustrous, colorless crystals, difficultly soluble in cold water, more easily soluble in hot water; fuses at 110°; combines with acids by heating, water being eliminated and bodies, like ethers, being formed.

Piperic acid, $C^{12}H^{10}O^4$. The potassium salt is formed together with piperidin by boiling piperin (see Alkaloids) with alcoholic potassa. Hydrochloric acid throws down the free acid from the solution of the salt.—Crystallizes from alcohol in long, capillary needles; almost insoluble in water, but slightly soluble in cold alcohol, more easily in hot alcohol; fuses at 216–217° and sublimes, at the same time undergoing partial decomposition. Monobasic acid; yields protocatechuic (p. 356), acetic, and oxalic acids when fused with potassium hydroxide. Potassium bichromate and dilute sulphuric acid destroy it completely with the aid of gentle heat, forming carbonic anhydride and water. Potassium hypermanganate converts it into piperonal (p. 324).

Hydropiperic acid, $C^{12}H^{12}O^4$. Is formed from piperic acid by direct combination with nascent hydrogen.—Long, colorless, fine needles, fusing at 70–71°.

FOURTH GROUP (INDIGO-GROUP).

Indigo-blue, C^8H^5NO or $C^{16}H^{10}N^2O^2$. Indigo is obtained from the various species of *Indigofera* of East India and South America, from *Isatis tinctoria*, *Nerium tinctorium*, *Polygonum tinctorium*, and other plants. The blossoming plants are cut off and allowed to stand under water from twelve to fifteen hours. The liquid is then drawn off, and, by means of beating with wooden shovels, etc., brought in contact with the air as much as possible. The indigo, which separates during this process, is separated from the brown liquid, boiled with water, and dried.—Indigo is not contained, ready formed, in the plants. From what compound and by what decomposition it is formed during the process of preparation, is not positively known.

A substance yielding indigo (probably indol, described below), is sometimes contained in human urine and blood. The conversion of this substance into indigo-blue is the cause of the lilac or blue color frequently noticed in urine on the rapid addition of sulphuric acid.

In order to prepare indigo-blue in a pure condition from commercial indigo, which often contains foreign substances mixed with it in large quantities, the latter is finely powdered; mixed with calcium hydroxide and iron vitriol; the mixture put in a flask, that can be closed; this filled completely with boiling hot water and hermetically closed.* In this operation the real indigo-blue, by the action of ferrous hydroxide which becomes ferric hydroxide, takes up hydrogen and is converted into indigo-white, which dissolves in combination with lime (indigo vat of dyers). After the transformation is completed and this solution has turned a clear, deep yellow, it is allowed to pour through a siphon into a vessel containing very dilute hydrochloric acid, the indigo-blue, in consequence of the access of air, being regenerated and separating in the form of a deep-blue powder, after violent shaking with air. This powder is then filtered off, washed out and dried.

Or indigo is mixed with an equal weight of grape-sugar; hot alcohol and 1½ part of the most concentrated soda-ley poured upon it in a large flask; the flask then completely filled with hot alcohol, and allowed to stand for some time. The clear liquid, being thereupon poured off, gradually deposits indigo-blue in crystalline form when allowed to remain in contact with air.

Indigo is obtained artificially in very small quantity when liquid nitro-acetophenone (p. 336) is converted into a solid resinous mass by being heated alone,

* Three parts indigo, the hydrate of 6 parts lime, 4 parts iron vitriol, and about 450 parts water.

and this mass then heated with zinc-dust and soda-lime.

Properties. Deep blue, approaching purple; pressure gives it a copper color and a half metallic lustre. Tasteless, inodorous; completely insoluble in water, alcohol, ether, dilute sulphuric acid, hydrochloric acid, and alkalies; soluble in anilin. At about 300°, it is transformed into a purple vapor, which condenses in the form of lustrous, deep copper-colored prisms; this property can also be made use of for the purpose of preparing pure indigo, though it involves loss in consequence of decomposition and carbonization. Distilled with potassium hydroxide, it is resolved into anilin and carbonic acid.—When boiled for a long time with potassa-ley and finely-divided black oxide of manganese, it is converted into anthranilic acid (p. 330).

Indigo-white, $C^{16}H^{12}N^{2}O^{2}$. Is produced from indigo-blue when this comes in contact with nascent hydrogen or with any other reducing agents in the presence of a base. It is contained in the solutions described above, which are not colored blue, and can be obtained from them in an isolated condition when they are allowed to flow directly into boiled, dilute hydrochloric acid by means of a siphon, care being taken that they do not come in contact with air. The indigo-white is thus separated in the form of white, glittering flocks.

After being filtered off and washed with water, that has been boiled for a long time, it must be dried either in a vacuum or in a current of hydrogen.—White, fine, crystalline powder; inodorous and tasteless; insoluble in water. In contact with air, particularly when in a moist condition or in water containing air, it is soon reconverted into indigo-blue. It is a weak acid, and is readily dissolved by alkalies, forming yellow solutions. These solutions, as well as its salts formed by double decomposition, are exceedingly unstable, take up oxygen rapidly from the air, and deposit indigo-blue.

Sulphindigotic acid (Sulphocœrulic acid), $C^8H^4NO.SO^2.OH$, is formed when 1 part indigo is digested for three days, at 30–40°, with 15 parts concentrated sulphuric acid. Pure wool is then placed in the diluted solution. Upon this the acid formed is deposited, there remaining in the liquid only the excess of free sulphuric acid. The wool, which is dyed blue, is now well washed with water; and the acid extracted by means of ammonium carbonate; the solution evaporated at as low a temperature as possible; and the residue washed with alcohol for the purpose of removing another acid, *hyposulphindigotic acid*, which is formed, together with sulphindigotic acid, particularly when indigo is dissolved in fuming sulphuric acid; thereupon the substance is dissolved in water; precipitated with lead acetate; and the lead salt, suspended in water, decomposed by sulphuretted hydrogen. On evaporating the filtered solution at a but slightly elevated temperature, the acid remains behind in the form of a blue, amorphous mass, easily soluble in water and alcohol.

Its salts are amorphous. The *potassium salt*, $C^8H^4NO.SO^3K$, and the *sodium salt* occur in commerce under the name of *indigo-carmine*, and are prepared on the large scale by adding potassium acetate, or Glaubers salt, to a diluted solution of indigo-blue in sulphuric acid; washing out the blue precipitate with solutions of the salts employed; and pressing. They form copper-colored masses, which appear blue in a finely-divided condition, and dissolve with blue color in pure water.

If, in the preparation of sulphindigotic acid, less (only 8 parts) sulphuric acid is employed, on subsequently diluting with water, a blue precipitate is thrown down, consisting of

Sulphophœnicic acid (Sulphopurpuric acid), $C^{16}H^9N^2O^2.SO^2.OH$, which dissolves in pure water, free of acids, and forms purplish-red salts with bases; these salts are soluble in water, the solutions having a blue color.

Isatin, $C^8H^5NO^2$, is formed by the oxidation of indigo-blue. Finely powdered indigo is heated with water to boiling, and to the liquid concentrated nitric acid is added, until the blue color has completely disappeared. By repeatedly boiling the mass with water, the isatin formed is dissolved, and, on cooling, it gradually crystallizes out. It may now be purified by dissolving in potassa, precipitating with hydrochloric acid and recrystallizing.

Yellowish-red prisms, of a strong lustre; soluble in alcohol, forming a brown-red solution; in cold potassa-ley forming a violet solution; fusible; partially sublimable without decomposition. Combines with the alkaline bisulphites, forming crystallizing compounds.

When distilled with concentrated potassa-ley, anilin passes over, hydrogen being at the same time set free. Suspended in water, and treated with nitrous acid, it is converted into nitrosalicylic acid (p. 346), a gas being evolved at the same time; treated with ammonia in an alcoholic solution, it yields a large number of crystallizing compounds, the composition of which shows that they have resulted from isatin by the addition of ammonia and the elimination of water.

Chlorisatin, $C^8H^4ClNO^2$. Is produced by the action of chlorine gas on a boiling-hot solution of isatin in water, it being thrown down under these circumstances as a yellow, flocky precipitate; further, together with secondary products, by conducting chlorine into pure indigo mixed with water. From the crude product thus obtained, the chlorisatin is extracted by means of boiling water, and separated by means of crystallization from *dichlorisatin*, $C^8H^3Cl^2NO^2$, which is formed at the same time, and is more easily soluble in water. —Orange-yellow, transparent, four-sided prisms, of bitter taste; inodorous; scarcely soluble in cold water, soluble in alcohol; partially sublimable.—Towards bromine it conducts itself in the same way. The substitution-products of isatin are decomposed by fusing caustic potassa like isatin, substitution-products of anilin being formed.

Isatosulphuric acid, $C^8H^4NO^2.SO^2.OH$, is formed by the action of potassium bichromate and sulphuric acid on sulphindigotates (indigo-carmine).—Difficultly crystallizable, very easily soluble acid; monobasic. Its *barium salt*, $(C^8H^4NSO^5)^2Ba + 4H^2O$, forms brass-red, strongly lustrous, crystalline scales, but slightly soluble in cold water.

Trioxindol (Isatic acid), $C^8H^7NO^3$. The violet solution of isatin in potassa-ley becomes yellow when boiled, and then contains potassium isatate. The free acid is exceedingly unstable; when the attempt is made to set it free by means of another acid, it breaks up into isatin and water.

The substitution-products of isatin conduct themselves towards caustic potassa in like manner. They yield chlorinated or brominated isatic acids, which are likewise exceedingly unstable in a free condition.

Dioxindol (Hydrindic acid), $C^8H^7NO^2$, is formed by the action of sodium-amalgam on isatin, to which is added water, by reduction of the isatic acid, which is at first formed.—Transparent, rhombic prisms; easily soluble in water and alcohol; fuses at 180°, and decomposes at 195°, anilin being formed. The aqueous solution in contact with air turns red, oxygen being taken up and isatin formed. Combines with acids and bases, forming salts. Treated with chlorine or bromine it yields crystallizing substitution-products. Treated with nitrous acid in an alcoholic solution, it is at first converted into a crystalline substance, *nitrosodioxindol*, $C^8H^6(NO)NO^2$, insoluble in water; fusing at 300–310°; further action converts it into ethyl benzoate and other products. It yields oil of bitter almonds when gently warmed with nitric acid or silver oxide.

Oxindol, C^8H^7NO. Is formed by further reduction of dioxindol with tin and hydrochloric acid or with sodium-amalgam in a dilute solution, kept constantly acid.—Long, colorless needles or feathery groups. Difficultly soluble in cold water, easily soluble in hot water and in alcohol; fuses at 120°; and in small quantities, it can be distilled without undergoing decomposition. When its aqueous solution is evaporated in contact with air, it becomes partially oxidized again, forming dioxindol. Like dioxindol, it yields crystallizing salts with acids, as well as bases. Nitrous acid transforms it in very dilute aqueous solutions into *nitroso-oxindol*, $C^8H^6(NO)NO$, a substance that crystallizes in long, golden needles, difficultly soluble in water.

Indol, C^8H^7N. Is formed when the vapors of oxindol are conducted over heated zinc-dust; or when indigo-blue is boiled with zinc and hydrochloric acid until it is converted into a brownish-yellow powder, and this then distilled with an excess of zinc-dust. It is also formed in small quantity when nitrocinnamic acid is fused with potassium hydroxide with an addition of iron filings.—Large, colorless laminæ similar to benzoic acid; fusing point, 52°; not distillable alone, but very well with water vapor. Very weak base. With hydrochloric acid, it forms a difficultly soluble salt, which, when boiled with water, yields free indol.

Isatyde, $C^{16}H^{12}N^2O^4$, is formed by heating isatin with dilute sulphuric acid; or when its warm saturated alcoholic solution is mixed with ammonium sulphydrate in a closed flask, and allowed to stand for some time, it being deposited gradually in the form of crystalline scales. It bears the same relation to isatin, as indigo-white bears to indigo-blue.—Colorless, fine crystalline inodorous and tasteless substance, insoluble in water, but slightly in alcohol.

Sulphisatyde, $C^{16}H^{12}N^2O^2S^2$. When sulphuretted hydrogen is conducted into an alcoholic solution of isatin, a mixture of sulphur and isatyde is thrown down, and the solution contains sulphisatyde, which is precipitated when the solution is allowed to drop in water.—Grayish-yellow, pulverous substance; becomes soft in hot water, soluble in alcohol, not crystallizable.

Indin, $C^{16}H^{10}N^2O^2$, isomeric with indigo-blue; is formed when sulphisatyde is well mixed with alcohol, and a solution of potassa gradually added; and the mass, when it has become red, washed out with water.—Beautiful rose-colored, crystalline powder; insoluble in water, but slightly soluble in alcohol. It dissolves, when warmed with alcoholic potassa, and, on cooling, *indin-potassium*, $C^{16}H^9N^2O^2K$, is deposited in small black crystals.

III. NAPHTHALENE-DERIVATIVES.

The bodies of this group are derived from naphthalene $C^{10}H^{8}$, in the same way as the aromatic compounds are derived from benzene. Naphthalene is constituted very similarly to benzene; it consists of two benzene-groups, which are so united that they have two carbon atoms in common:—

$$\begin{array}{c} CH{:}CH.C.CH{:}CH \\ \dot{C}H{:}CH.\ddot{C}.CH{:}\dot{C}H \end{array}$$

A. Hydrocarbons, $C^{n}H^{2n-12}$.

1. *Naphthalene.*
$C^{10}H^{8}$.

Formation. By the dry distillation of a great many organic substances at a high temperature, particularly when the distillation-products are conducted through a red-hot tube. It is hence contained in coal-tar and wood-tar. It is also formed from alcohol, acetic acid, and a number of other substances, when their vapors are passed through red-hot tubes.

Preparation. Most advantageously from coal-tar oil by partial distillation and strong cooling of the distillate between 180 and 220°. The crude naphthalene thus separated is purified by recrystallization from hot alcohol, or, better, by means of sublimation.

Properties. Large, lustrous, colorless crystalline laminæ of peculiar odor and burning taste. Fuses at 80°; boils at 218°, and sublimes at a lower temperature; insoluble in water, but slightly in cold alcohol, easily in hot alcohol and in ether. Distils over readily with water. Burns with a luminous sooty flame. Combines with picric acid, forming a compound, $C^{10}H^8 + C^6H^3(NO^2)^3O$, which crystallizes in stellate, yellow needles. When oxidized with nitric acid, it yields phtalic acid (p. 362). When heated with phosphonium iodide to 170–190°, it yields a liquid hydrocarbon, $C^{10}H^{12}$; boiling point, 201°.

With chlorine it forms products of addition and substitution.

Naphthalene dichloride, $C^{10}H^8.Cl^2$, pale yellow oil, heavier than water, and insoluble in it.

Naphthalene tetrachloride, $C^{10}H^8.Cl^4$, transparent rhombohedral crystals; fusing point, 182°; difficultly soluble in alcohol and ether.

Chlornaphthalene tetrachloride, $C^{10}H^7Cl.Cl^4$, klinorhombic prisms; fusing point, 128–130°.

Dichlornaphthalene tetrachloride, $C^{10}H^6Cl^2.Cl^4$, klinorhombic prisms; fusing point, 172°.

When these chlorine compounds are boiled with alcoholic potassa, hydrochloric acid is given off and chlorine-substitution-products of naphthalene are formed. These, when further subjected to the action of chlorine, again form addition-products and substitution-products, containing more chlorine.

Monochlornaphthalene, $C^{10}H^7Cl$. Colorless liquid; boiling point, 250–252°.

α-Dichlornapthalene, $C^{10}H^6Cl^2$. Crystalline mass; fusing point, 35–36°; boiling point, 282°.

β-Dichlornaphthalene, $C^{10}H^6Cl^2$. Colorless prisms; fusing point, 68°; boiling point, 281–283°.

Trichlornaphthalene, $C^{10}H^5Cl^3$. Brittle prisms; fusing point, 81°.

Tetrachlornaphthalene, $C^{10}H^4Cl^4$. Colorless needles; fusing point, 130°.

Enneachlordinaphthalene, $C^{20}H^7Cl^9$. The end-product of the action of chlorine on heated chlorinated naphthalene.—White, delicate needles; fusing point, 156–158°.

Pentachlornaphthalene, $C^{10}H^3Cl^5$. Is formed by the action of phosphorus chloride on dichlornaphthoquinone and chloroxynaphthalenic acid.—Colorless needles; fusing point, 168.5°.

Perchlornaphthalene, $C^{10}Cl^8$. Prisms; fusing point, 135°.

Naphthalene yields substitution-products with bromine, but does not combine directly with it.—*Monobromnaphthalene*, $C^{10}H^7Br$. Colorless liquid; boiling point, 277°.—α-*Dibromnaphthalene*, $C^{10}H^6Br^2$. Long needles of a silky lustre; fusing point, 81°.—β-*Dibromnaphthalene*, $C^{10}H^6Br^2$, is formed together with the α-compound when bromine acts upon α-sulphonaphthalic acid. Needles; fusing point, 126–127°.—*Tribromnaphthalene*, $C^{10}H^5Br^3$. Colorless needles; fusing point, 75°.—*Tetrabromnaphthalene*, $C^{10}H^4Br^4$. Colorless prisms, but slightly soluble in alcohol.—*Pentabromnaphthalene*, $C^{10}H^3Br^5$. Colorless granular crystals, insoluble in alcohol.

α-**Cyannaphthalene,** $C^{10}H^7.CN$, is formed by the distillation of a mixture of potassium α-sulphonaphthalate with potassium cyanide. Is also formed when naphthylamine oxalate is distilled, and the distillate, which contains a great deal of naphthylformamide, heated with concentrated hydrochloric acid.—Colorless, broad needles; insoluble in water, easily soluble in alcohol. Fuses at 37.5°; has a strong tendency to remain liquid, and boils without undergoing decomposition at 297–298°.—β-*Cyannaphthalene*, $C^{10}H^7.CN$. Is obtained in the same way from potassium β-sulphonaphthalate.—Colorless laminæ; fusing point, 66.5°; boiling point, 304–305°.

Nitronaphthalene, $C^{10}H^7(NO^2)$, is formed by the action of concentrated nitric acid on naphthalene; slowly at the ordinary temperature, rapidly by boiling. —Crystallizes from alcohol in sulphur-colored prisms; fusing point, 58.5°, and sublimes when carefully heated; insoluble in water, easily soluble in alcohol and ether.

α-**Dinitronaphthalene** and *β-Dinitronaphthalene*, $C^{10}H^6(NO^2)^2$, are produced simultaneously when the preceding compound, or naphthalene, is boiled with nitric acid until no oily body (melted nitronaphthalene) can be detected on the surface of the liquid. The two compounds may be separated by boiling with alcohol, in which the α-compound is more easily soluble; and crystallizing from chloroform. α-Dinitronaphthalene crystallizes in four- or six-sided rhombic plates, that detonate when heated; fusing point, 170°; β-dinitronaphthalene crystallizes in colorless, sublimable prisms, that fuse at 214°.—If the boiling with nitric acid is continued for several days *trinitronaphthalene*, $C^{10}H^5(NO^2)^3$, is formed; small, monoclinate prisms, fusing at 214°. When this is heated for a long time with fuming nitric acid in sealed tubes at 100°, it is converted into *tetranitronaphthalene*, $C^{10}H^4(NO^2)^4$, which crystallizes in fine needles, resembling asbestos; fusing point, 200°.

Naphthylamine (Naphthalidine), $C^{10}H^7.NH^2$, is produced from nitronaphthalene in the same way that anilin is produced from nitrobenzene (*cf.* p. 258).—Fine, colorless prisms, of an unpleasant odor; almost insoluble in water, easily soluble in alcohol; fuses at 50°; sublimes easily, and boils at 300°. Turns gradually red in contact with the air. Combines with acids, forming crystallizing and, for the greater part, easily soluble salts. Oxidizing agents, iron chloride, silver nitrate, chromic acid, tin chloride produce a blue precipitate in the solutions of these salts, which is rapidly converted into a purple-red, amorphous powder, *oxy-naphthylamine*, $C^{10}H^{9}NO$.

When nitrous acid is allowed to act on naphthylamine, diazocompounds are formed as in the case of anilin.

Sulphonaphthalic acids, $C^{10}H^{7}.SO^{2}.OH$. When naphthalene is carefully heated with sulphuric acid, two isomeric sulphonaphthalic acids are formed, which may be separated by partial crystallization of the lead or barium salts. Both salts of α-sulphonaphthalic acid are much more easily soluble in water and alcohol than those of the β-acid. The α-acid, when heated, is converted into the β-acid, and hence, when naphthalene is treated with sulphuric acid at an elevated temperature (160°), the product consists almost entirely of β-sulphonaphthalic acid.

Barium α-sulphonaphthalate, $(C^{10}H^{7}.SO^{3})^{2}Ba + H^{2}O$. Colorless laminæ; soluble in 87 parts water and 350 parts alcohol (of 85 per cent.) at 10°. The *lead salt*, $(C^{10}H^{7}SO^{3})^{2}Pb + 3H^{2}O$, forms lustrous, colorless laminæ; soluble in 27 parts water and 11 parts alcohol at 10°.

Barium β-sulphonaphthalate, $(C^{10}H^{7}.SO^{3})^{2}Ba + H^{2}O$. Colorless laminæ; soluble in 290 parts water and 1950 parts alcohol at 10°.—The *lead salt* crystallizes in small, hard scales, with varying amounts of water of crystallization; soluble in 115 parts water and 305 parts alcohol at 10°.

Naphthalene sulphochlorides, $C^{10}H^{7}.SO^{2}Cl$, are obtained by gently heating potassium α- and β-sulphonaphthalate with phosphorus chloride. The *α-chloride* forms lustrous laminæ; easily soluble in ether; fusing at 66°; the *β-chloride* is more difficultly soluble in ether, and fuses at 76°.

Disulphonaphthalic acid, $C^{10}H^{6}(SO^{2}.OH)^{2}$, is formed by continued heating of naphthalene with an excess of sulphuric acid.—Bibasic acid. The *barium salt*, $C^{10}H^{6}S^{2}O^{6}Ba$, is much less easily soluble in water, and particularly in alcohol, than the sulphonaphthalates.

Mercurynaphthyl, $(C^{10}H^{7})^{2}Hg$. Is formed by continued boiling of a solution of monobromnaphthalene in benzene with sodium-amalgam.—Small, colorless, rhombic, columnar crystals. Insoluble in water, difficultly soluble in boiling alcohol, easily in carbon bisulphide and in chloroform. Fuses at 243°; not volatile without decomposition; combines directly with iodine; and, when heated with hydriodic, hydrobromic, or hydrochloric acids, it yields naphthalene and mercury iodide, bromide, or chloride. Conducts itself exactly like mercuryphenyl (p. 272).

Dinaphthyl, $C^{20}H^{14} = (C^{10}H^{7})^{2}$, is formed by the decomposition of monobromnaphthalene with sodium; and by heating naphthalene with black oxide of manganese and sulphuric acid.—Colorless laminæ, of a mother-of-pearl lustre. But slightly soluble in cold alcohol, easily soluble in ether; fuses at 154°; and is sublimable without decomposition. When further oxidized with black oxide of manganese and sulphuric acid, it is converted into phtalic acid (p. 362).

2. *Methylnaphthalene.*
$C^{11}H^{10} = C^{10}H^{7}.CH^{3}$.

Is obtained by the action of sodium on a mixture of monobromnaphthalene and methyl iodide, diluted with ether.—Colorless, clear, somewhat viscid liquid; specific gravity, 1.0287; boiling point, 231–232°; does not congeal at —18°.

3. *Ethylnaphthalene*, $C^{12}H^{12} = C^{10}H^{7}.CH^{2}.CH^{3}$. Is formed like methylnaphthalene.—Colorless, clear liquid; specific gravity, 1.0184; boiling point, 251–252°; still liquid at —14°.

B. Phenols.

1. *Naphthol* (*α-Naphthol*).

$C^{10}H^8O = C^{10}H^7.OH$.

Is formed by heating potassium α-sulphonaphthalate with potassium hydroxide.—Colorless, monoclinate prisms; fusing point, 94°; boiling point, 270–280°; almost insoluble in cold water, somewhat soluble in hot water, easily in alcohol and ether. Towards alkalies it conducts itself like phenol (p. 290).

Naphthol-ethylether, $C^{10}H^7.O.C^2H^5$. By heating naphthol-potassium with ethyl iodide.—Colorless liquid; boiling point, 272°; does not congeal at —5°.

Naphthol-acetate, $C^{10}H^7.O.C^2H^3O$. By the action of acetyl chloride on naphthol.—Yellowish liquid, insoluble in water.

Nitronaphthol, $C^{10}H^6(NO^2).OH$. Is formed, when 1 part nitronaphthalene is heated in a current of air for a long time at 140°, intimately mixed with 1 part potassium hydroxide and 2 parts calcium hydroxide, and the aqueous extract from the mass decomposed with hydrochloric acid.—Bright-yellow, light mass; crystallizing from acetic acid or acetone in golden-yellow prisms; fusing point, 151–152°.

Dinitronaphthol, $C^{10}H^5(NO^2)^2.OH$. Cannot be prepared directly from naphthol. Is, however, readily obtained by pouring nitric acid (specific gravity, 1.35) upon naphthylamine; and by gently heating a solution of sulphonaphthalic acid, to which is added nitric acid. Is also formed by boiling diazonaphthalene hydrochlorate (from naphthylamine hydrochlorate with nitrous acid) with nitric acid.—Lustrous sulphur-colored crystals; fusing point, 138°; almost insoluble in boiling water, difficultly soluble in alcohol and ether, more easily in chloroform. With bases it yields salts; and liberates carbonic acid from its salts. The sodium and calcium salts are excellent yellow dyes (naphthalene yellow).

Diamidonaphthol, $C^{10}H^5(NH^2)^2.OH$. Cannot be isolated and obtained in a free condition. Its *double salt with stannous chloride*, $C^{10}H^5(NH^2)^2.OH + 2HCl + SnCl^2 + 2H^2O$, is obtained by heating dinitronaphthol with tin and hydrochloric acid. It crystallizes in large, lustrous, monoclinate prisms. When its solution is precipitated with sulphuretted hydrogen, a solution of diamidonaphthol hydrochlorate is obtained, which, in contact with air, and rapidly on the addition of iron chloride, yields *diimidonaphthol hydrochlorate* $C^{10}H^6(OH)\left\{\begin{matrix}HN\\HN\end{matrix}\right> + HCl$. This salt crystallizes in large columnar or tabular crystals of a metallic lustre, which in transmitted light are dark red; in reflected light, green. With ammonia it yields *diimidonaphthol*, $C^{10}H^5(OH)(HN)^2$, a yellow crystalline body, almost insoluble in water.

Naphtholsulphuric acid, $C^{10}H^6\left\{\begin{matrix}OH\\SO^2.OH.\end{matrix}\right.$ By heating naphthol with double its weight of concentrated sulphuric acid. The free acid, separated from the lead salt, forms long, colorless, deliquescent needles; fusing point, 101°; its solution is colored deep blue on the addition of iron chloride; if heated it becomes green.

Naphthyl sulphydrate, $C^{10}H^7.SH$. Is formed by the action of zinc and dilute sulphuric acid on α-naphthyl sulphochloride.—Colorless liquid, insoluble in water; boiling point, 285°; volatile with water vapor; yields salts with bases.

Naphthyl disulphide, $(C^{10}H^7)^2S^2$. Is formed by the spontaneous evaporation of an ammoniacal alcoholic solution of the sulphydrate in contact with air.—Yellowish, transparent crystals; fusing point, 85°.

2. *Isonaphthol (β-Naphthol).*
$C^{10}H^8O = C^{10}H^7.OH.$

Is obtained, like naphthol, from potassium β-sulpho-naphthalate.—Small, colorless, rhombic plates; fusing point, 122°; boiling point, 285–290°; easily sublimable. Difficultly soluble in boiling water; easily in alcohol and ether.

The derivatives of isonaphthol are prepared like those of naphthol.

Isonaphthol-ethylether, $C^{10}H^7.O.C^2H^5$. Colorless, crystalline mass; fusing point, 33°.

Isonaphthol-acetate, $C^{10}H^7.O.C^2H^3O$. Small, colorless needles; fusing point, 60°.

Dinitro-isonaphthol, $C^{10}H^5(NO^2)^2.OH$. Is obtained by warming an alcoholic solution of isonaphthol with dilute nitric acid.—Lustrous, bright-yellow needles; fusing point, 195°.

Isonaphtholsulphuric acid, $C^{10}H^6 \begin{cases} OH \\ SO^2.OH. \end{cases}$ Small, colorless laminated crystals; fusing point, 125°; not deliquescent, but easily soluble in water and alcohol. The aqueous solution turns slightly green on the addition of iron chloride, and, when heated with it, it deposits brown flocks.

Isonaphthyl sulphydrate, $C^{10}H^7.SH$. Small, lustrous scales; fusing point, 136°; not volatile with water vapor; insoluble in water; easily soluble in ether and alcohol.

3. *Dioxynaphthalene.*
$C^{10}H^8O^2 = C^{10}H^6(OH)^2.$

Is obtained by melting potassium disulphonaphthalate with potassium hydroxide.—Colorless needles, which become violet in the air; difficultly soluble in water, easily in alcohol and ether; sublimable; in an alkaline solution it absorbs oxygen rapidly from the air, and turns black.

4. *Trioxynaphthalene.*

$$C^{10}H^{8}O^{3} = C^{10}H^{5}(OH)^{3}.$$

Is formed by the action of tin and hydrochloric acid on oxynaphthoquinone, and after the solution has been freed from tin by sulphuretted hydrogen, it can be extracted by means of ether.—Yellow needles; soluble in water, alcohol, and ether; the solutions, which are at first colorless, turn yellow and brown in the air. Is a strong reducing agent, especially in alkaline solution.

C. Quinones.

Naphthoquinone, $C^{10}H^{6}\left\{\begin{matrix}O\\O\end{matrix}\right>$, is as yet not known. Only substitution-products and other derivatives of it have been discovered.

Dichlornaphthoquinone (Chloroxynaphthalene chloride), $C^{10}H^{4}Cl^{2}\left\{\begin{matrix}O\\O\end{matrix}\right>$. Is produced by the action of nitric acid on chlornaphthalene chloride. Can be most easily prepared by treating naphthol or commercial naphthalene yellow (see Dinitronaphthol p. 397) with hydrochloric acid and potassium chlorate, or by the addition of chromium oxichloride to a solution of naphthalene in concentrated acetic acid.—Golden-yellow needles; fusing point, 189°; insoluble in water, but slightly in cold alcohol and in ether, more readily in hot alcohol; easily sublimable. Hot concentrated nitric acid converts it into phtalic acid. Sulphurous acid and hydriodic acid convert it into *dichlordioxynaphthalene*, $C^{10}H^{4}Cl^{2}(OH)^{2}$, which crystallizes in reddish-colored needles, that fuse at 135–140°, and are reconverted into dichlornaphthoquinone by iron chloride.—Heated with two molecules phosphorus chloride at 180–200°, dichlornaphthoquinone is transformed into pentachlornaphthalene (p. 393).

Oxynaphthoquinone (Naphthalic acid), $C^{10}H^{5}(OH)\left\{\begin{matrix}O\\O\end{matrix}\right>$. Is most readily obtained by heating diimidonaphthol hydrochlorate (p. 398) with dilute hydrochloric or sulphuric acid at 120°.—Bright yellow, electric powder, or yellow needles; sublimes partially when carefully heated, condensing in reddish-yellow needles. Almost insoluble in cold water, somewhat soluble in boiling water, easily in alcohol and ether. It combines with nascent hydrogen, forming trioxynaphthalene.—It conducts itself as a moderately strong monobasic acid towards bases, and liberates carbonic acid from its salts. The *alkaline salts* are blood-red and easily soluble in water.

Chloroxynaphthoquinone (Chloroxynaphthalic acid), $C^{10}H^5ClO^3 = C^{10}H^4Cl(OH)\left\{\begin{matrix}O\\O\end{matrix}\right>$. The *potassium salt*, $C^{10}H^4Cl(OK)O^2$, is obtained, when dichlornaphthoquinone is placed under alcohol, and concentrated potassa-ley then added. It forms cherry-colored needles, which are but slightly soluble in cold water, easily soluble in hot water; hydrochloric acid throws down the free acid from this solution.—Straw-colored, crystalline powder; fuses somewhat above 200°, and sublimes in needles; but slightly soluble in cold water, moderately in boiling, more easily in alcohol and ether. Strong monobasic acid; its salts, when heated, give a sublimate of phtalic anhydride. Heated with phosphorus chloride, pentachlornaphthalene is formed.

Dioxynaphthoquinone (Naphthazarin), $C^{10}H^6O^4 = C^{10}H^4(OH)^2\left\{\begin{matrix}O\\O\end{matrix}\right>$. Is obtained by simultaneously adding β-dinitronaphthalene and zinc in small quantities to concentrated sulphuric acid heated to 200°. Subsequently the mass is diluted with water, boiled, filtered boiling hot and the gelatinous mass, that separates on cooling, purified, when dried, by means of sublimation.—Long needles with a beautiful green metallic lustre. But slightly soluble in boiling water, more easily in alcohol, the solution having a red color. It dissolves in ammonia forming a sky-blue solution, which turns reddish-violet on standing. Its solution gives beautiful violet precipitates with baryta and lime-water. Excellent dye, very similar to alizarin.

D. Acids.

1. *Naphthoic Acid* (*Menaphthoxylic Acid*).

$C^{11}H^8O^2 = C^{10}H^7.CO.OH$.

Formation. From α-cyannaphthalene (p. 393), by boiling with alcoholic potassa, and decomposing the potassium salt thus formed with hydrochloric acid. Its ether is also formed by the action of sodium-amalgam on a mixture of monobromnaphthalene and ethyl chlorcarbonate; its potassium salt, by fusing a mixture of potassium α-sulphonaphthalate with sodium formate.

Properties. Colorless crystalline needles; fusing point, 160°; difficultly soluble in boiling water, easily soluble in boiling alcohol. Heated with baryta, it is resolved into naphthalene and carbonic acid.

Barium naphthoate, $(C^{11}H^7O^2)^2Ba + 4H^2O$, and **Calcium naphthoate** $(C^{11}H^7O^2)^2Ca + 2H^2O$, are difficultly soluble in water (the calcium salt in 93 parts at 15°); and crystallize in colorless needles.

Ethyl naphthoate, $C^{10}H^7.CO.O.C^2H^5$. Liquid of an aromatic odor; boiling point, 309°.

Naphthoyl chloride, $C^{10}H^7.COCl$. By the action of phosphorus chloride on naphthoic acid.—Liquid; boils at 297.5°; congeals at a low temperature.

Naphthoylamide, $C^{10}H^7.CO.NH^2$. Is obtained by the action of ammonia on the chloride; and by dissolving α-cyannaphthalene in alcoholic soda-ley and precipitating with water.—Colorless needles; fusing point, 204°; insoluble in water, difficultly soluble in alcohol; sublimable.

2. *Isonaphthoic Acid* (β-*Naphthoic Acid*).

$C^{11}H^8O^2 = C^{10}H^7.CO.OH$.

Is obtained from β-cyannaphthalene like naphthoic

acid.—Long, colorless needles; fusing point, 182°; boils above 300° without undergoing decomposition. But slightly soluble in boiling water, easily soluble in alcohol and ether. Heated with barium hydroxide, it is resolved like naphthoic acid into carbonic acid and naphthalene.

Barium isonaphthoate $(C^{11}H^{7}O^{2})^{2}Ba + 4H^{2}O$, and **Calcium isonaphthoate** $(C^{11}H^{7}O^{2})^{2}Ca + 3H^{2}O$, crystallize in needles and are insoluble in cold water (in 1400–1800 parts at 15°).

3. *Oxynaphthoic Acid* (*Carbonaphtholic Acid*).

$$C^{11}H^{8}O^{3} = C^{10}H^{6}\left\{\begin{matrix} OH \\ CO.OH. \end{matrix}\right.$$

The sodium salt is produced by the simultaneous action of sodium and carbonic anhydride on α-naphthol. Hydrochloric acid precipitates the acid from the solution of this salt.—Small, stellate, colorless needles; fusing point, 185–186°; but slightly soluble in water even at boiling temperature, easily soluble in alcohol and ether. Its salts are, for the greater part, difficultly soluble in water. These solutions are turned deep blue by iron chloride.

β-*Naphthol* (p. 399), when treated in the same way, yields with difficulty an isomeric oxyacid of similar properties.

IV. ANTHRACENE-DERIVATIVES.

ANTHRACENE, the substance from which the bodies of this group are derived, has a chemical constitution similar to that of benzene and naphthalene. It bears the same relation to naphthalene that the latter bears to benzene. It may be considered as a combination of three benzene-rings, of which each one has two carbon atoms in common with one or both the others:—

CH:CH.C.CH:C.CH:CH
ĊH:CH.C̈.CH:Ċ.CH:ĊH

Anthracene.
$C^{14}H^{10}$.

Formation. By dry distillation of anthracite coal; hence contained in coal-tar. By heating benzyl chloride (p. 274) with water at 190°, together with liquid ditolyl (p. 282) and benzylic ether (p. 313).

Preparation. From those portions of coal-tar, that boil at high temperatures, by means of repeated distillations, pressing, and recrystallizing from benzene. To obtain it perfectly pure and colorless, it must be sublimed at as low a temperature as possible, best by heating it until it begins to boil, and then blowing a strong current of air over it by means of a pair of bellows. Or the solution in hot benzene is bleached in direct sunlight.

Properties. Colorless, monoclinate plates; when perfectly pure exhibiting a beautiful blue fluorescence; fusing point, 213°; boiling point, somewhat above

360°. Insoluble in water, difficultly soluble in alcohol and ether, easily in boiling benzene, less soluble in cold benzene. Heated with picric acid and benzene, it yields a compound, $C^{14}H^{10} + 2C^{6}H^{3}(NO^{2})^{3}O$, that crystallizes in red needles.

Paranthracene, $C^{14}H^{10}$. When a cold saturated solution of anthracene in benzene is exposed to direct sunlight, tabular crystals of this compound, which is isomeric or polymeric with anthracene, are deposited. It is almost insoluble in benzene, and is much more stable than anthracene; it is attacked neither by bromine nor hot concentrated nitric acid. It fuses at 244°, and is at this temperature reconverted into anthracene.

Anthracene dihydride, $C^{14}H^{12}$. Is formed by heating anthracene with hydriodic acid and a little phosphorus at 160–170°; and by gently heating it for a long time with alcohol and sodium-amalgam.—Small, colorless, monoclinate plates; fusing point, 106°; boiling point, 305°; sublimes readily in the form of needles. Easily soluble in alcohol and ether. When conducted in the form of vapor through a tube heated to low red-heat, it is resolved into anthracene and hydrogen. Heated with concentrated sulphuric acid, it yields sulphurous anhydride and anthracene; with bromine and oxidizing agents, the same products as anthracene.

Anthracene hexahydride, $C^{14}H^{16}$. Is obtained by heating the preceding compound for a long time with hydriodic acid and a little phosphorus at 200–220°.—Colorless laminæ; fusing point, 63°; boiling point, 290°. Very easily soluble in alcohol, ether, and benzene. At red-heat it is broken up, like the dihydride, into anthracene and hydrogen.

Anthracene forms addition- and substitution-products with chlorine and bromine.

Anthracene dichloride, $C^{14}H^{10}Cl^{2}$. Long, radiating needles; easily soluble in alcohol, but slightly soluble in ether.

Monochloranthracene, $C^{14}H^{9}Cl$. Is obtained directly from anthracene in a current of chlorine gas; and by decomposing the dichloride with alcoholic potassa.—Small, hard, scaly crystals.—*Dichloranthracene*, $C^{14}H^{8}Cl^{2}$. Yellow laminæ or needles; fusing point, 205°; sublimable.—*Tetrachloranthracene*, $C^{14}H^{6}Cl^{4}$. Stellate, gold-colored needles; fusing point, 220°.

Dibromanthracene, $C^{14}H^{8}Br^{2}$, is formed alone when bromine is added to a solution of anthracene in carbon bisulphide.—Gold-colored needles; fusing point, 221°. Heated with alcoholic ammonia at 160–170°, it is reconverted into anthracene.—*Dibromanthracene tetrabromide*, $C^{14}H^{8}Br^{2}.Br^{4}$, is formed when bromine vapor is allowed to act on finely divided anthracene or dibromanthracene.—Hard, thick, colorless plates; fuses at 170–180°, undergoing decomposition.—*Tribromanthracene*, $C^{14}H^{7}Br^{3}$, by heating the preceding compound to 200.°—Yellow needles; fusing point, 169°; sublimable.—*Tetrabromanthracene*, $C^{14}H^{6}Br^{4}$. From dibromanthracene tetrabromide with alcoholic potassa.—Yellow crystals; fusing point, 254°.

Nitroanthracene, $C^{14}H^{9}(NO^{2})$. Is obtained by heating an alcoholic solution of anthracene with nitric acid.—Stellate, red needles. Insoluble in cold alcohol and benzene, difficultly soluble in the hot liquids. Sublimable.

Anthraquinone (Oxanthracene).

$$C^{14}H^{8}\left\{\begin{matrix} O \\ O \end{matrix}\right> .$$

Formation and preparation. By the oxidation of anthracene, dichlor-, or dibromanthracene with nitric or chromic acid. Can be most readily prepared by adding a solution of chromic acid in glacial acetic acid, or finely powdered potassium bichromate, to a hot solution of anthracene in glacial acetic acid.

Properties. Purified by sublimation, it forms lustrous, yellow needles; fusing point, 273°; insoluble in water, but slightly soluble in alcohol, ether, and cold benzene, more easily in hot benzene. Very stable; alcoholic potassa-ley produces no effect upon it. Heated with hydriodic acid at 150°, or with zinc-dust, it is converted into anthracene. Fused with caustic potassa it yields benzoic acid.

Dichloranthraquinone, $C^{14}H^6Cl^2O^2$. Is obtained by oxidizing tetrachloranthracene like anthraquinone. —Yellow needles.

Monobromanthraquinone, $C^{14}H^7BrO^2$. By oxidation of tribromanthracene.—Bright-yellow needles; fusing point, 187°; sublimable.—*Dibromanthraquinone*, $C^{14}H^6Br^2O^2$. By heating anthraquinone with two molecules bromine at 100°; more readily by oxidizing tetrabromanthracene.—Bright-yellow needles; sublimable.

Dinitroanthraquinone, $C^{14}H^6(NO^2)^2O^2$. Is formed together with anthraquinone, by heating anthracene with dilute nitric acid. From the solution of the product in a great deal of hot alcohol, it separates first on cooling. It is more readily obtained by the action of nitric-sulphuric acid on anthraquinone.—Small, bright-yellow, monoclinate crystals; difficultly soluble in alcohol, ether, and benzene; sublimes in the form of yellow needles, at the same time undergoing partial decomposition. Combines with hydrocarbons, the same as picric acid, forming very characteristic compounds.

Diamidoanthraquinone, $C^{14}H^6(NH^2)^2O^2$. Is obtained from dinitroanthroquinone by warming with tin and hydrochloric acid, or with a solution of sodium sulphydrate.—Small, cinnabar-colored needles; fusing point, 236°. Scarcely soluble in water, soluble in alcohol, ether, and concentrated acids. Sublimes in garnet-colored, flat needles. Very weak base. From its solutions in acids it is thrown down in a free condition on the addition of water.

Anthraquinonedisulphuric acid, $C^{14}H^6O^2(SO^2.OH)^2$. Dichlor- and dibromanthracene dissolve readily in fuming sulphuric acid with the aid of gentle heat, forming dichlor- or dibromanthracenedisulphuric acids, $C^{14}H^6Cl^2(SO^2.OH)^2$, which when treated with oxidizing agents, and also when heated with concentrated sulphuric acid, are easily converted into anthraquinonedisulphuric acid. The *barium salt*, $C^{14}H^6O^2.(SO^3)^2Ba$, is difficultly soluble in water.

Oxyanthraquinone, $C^{14}H^8O^3 = C^{14}H^7(OH)O^2$. Is formed by fusing potassium anthraquinonedisulphate with potassium hydroxide, when the action is moderated by the addition of indifferent bodies (sodium chloride, chalk).—Yellow laminæ or needles, sublimable. Soluble in alkalies and baryta-water forming reddish-brown solutions.

Alizarin (Dioxyanthraquinone), $C^{14}H^8O^4 = C^{14}H^6(OH)^2O^2$. Is contained in old madder, and is obtained from rubianic acid (see Glucosides, p. 418), which is contained in fresh madder, by boiling with acids or alkalies. It can be artificially prepared by heating dichloranthraquinone, mono- or dibromanthraquinone, oxyanthraquinone and potassium anthraquinonedisulphate with potassium hydroxide at 250–270°. The mass is then dissolved in water, precipitated with hydrochloric acid, and the precipitate purified by recrystallization from alcohol, or, better, by sublimation.—Long, orange-red needles. Carefully heated, sublimable without decomposition. Almost insoluble in cold water, more easily in boiling water, in alcohol and ether.

Towards bases it conducts itself like a weak bibasic acid; soluble in alkalies, forming purple solutions. The alcoholic solution gives blue precipitates, $C^{14}H^6(O^2Ba)O^2 + H^2O$ and $C^{14}H^6(O^2Ca)O^2 + H^2O$, with baryta- or lime-water; the solution in alkalies gives a beautiful red precipitate (madder lake) with a solution of alum.

When heated with zinc-dust, alizarin is converted into anthracene; when oxidized with nitric acid, phtalic acid is the product. Excellent dye.

Chrysophanic acid (Parietic acid, Rheic acid), $C^{14}H^6(OH)^2O^2$ or $C^{14}H^8(OH)^2O^2$ (isomeric with alizarin or an analogous derivative of anthracene dihydride). Is contained in the lichen *Parmelia parietina;* in rhubarb (the root of various species of *Rheum*); and in senna leaves (from *Cassia lanceolata* and *Cassia obovata*). Can be readily obtained from these plants by extracting with caustic potassa, precipitating with hydrochloric acid and recrystallizing from chloroform.—Yellow, lustrous prisms; fusing point, 162°; partially sublimable; almost insoluble in water, but slightly in alcohol, easily soluble in ether. Soluble in alkalies, the solutions being red. Heated with zinc-dust it is converted into anthracene.

Chrysammic acid (Tetranitro- Dioxyanthraquinone), $C^{14}H^2(NO^2)^4(OH)^2O^2$. Is formed by warming crysophanic acid and aloes (see Aloin) with concentrated nitric acid.—Golden yellow, lustrous laminæ, very similar to lead iodide. But slightly soluble in water. Strong bibasic acid. Reducing substances (hydriodic acid, zinc and dilute sulphuric acid, potassium sulphide) convert it into *hydrochrysamide*, $C^{14}H^2(NH^2)^3(NO^2)(OH)^2O^2$, a body that forms indigo-colored needles; sublimable, when carefully heated; insoluble in water.

Purpurin (Trioxyanthraquinone), $C^{14}H^8O^5 = C^{14}H^5(OH)^3O^2$. Is contained in old madder, and is also sometimes formed as a by-product in the artificial preparation of alizarin.—Reddish-yellow prisms. Easily fusible and sublimable. Somewhat more easily soluble in water than alizarin, easily soluble in alcohol, ether, and alkalies, the solutions having a red color. It gives purplish-red precipitates with lime- and baryta-water. When heated with zinc-dust it is converted into anthracene.

Anthracenecarbonic Acid.

$$C^{15}H^{10}O^{2} = C^{14}H^{9}.CO.OH.$$

Preparation. By heating anthracene with phosgene in sealed tubes for twelve hours at 200°, dissolving the product in a solution of sodium carbonate, and precipitating with hydrochloric acid.

Properties. Long bright-yellow needles of a silky lustre. Fuses at 206°, with decomposition. Almost insoluble in cold water, difficultly in boiling water, easily soluble in alcohol. Heated alone or with soda-lime, it is resolved into anthracene and carbonic acid. When its solution in glacial acetic acid containing chromic acid is gently warmed it is converted into anthraquinone.

Most of its salts are soluble in water and alcohol.

In connection with this group, a few hydrocarbons that are not so well known, will here be described.

Pyrene, $C^{16}H^{10}$ (isomeric with diacetenylphenyl, p. 379). Is contained in those portions of coal-tar that boil at a high temperature. Those hydrocarbons that boil higher than anthracene are extracted by means of carbon bisulphide. Crude chrysene (p. 411) is thus left behind, while pyrene and other hydrocarbons are dissolved. In order to purify the pyrene, the carbon bisulphide is distilled off, the residue dissolved in alcohol, and mixed with an alcoholic solution of picric acid. Red crystals of a compound of pyrene with picric acid are deposited, which, after repeated recrystallizations from alcohol, are decomposed with ammonia.—Colorless laminæ; fusing point, 142°; but slightly soluble in cold alcohol, more readily in hot alcohol, very easily soluble in benzene, ether, and carbon bisulphide.

Nitric acid readily converts it into substitution-products; bromine yields substitution- and addition-products. Heated with potassium bichromate and dilute sulphuric acid, it is converted into *pyrenequinone*,

$C^{16}H^8O^2$, a brick-red powder, which, when heated sublimes partially in red needles and decomposes partially.

Chrysene, $C^{18}H^{12}$. That portion of the high-boiling hydrocarbons of coal-tar (see Pyrene), which is insoluble in carbon bisulphide, is repeatedly recrystallized from benzene.—Small, yellow laminæ; fusing point, 245–248°; difficultly soluble in alcohol, ether, and carbon bisulphide; more easily in hot benzene. Treated with picric acid in boiling benzene, it yields a compound, $C^{18}H^{12} + C^6H^3(NO^2)^3O$, that crystallizes in brown needles.—Nitric acid and bromine yield substitution-products. Heated with glacial acetic acid and chromic acid, it is converted into *chrysoquinone*, $C^{18}H^{10}O^2$, which crystallizes in beautiful red needles, dissolves in cold concentrated sulphuric acid forming a deep indigo-blue solution, and is reprecipitated from this solution, unchanged, by the addition of water.

Retene, $C^{18}H^{18}$. Is contained in the tar from very resinous pine- and fir-wood; and is formed together with benzene, cinnamene and other hydrocarbons by heating acetylene.—White laminæ of a mother-of-pearl lustre; fusing point, 98–99°; difficultly soluble in alcohol, easily in ether and benzene. Combines with picric acid forming a compound, $C^{18}H^{18} + C^6H^3(NO^2)^3O$, that crystallizes in orange-yellow needles. It dissolves in concentrated sulphuric acid, a crystalline *disulpho-acid*, $C^{18}H^{16}(SO^2.OH)^2$, being formed, the barium salt of which crystallizes in colorless needles. When treated with potassium bichromate and dilute sulphuric acid, it yields carbonic anhydride, acetic and phtalic acids, and a brick-red powder, *dioxyretistene*, $C^{16}H^{14}O^2$, which crystallizes in long, flat, orange-colored needles; fuses at 194–195°; and, when heated with zinc dust, yields a solid hydrocarbon *retistene*, $C^{16}H^{14}$.

Fichtelite in old pine trunks, *idrialin* in the mercury-ore of Idria, *scheererite* in beds of bituminous coal, are similar hydrocarbons, the chemical character of which is but little understood.

V. GLUCOSIDES.

A NUMBER of natural substances possess the common property of breaking up into sugar and other bodies by the action of certain agents (ferments, acids, alkalies). Neither the sugar nor the other bodies exist ready formed in them, but are formed during the process of decomposition, water being assimilated. With very few exceptions, the variety of sugar that results from the glucosides is grape-sugar; the other bodies, however, which make their appearance, are of very various character. The glucosides are to be considered as complicated ether-like compounds of grape-sugar. They still contain a number of hydroxyl-groups, the hydrogen of which is readily displaced by acid radicles.

1. *Amygdalin.*
$C^{20}H^{27}NO^{11}$.

Occurrence. In bitter almonds; in the leaves and berries of *Prunus laurocerasus;* in the blossoms, bark, and fruit kernels of *Prunus padus;* in the bark and young shoots and leaves of *Sorbus aucuparia;* in the fruit kernels of cherries, apricots, peaches, and in a great many other plants of the orders *Amygdaleæ* and *Pomaceæ.*

Preparation. The fatty oil is pressed from the paste of bitter almonds, and the mass then boiled repeatedly with fresh quantities of alcohol, the liquid being filtered each time boiling hot; and then about three-fourths of the alcohol distilled off from the mixed solutions. The amygdalin separates from the residue after being allowed to stand for several days in a cool place, in the form of a stellate, crystalline mass. By maceration with ether and subsequent recrystallization from alcohol it is freed of fatty oil.

Properties. Crystallized from alcohol, it forms colorless, fine crystalline scales, of a pearly lustre, without odor, of a slightly bitter taste. Easily soluble in water, from which it crystallizes in large, transparent prisms with three molecules of water. Not volatile. When heated with acetic anhydride, it is converted into *heptacetyl-amygdalin*, $C^{20}H^{20}NO^{4}(O.C^{2}H^{3}O)^{7}$, which crystallizes in long needles, of a silky lustre, insoluble in water, soluble in alcohol and ether.

Decompositions. It is resolved by treatment with dilute acids, or when in contact with *emulsin* (an albuminous body contained in almonds) into *sugar*, *hydrocyanic acid*, and *oil of bitter almonds* (p. 317), two molecules of water being taken up. Boiled with potassium or barium hydroxide, it is decomposed, forming ammonia and a white, crystalline, deliquescent acid, *amygdalic acid*, $C^{20}H^{28}O^{13}$.

2. *Solanin.*

$C^{43}H^{71}NO^{16}$(?).

Occurrence. In the various species of *Solanum*, particularly in the young sprouts of old potatoes.

Preparation. Potato sprouts are macerated with water containing a little sulphuric acid, the quickly filtered solution mixed warm with ammonia, the precipitate filtered off after prolonged standing, thoroughly dried, and repeatedly boiled with alcohol. On the cooling of the boiling-hot filtered solution, the greater part of the solanin separates, and, by recrystallization from alcohol, is now thoroughly purified.

Properties. Fine prisms, of a silky lustre, almost insoluble in water, but slightly soluble in cold alcohol, more easily in hot; fuses at 235°. Acts poisonously. It is a weak base, possesses a weak alkaline reaction, dissolves readily in acids, and yields with them gummy, uncrystalline salts, which can be precipitated from their solutions in alcohol by ether.

Decompositions. By boiling with dilute hydrochloric or sulphuric acid, it is resolved into sugar and solanidin, with assimilation of three molecules of water. On cooling, the solanidin is deposited in the form of a sulphate or hydrochlorate, from the solutions of which in alcohol, solanidin is precipitated by means of ammonia.

Solanidin, $C^{25}H^{41}NO$(?). Fine needles, of silky lustre; but slightly soluble in water, in alcohol more easily soluble. It fuses above 200°, and sublimes by rapid heating almost without decomposition. A stronger base than solanin; gives with acids easily crystallizing salts; difficultly soluble in water.

Solanidin hydrochlorate, $C^{25}H^{41}NO.HCl$, forms rhombic columns; can be sublimed undecomposed.

Solanin, in contact with concentrated cold acids, yields sugar, but no solanidin, but two other, still but slightly known, bases, which are also formed from solanidin when it is heated with concentrated acids.

3. *Salicin.*

$C^{13}H^{18}O^7$.

Occurrence. In the bark and leaves of most willows (*Salix* species) and of some poplar species.

Preparation. The bark is cut up and boiled with water, the liquid concentrated and boiled with litharge until decolorized, by which means gums, tannic acid, etc., are thrown down. The dissolved lead combined with salicin is at first precipitated with sulphuric acid, afterwards completely with sulphuretted hydrogen or barium sulphide; the solution of salicin, filtered from the precipitate, is evaporated to the point of crystallization.

Properties. Small, colorless, lustrous prisms or laminæ, of a bitter taste; fusible at 198°; easily soluble in hot water, difficultly soluble in cold water, soluble in alcohol.—In contact with emulsin or saliva, it assimilates one molecule of water, and is resolved into *sugar* and *saligenin* (p. 315). When heated with dilute hydrochloric or sulphuric acid, it yields sugar and *saliretin* (p. 315).

Tetracetyl-salicin, $C^{13}H^{14}(C^2H^3O)^4O^7$. Is obtained by heating salicin with acetyl chloride or acetic anhydride.—Colorless, lustrous needles; but slightly soluble in water, ether, and cold alcohol, easily soluble in hot alcohol.

Benzoyl-salicin (Populin), $C^{20}H^{22}O^8 + 2H^2O = C^{13}H^{17}(C^7H^5O)O^7 + 2H^2O$. Is contained in the bark and leaves of *Populus tremula*, from which it may be prepared in the same manner as salicin. It is formed together with di- and tribenzoyl-salicin by the action of benzoyl chloride on salicin; and by fusing salicin and benzoic anhydride together.—Small, colorless prisms of a sweetish taste, difficultly soluble in cold water, more easily soluble in hot water and in alcohol.—When boiled with baryta-water or milk of lime, it yields benzoic acid and salicin. Dilute acids (but not emulsin), resolve it into sugar, saliretin, and benzoic acid.

Dibenzoyl-salicin, $C^{13}H^{16}(C^7H^5O)^2O^7$, and *tribenzoyl-salicin*, $C^{13}H^{15}(C^7H^5O)^3O^7$, are formed from salicin together with populin. They are white powders, insoluble in water, scarcely crystalline.

Helicin, $C^{13}H^{16}O^7$. Is formed together with nitro-salicylic acid by the action of nitric acid (containing hyponitric acid) on salicin.—Small, white needles, difficultly soluble in water, more easily in alcohol; fusing point, 175°. Ferments, dilute acids, and alkalies resolve it into sugar and salicylic aldehyde (p. 322).

4. *Æsculin.*

$C^{15}H^{16}O^9 + 2H^2O$.

Occurrence. In the bark of *Æsculus hippocastanum*, and several other trees.

Preparation. The bark of horsechestnut-trees is

cut up into small pieces, boiled with water, foreign substances precipitated by means of lead acetate, the excess of lead removed from the filtered solution by means of sulphuretted hydrogen, and the filtrate evaporated to a syrup from which the æsculin gradually crystallizes.

Properties. Colorless, fine prisms, of a slightly bitter taste, but little soluble in water. Even an exceedingly dilute solution is very fluorescent, the reflected light being of a bright-blue color. The fluorescence disappears in the presence of acids, reappears on the addition of alkalies. Difficultly soluble in alcohol. Dilute acids resolve it into sugar and æsculetin.

Hexacetyl-æsculin, $C^{15}H^{10}(C^2H^3O)^6O^9 + H^2O$. Is formed by the action of acetyl chloride or acetic anhydride on æsculin.—Small, colorless needles, that give up water at 130°. Brought in contact with anilin at the boiling temperature æsculin yields *trianil-æsculin*, $C^{33}H^{31}N^3O^6 = C^{15}H^{16}O^9 + 3C^6H^7N - 3H^2O$.

Æsculetin, $C^9H^6O^4 + H^2O$. Exists ready formed in the bark of the horsechestnut. If æsculin is digested with dilute sulphuric acid, it dissolves, the solution having a yellow color, and æsculetin is deposited in its place in crystals.—Fine, colorless needles and laminæ, very sparingly soluble in water, but slightly in alcohol, very easily soluble in alkalies, the solutions being yellow. Is decomposed by heating with caustic potassa into formic acid, oxalic acid, and protocatechuic acid (p. 356), or an acid isomeric with the latter, *æscioxalic acid.*

5. *Phlorizin.*

$$C^{21}H^{24}O^{10} + 2H^2O.$$

Occurrence. In the bark, especially the root-bark, of the apple, cherry, pear, and plum-tree, from which it can be extracted by means of boiling water or warm dilute alcohol. It is purified by recrystallizing from hot water.

Properties. Fine, silky prisms of a bitter taste; easily soluble in boiling water and alcohol, difficultly soluble in cold water. It loses its water of crystallization at 100°; fuses at 106–109°; solidifies again at 130°; and appears to be converted into another modification at this temperature, which does not fuse again below 160°, is less soluble in water, and is deposited in an amorphous condition from this solution, gradually passing into the crystalline modification. Treated with acetic anhydride, it yields acetyl-compounds (with 1, 3, and 5 times the group C^2H^3O), similar to those of salicin.

Decompositions. Under a bell-jar filled with ammonia vapors and moist air, phlorizin deliquesces, forming a thick, dark syrup, from which, by means of careful evaporation and washing with alcohol, is obtained *phlorizein*, $C^{21}H^{30}N^2O^{13}$, a red, amorphous body, easily soluble in hot water, very sparingly soluble in alcohol. Boiled for a long time with dilute hydrochloric or sulphuric acid, two molecules of water are assimilated and phlorizin is resolved into grape-sugar and

Phloretin, $C^{15}H^{14}O^5$, which separates from the solution on cooling.—Small, colorless laminæ, very slightly soluble in water, easily soluble in alcohol, dissolves also easily in alkalies, but on evaporating this solution, it is decomposed into *phloretic acid* (p. 353), and *phloroglucin* (p. 311).

6. *Quercitrin.*
$C^{33}H^{30}O^7$ (?).

Occurrence. In the bark of *Quercus tinctoria* (which occurs in commerce under the name of quercitron, and is used as a yellow dye); and the blossoms of *Æsculus hippocastanum;* and is prepared from these sources in the manner described in connection with phlorizin.

Properties. Yellow, crystalline powder, difficultly soluble in water even at boiling temperature. Treated with acids, it is resolved into a crystallizing, unfermentable, saccharine body, *isodulcite*, and into

Quercitin, $C^{27}H^{18}O^{12}$, which also occurs ready formed in *Calluna vulgaris*, in tea, in the root-bark and trunk-bark of the apple-tree and other plants.—Yellow, crystalline powder, sublimes in large yellow needles with partial decomposition. But slightly soluble in water, easily soluble in alcohol. Fusing potassa decomposes it, like phloretin, into *phloroglucin* and *quercetic acid*, $C^{15}H^{10}O^{7}$, which crystallizes in fine, silky prisms; sparingly soluble in cold water. By further treatment with fusing potassa, it yields *protocatechuic acid* (p. 356), *quercimeric acid*, $C^{8}H^{6}O^{5} + H^{2}O$ and *paradatiscetin*, $C^{15}H^{10}O^{6}$.

Rutin is a glucoside, very similar to quercitrin, but not identical with it, contained in *Ruta graveolens*—the loppers (blossom-buds) of *Capparis spinosa*. Yields, when treated with acids, quercitin and an unfermentable sugar, which appears to be different from isodulcite.

---

7. *Frangulin.*
$C^{20}H^{20}O^{10}$.

In the bark of *Rhamnus frangula*.—Yellow, crystalline mass; fusing point, 226°. Almost insoluble in cold water, difficultly soluble in cold alcohol and ether, easily in hot alcohol. Soluble in alkalies, forming red solutions. Acids resolve it into sugar and *frangulic acid*, $C^{14}H^{10}O^{5} + H^{2}O$, which forms an orange-yellow, loose crystalline mass, but slightly soluble in water, easily soluble in alcohol; fusing point, 246–248°.

8. *Rubianic Acid.*
$C^{20}H^{22}O^{11}$ (?).

In madder (the root of *Rubia tinctorum*). In order to prepare it, fresh madder-root is boiled with water, foreign substances removed from the solution by means of neutral lead acetate, the liquid filtered, the rubianic acid precipitated from it by means of basic lead acetate, and the red precipitate then decomposed. The acid is thrown down with the lead sulphide, and separated from this by extracting with alcohol.—Yellow prisms, sparingly soluble in cold water, easily in hot water, alcohol, and ether.—By boiling with acids and alkalies, as well as by contact with a ferment contained in madder, it is resolved into sugar and alizarin. In old madder, as it is used in dying, this decomposition has already partially taken place; it is accelerated by treating the madder with sulphuric acid (*Garancin*, a commercial product, is madder which has been treated in this way).

Morindin, a body contained in the root-bark of *Morinda citrifolia*, is probably identical with rubianic acid; and the dye *morindon*, prepared from it by means of sublimation, appears to be alizarin.

---

9. *Arbutin.*

$C^{12}H^{16}O^7$.

In the leaves of the bearberry (*Arbutus uva ursi*), and of winter-green (*Pyrola umbellata*).—Long, colorless, bitter tasting needles, which fuse at 170°, the solution of which is colored deep blue by iron chloride. In contact with emulsin, and by boiling with dilute sulphuric acid, it is resolved into sugar and *hydroquinone* (p. 303), which is also formed by the dry distillation of arbutin. Concentrated nitric acid converts it into bright-yellow needles of *dinitroarbutin*, $C^{12}H^{14}(NO^2)^2O^7 + 2H^2O$. When chlorine is conducted into a watery solution of arbutin, substitution-products of quinone (p. 301) are formed.

10. *Fraxin.*

$C^{32}H^{36}O^{20}$.

In the bark of *Fraxinus excelsior* and *Æsculus hippocastanum*.—Fine, fascicular needles; slightly soluble in cold water, easily soluble in alcohol; fuses at 190°. With dilute acids it yields sugar and *fraxetin*, $C^{10}H^8O^5$.

11. *Phillyrin.*
$C^{27}H^{34}O^{11}$.

Contained in the bark of *Phyllyrea latifolia.*—Colorless crystals; difficultly soluble in cold water; fusing at 160°. Dilute acids resolve it into sugar and *phillygenin*, $C^{21}H^{24}O^6$.

12. *Daphnin.*
$C^{31}H^{34}O^{19} + 4H^2O$.

In the bark of *Daphne alpina* and *Daphne mezereum.*—Colorless, transparent prisms; fusing at 200°; insoluble in cold water and in ether, easily soluble in hot water and alcohol. Emulsin or dilute acids resolve it into sugar and *daphnetin*, $C^{19}H^{14}O^9$.

13. *Myronic Acid.*
$C^{10}H^{19}NS^2O^{10}$.

In the seed of black-mustard in the form of the potassium salt. This can be extracted from the residue by means of water after the powdered seed has been boiled with alcohol.—Small, silky needles; easily soluble in water. In contact with *myrosin*, a ferment contained in mustard seed, and heated with baryta-water, it is decomposed into allyl mustard-oil (p. 215) and potassium bisulphate. Its solution gives a white precipitate with silver nitrate, $C^4H^5NSO^4Ag^2$, which, when treated with sulphuretted hydrogen, yields silver sulphide, sulphur, free sulphuric acid, and allyl cyanide (p. 120).

14. *Convolvulin* (*Rhodeoretin*).
$C^{31}H^{50}O^{16}$.

In jalap root (of *Convolvulus schiedeanus*). The root is first thoroughly exhausted with boiling water, then treated with alcohol; the alcoholic solution decolorized with animal charcoal; evaporated; the crude convolvulin dissolved in alcohol, and reprecipitated with ether.—Colorless, resinous mass; inodorous and tasteless; fuses at 150°; but slightly soluble in water, easily in alcohol. It exerts a purgative action. Dissolves in alkalies, and is converted by them into an easily soluble, amorphous substance, *convolvulic acid* (rhodeoretic acid), $C^{31}H^{52}O^{17}$(?), water being assimilated in the reaction. Convolvulin, as well as convolvulic acid, in contact with emulsin, or when treated with dilute acids, is decomposed into sugar and *convolvulinol*, $C^{13}H^{24}O^{3} + \frac{1}{2}H^{2}O$, which dissolves in alkalies, forming *convolvulinolic acid*, $C^{13}H^{26}O^{4}$.

15. *Jalapin.*

$C^{34}H^{56}O^{16}$.

Homologous with convolvulin. In jalap-root (of *Convolvulus orizabensis*) and scammony-resin (the hardened sap of *Convolvulus scammonia*).—Very similar to convolvulin. Is decomposed by emulsin or acids into sugar and *jalapinol*, $C^{16}H^{30}O^{3} + 1\frac{1}{2}H^{2}O$; and conducts itself towards alkalies like convolvulin.

Turpethin, a resinous glucoside, isomeric with jalapin, is contained in turpeth-resin (from the root of *Ipomœa turpethum*). It yields, when treated with baryta-water, amorphous *turpethic acid*, $C^{34}H^{60}O^{18}$, and is decomposed by mineral acids into sugar and white, amorphous *turpetholic acid*, $C^{16}H^{32}O^{4}$.

16. *Saponin.*

$C^{32}H^{54}O^{18}$.

In the root of a number of plants (*Saponaria officinalis*, *Gypsophila struthium*, *Polygala senega*, *Agrastemma githago*).—White, amorphous powder, which causes sneezing; poisonous; easily soluble in hot water. This solution foams like soap-water, even when very dilute. Treated with hydrochloric acid gas or fuming hydrochloric acid, it yields an uncrystalline, saccharine body, and *sapogenin*, $C^{14}H^{22}O^{4}$, white crystals, sparingly soluble in water and alcohol.

17. *Caïncin* (*Caïncic Acid*).
$C^{40}H^{64}O^{18}$.

In the root of *Chiocca racemosa.*—Fine, colorless prisms; sparingly soluble in cold water, easily soluble in alcohol. Is resolved by hydrochloric acid gas into an uncrystalline sugar, and crystalline *caïncetin*, $C^{22}H^{34}O^{3}$, which, treated with fusing potassa, is decomposed into butyric acid and *caïncigenin*, $C^{14}H^{24}O^{2}$.

18. *Quinovin.*
$C^{30}H^{48}O^{8}$.

In cinchona-bark, particularly in a false bark, *China nova.*—White, amorphous substance; insoluble in water. When hydrochloric acid gas is conducted into its alcoholic solution, and when it is treated with sodium-amalgam, it is decomposed into a sugar, very similar to mannitan (p. 189), perhaps identical with it, and *quinovic acid*, $C^{24}H^{38}O^{4}$, which separates as a white, crystalline powder.

19. *Pinipicrin.*
$C^{22}H^{36}O^{11}$.

In the bark and needles of *Pinus sylvestris;* in the green portions of *Thuja occidentalis.*—Yellow, amorphous, bitter powder, soluble in water and alcohol. Is decomposed by heating with sulphuric acid into sugar and *ericinol*, $C^{10}H^{16}O$.

20. *Carminic Acid.*
$C^{17}H^{18}O^{10}$.

In the blossoms of *Monarda didyma*, and probably also in other plants. Most particularly, however, in cochineal (the female of the insect *Coccus cacti*), from which it is obtained by boiling with water, precipitating with lead acetate, and decomposing the lead precipitate with sulphuretted hydrogen.—Purple, amorphous mass. Easily soluble in water and alcohol. Combines with bases, forming colored salts. When boiled with dilute sulphuric acid, it is decomposed, yielding a peculiar uncrystalline, unfermentable sugar, which is optically inactive; and *carmine red*, $C^{11}H^{12}O^7$, dark-purple, shiny mass; soluble in water and alcohol, the solution formed being of a red color. Weak acid.

Fused with potassium hydroxide, carminic acid yields oxalic, succinic, and acetic acids, and a yellow, crystalline substance, *coccinin*, $C^{14}H^{12}O^5$; heated with concentrated nitric acid: oxalic acid and *nitrococcusic acid*, $C^8H^5(NO^2)^3O^3 + H^2O$.

21. *Helleboreïn.*

$C^{26}H^{44}O^{15}$.

In the root of *Helleborus niger*, and in smaller quantity in that of *Helleborus viridis*.—Colorless nodules, consisting of microscopical needles. Easily soluble in water, but slightly in alcohol. Has a narcotic action. Is resolved, by boiling with dilute acids, into sugar and amorphous *helleboretin*, $C^{14}H^{20}O^3$, which is deposited as a dark-violet precipitate, that, however, becomes grayish-green by drying.

22. *Helleborin.*

$C^{36}H^{42}O^6$.

In the root of *Helleborus viridis*, and in traces in that of *Helleborus niger*.—Shiny, colorless needles, arranged concentrically. Insoluble in cold water, easily soluble in boiling alcohol. Is colored an intense red by concentrated sulphuric acid. Has a stronger narcotic action than helleboreïn. When heated with dilute acids, it is resolved into sugar and an amorphous, resinous substance, *helleboresin*, $C^{30}H^{38}O^4$.

23. *Glycyrrhizin.*

$C^{24}H^{36}O^9$ (?).

In liquorice root (from *Glycyrrhiza glabra*), and in the extract prepared from it.—Amorphous, yellowish-white powder, easily soluble in hot water and in alcohol. By boiling with dilute acids, it yields sugar and a yellowish resin *glycyrrhetin*, $C^{18}H^{26}O^{4}$ (?).

24. *Digitalin.*

In *Digitalis purpurea.*—Small colorless crystals; sparingly soluble in water, easily soluble in alcohol, of an intensely bitter taste. Exceedingly poisonous. Very difficult to obtain in a pure state, and hence but little known as yet. Is resolved by sulphuric acid into sugar and amorphous *digitalretin.*

25. *Tannic Acids.*

By the name tannic acids is understood a class of weak acids, which are widely distributed in the vegetable kingdom, and which bear a close relation to each other, as regards their properties, as well as their composition; the composition is, however, not yet determined with certainty for all of them. Most of the tannic acids have been shown to be glucosides. In general they are characterized by a sharp astringent taste; by the property of giving bluish-black or green compounds with iron salts; of precipitating solutions of gelatin; and by the ability to *tan* animal hides; *i. e.* to convert them into leather. Their important uses in dyeing, in the preparation of ink, and dressing of leather, depend upon these properties. They also constitute the active principles of a number of plants employed in medicine.

Gallotannic acid (Tannin), $C^{27}H^{22}O^{17}$. Occurs particularly in gall-nuts, the excrescences found on the young branches of *Quercus infectoria*, caused by the punctures of the gall-wasp; these contain about half their weight of tannic acid; in still larger quantity in Chinese gall-nuts, formed in a similar manner; also in the various species of sumach (the branches of *Rhus coriaria*); and probably in still other plants.

Eight parts powdered gall-nuts (most profitably

from Chinese gall-nuts) are macerated with 12 parts ether and 3 parts alcohol for two days, the mixture being frequently shaken; the solution is then poured off, and the residue again treated in the same way with the same quantity of ether and alcohol. To the united extracts 12 parts of water are added; the alcohol and ether distilled off over a water bath; the solution filtered; and the filtrate evaporated to dryness.

Colorless amorphous mass, of a purely astringent taste; inodorous; easily soluble in water; reddens litmus. It forms bluish-black precipitates with solutions of ferric salts. It is thrown down from its solution by mineral acids and a number of alkaline salts (sodium and potassium chlorides, not by saltpetre and sodium sulphate); most thoroughly by a solution of gelatin and by animal membranes. Further, it precipitates most organic bases, starch, albumen.—Tribasic acid. Its salts are amorphous and difficult to obtain of constant composition. The solutions of the alkaline salts become colored red or brown rapidly in the air, oxygen being taken up and the acid decomposed. The free acid in an aqueous solution is also decomposed in the air. If a concentrated extract of gall-nuts is allowed to stand in contact with the air, gallic acid is deposited from it, mixed with another crystalline acid, *ellagic acid*, $C^{14}H^6O^8 + 2H^2O$. This is very difficultly soluble in water; it is also formed by heating two molecules gallic acid with one molecule arsenic acid in aqueous solution, and is the principal ingredient of a known intestinal concretion, *bezoar*, found in a species of goat of Persia.

By boiling with dilute acids, it is resolved into sugar and gallic acid (p. 360); also by boiling with alkalies (only in the latter case the sugar undergoes further decomposition); and also by the action of yeast, emulsin, or a ferment contained in gall-nuts.

Heated alone it yields pyrogallic acid (p. 310).

Catechutannic acid. In catechu, a dark or light brown extract prepared from *Acacia catechu*, *Areca catechu*, and *Nauclea gambir*.—Very similar to gallotannic acid. With iron salts, however, it does not give a bluish-black, but a dirty green precipitate; it can also not be converted into gallic acid. Composition unknown.

Catechin (Catechuic acid), $C^{19}H^{18}O^{8}$ (?). Occurs together with tannic acid in catechu, more especially in the cubical variety from *Nauclea gambir*.

Powdered catechu is macerated with cold water; the brown tannic acid solution filtered from the undissolved catechin; this pressed and dissolved in boiling water, from which it is deposited slowly on cooling, generally not yet quite white. It is purified by recrystallization.

Colorless mass consisting of interwoven fine crystalline scales; almost tasteless; fusible at 217°; difficultly soluble in cold water, easily in boiling water and in alcohol. Turns a reddish color in the air, finally brown. Ferric salts are colored green by it; solutions of salts of the noble metals are reduced. Very weak acid, does not expel carbonic acid from it salts.—When heated, it yields pyrocatechin (p. 305); when fused with potassa, protocatechuic acid (p. 356) and phloroglucin (p. 311).

Kinotannic acid. Forms the principal ingredient of kino, a brittle reddish-brown extract, which is prepared in West India from *Coccoloba uvifera*, in Africa, from *Pterocarpus erinaceus*. The tannic acid contained in it has been but little investigated; it is not yet known in a pure condition. It colors ferric salts blackish-green. Fused with potassa, it yields phloroglucin.

Morintannic acid (Maclurin), $C^{13}H^{10}O^{6} + H^{2}O$. In old fustic (of *Morus tinctoria*), from which it is obtained by boiling with water. On evaporation of the solution, morin is at first deposited and then morintannic acid.—Bright-yellow, crystalline powder, easily soluble in hot water and alcohol. Its solution gives with ferrous sulphate a blackish-green precipitate. Heated alone, it yields pyrocatechin; fused with potassa, phloroglucin and protocatechuic acid. Treated with zinc and sulphuric acid, it is resolved into phloroglucin and a white, crystalline substance, *machromin*, $C^{14}H^{10}O^{5}$, which is converted into an indigo-blue body by the action of light, heat, or oxidizing agents.

Morin (Moric acid), $C^{12}H^{8}O^{5}$. Is contained in old fustic, together with morintannic acid, and, being much less soluble in water than the latter, it can be easily separated from it.—Crystallizes from alcohol in almost colorless, shiny needles; almost insoluble in cold water, but sparingly soluble in boiling water. Treated with sodium-amalgam in an alkaline solution, and fused with potassa, it is converted into phloroglucin.

Quino-tannic acid. In the bark of the various species of cinchona, partially combined with bases also contained in the bark.—Very similar to gallotannic acid; precipitates ferrous salts, however, green or grayish-green. By boiling with acids it is resolved into sugar and *quino-red*, $C^{28}H^{22}O^{14}$, a reddish-brown, amorphous substance, with weak acid properties, which is itself contained ready formed in cinchona-bark, and can be extracted from it by means of ammonia. With fusing potassa it yields protocatechuic and acetic acids.

Oak-bark-tannic acid. In oak bark, together with a small quantity of gallotannic acid. The bark extract is subjected to partial precipitation with lead acetate; the dirty-brown precipitate, which is first formed, and that formed later, of a lighter color, are decomposed with sulphuretted hydrogen. On evaporating the filtrate, the tannic acid remains behind as an easily soluble, yellowish-brown, amorphous mass. Its solution is colored a deep blue by iron chloride. By boiling with dilute sulphuric acid, it is resolved into sugar and oak-red, a body very similar to *quino-red*, which, it appears, is also contained in oak bark. It yields, when fused with potassa, phloroglucin and protocatechuic acid.

Ratanhia-tannic acid, in ratanhia-root, *filix-tannic acid*, in fern-root, and *tormentill-tannic acid*, in tormentill-root, conduct themselves very similarly to quino-tannic acid and oak-bark-tannic acid. When boiled with dilute acids they are all resolved into sugar and reddish-brown bodies, which possess the greatest similarity with oak-red and quino-red; and when fused with potassa they yield phloroglucin and protocatechuic acid.

Caffetannic acid, $C^{15}H^{18}O^{8}$(?). In coffee. Coffee is boiled with alcohol; the acid precipitated by means of lead acetate; and the precipitate decomposed by sulphuretted hydrogen.—Gummy mass; easily soluble in water; colors ferric salts green.—With ammonia it becomes rapidly green in the air.—Subjected to dry distillation, it yields pyrocatechin (p. 305); when fused with potassa, protocatechuic acid. Heated with potassa-ley, it is decomposed, forming an uncrystalline sugar and caffeic acid (p. 378).

VI. VEGETABLE SUBSTANCES BUT LITTLE KNOWN.

THERE is a large number of compounds occurring in nature, whose chemical constitution and the relation they bear to other better known bodies have not yet been ascertained. Only the more important and better investigated of these will be here described.

A. ACIDS.

1. *Usnic acid*, $C^{18}H^{16}O^{7}$. In a great many lichens, particularly in the various species of *Usnea*, from which it is extracted by means of ether.—Sulphur-yellow, transparent prisms; insoluble in water, but sparingly soluble in alcohol; fusible at 202°. (A modification of usnic acid, from *Cladonia rangiferina*, called *beta-usnic acid*, fuses at 175°). Its solution, in an excess of alkali, becomes first red and then black in the air. Subjected to dry distillation, it yields *betaorcin* (p. 309).

2. *Cetraric acid*, $C^{18}H^{15}O^{8}$. In Iceland moss (*Cetraria islandica*). It can be obtained pure only with difficulty.—Very fine, white needles, of an intensely bitter taste; neither fusible nor volatile; scarcely soluble in water, easily soluble in alcohol. Dissolves in alkalies with yellow color, which is, however, rapidly converted into brown in the air, the acid undergoing decomposition. It suffers a similar rapid decomposition when boiled in alcohol or water, with access of air.

3. *Lichenstearic acid*, $C^{14}H^{24}O^3$. Together with cetraric acid in Iceland moss.—Fine, shiny crystalline laminæ; insoluble in water, easily soluble in alcohol and ether.

4. *Vulpic acid*, $C^{19}H^{14}O^5$. In the lichens, *Cetraria vulpina*, and a variety of *Parmelia parietina*, from which it can be extracted by lukewarm water and milk of lime; and then reprecipitated by hydrochloric acid.—Yellow crystals, very similar to usnic acid; insoluble in water, but slightly soluble in alcohol, more readily in ether. By boiling with barium hydroxide, it is decomposed into methyl alcohol, oxalic acid, and alpha-toluic acid (p. 340); by boiling with dilute caustic potassa, into methyl alcohol, carbonic acid, and *oxatolylic acid*, $C^{16}H^{16}O^3$. The latter crystallizes in colorless, four-sided columns, fusing at 154°; insoluble in water; in alcohol and ether more easily soluble; and is resolved, by continued boiling with concentrated potassa-ley, into oxalic acid and toluene.

5. *Meconic acid*, $C^7H^4O^7\left(= C^4 \left\{ \begin{matrix} OH \\ (CO.OH)^3 \end{matrix} (?)\right.\right)$. In the milky juice of the poppy (*Papaver somniferum*) and the opium prepared from this.—The crude calcium meconate, obtained in the preparation of morphine, is repeatedly treated with dilute, hot hydrochloric acid; the acid, which crystallizes out in a still impure condition on cooling, is dissolved in dilute, warm ammonia; the salt recrystallized several times from hot water, and finally the acid precipitated from the hot solution of the salt by means of hydrochloric acid.

Crystallizes from water in colorless, shiny laminæ or prisms, with three molecules of water of crystallization. Of a weak, sour taste; difficultly soluble in cold water, more easily in hot water and alcohol. Colors solutions of ferric salts a deep red. Tribasic acid.—When treated with sodium-amalgam, it yields an amorphous, deliquescent acid, difficultly soluble in alcohol, *hydromeconic acid*, $C^7H^{10}O^7$.—Heated to 220°, or boiled for a long time with water, particularly with dilute hydrochloric acid, meconic acid is resolved into carbonic anhydride and *comenic acid*, $C^6H^4O^5$, which consists of very hard and difficultly soluble granules. Comenic acid, in its turn, yields by distillation another, easily fusible, monobasic acid, subliming in shiny laminæ, *pyrocomenic acid*, $C^5H^4O^3$.

6. *Chelidonic acid*, $C^7H^4O^6$. In *Chelidonium majus*, particularly at the blossoming period of the plant.— The expressed, boiled, and filtered juice is acidified with nitric acid; and lead chelidonate precipitated with lead nitrate. This, when decomposed with sulphuretted hydrogen, yields impure chelidonic acid, which is purified by preparation of salts, and recrystallization.— Long, shiny needles. Difficultly soluble in cold water and alcohol, more easily in hot water; not volatile without decomposition. Strong acid; dissolves iron and zinc with evolution of hydrogen. Tribasic. Treated with bromine and water it is decomposed, forming bromoform, pentabromacetone (C^3HBr^5O), and oxalic acid.

B. Bases (Alkaloids).

In a large number of plants occur peculiar nitrogenized bases, combined with acids. Although present in but very small quantity, they form, as a rule, the active principle of these plants, which are mostly distinguished for poisonous or healing properties.

The majority of these bases are crystallizable and not volatile; only a few are liquid and distillable. Nearly all of them are sparingly soluble in water, easily soluble in alcohol, turn litmus-paper blue, and have a bitter taste.

Their preparation takes place usually in the following manner: the proper portions of the plants are exhausted with water or dilute hydrochloric acid, and the bases (if volatile) separated by distilling with an alkali, or (if not volatile) precipitated by means of a stronger, inorganic base. As in the latter case, however, a number of other substances are precipitated at the same time, it is necessary that the product be still subjected to various other purifying processes (preparation of salts, recrystallization and subsequent decomposition, etc.). Frequently the extract is mixed with neutral or basic lead acetate for the purpose of precipitating foreign substances; the filtrate is then freed of dissolved lead by sulphuretted hydrogen; and the alkaloid now precipitated by means of a stronger base.

All alkaloids are precipitated from their solutions by tannic acid, by phosphormolybdenic acid,* by potassio-mercuric iodide, potassio-cadmic iodide and potassio-bismuthic iodide, and can be set free from these precipitates by means of alkalies or barium hydroxide, and extracted by solvents (ether, benzene, amyl alcohol, chloroform, etc.).

1. *Conine.*

$$C^8H^{15}N = C^8H^{14}.NH.$$

Occurrence. In all parts of the hemlock (*Conium maculatum*), most abundantly in the ripe seeds.

Formation. When butyric aldehyde is treated with alcoholic ammonia a base, *butyraldin*, $C^8H^{17}NO$, not known in a pure state, is produced together with other substances. When this is subjected to dry distillation, it yields conine.

Preparation. The plant or the crushed seeds are distilled with dilute caustic potassa, in which process conine passes over dissolved in water. The distillate is saturated accurately with sulphuric acid, evaporated to a syrupy consistence, and distilled with concentrated caustic potassa, the conine now passing over as an oil, floating on a saturated solution in water. It is freed of ammonia in a vacuum.

Properties. Colorless, clear, oily liquid, specific gravity, 0.89; of a suffocating, unpleasant odor (somewhat resembling hemlock); and a very repulsive penetrating

* Prepared by precipitating ammonium molybdenate with sodium phosphate, dissolving the well-washed precipitate in hot sodium carbonate, evaporating, and then igniting the mass. The salt which remains behind is heated with ten parts of water; nitric acid added until the solution shows a strong acid reaction; and then filtered.

taste. Boils at 163.5°. Dissolves water, which is separated by the aid of heat; hence the property of conine, of becoming turbid even from the warmth of the hand. Soluble in 100 parts of water; miscible with alcohol and water. Strongly alkaline, and very poisonous. Monatomic.

Decomposition. On exposure to the air, conine, as well as the solutions of its salts, soon becomes brown, and is finally entirely destroyed, ammonia being evolved. Warmed with oxidizing substances, it yields butyric acid. By treatment with dry nitrous acid and subsequent addition of water, there is formed *azoconydrine*, $C^8H^{16}N^2O$, a bright-yellow liquid, insoluble in water, which, heated with phosphoric anhydride, is resolved into nitrogen, water, and a hydrocarbon, *conylene*, C^8H^{14} (homologous with acetylene, p. 131). Colorless, mobile liquid, boiling at 126°; combines directly with bromine, forming a liquid product, $C^8H^{14}Br^2$.

Methylconine, $C^8H^{14}.N.CH^3$, and **Ethylconine,** $C^8H^{14}.N.C^2H^5$, are colorless liquids, which are formed when conine is heated with methyl or ethyl iodide, and afterwards distilled with caustic potassa. The former is frequently contained in commercial conine. They both combine directly with another molecule of ethyl iodide, forming crystallizing iodides, which, when decomposed with silver oxide, yield bases, analogous to tetrethylammonium hydroxide; not volatile; very easily soluble in water.

Conhydrine, $C^8H^{17}NO$, occurs together with conine, particularly in the fresh blossoms, but also in the ripe seed of hemlock. Can be separated from conine by distillation in a current of hydrogen, the temperature being raised very slowly. At first conine passes over, and, at a higher temperature, crystals of conhydrine are deposited in the neck of the retort.—Colorless, iridescent, crystalline laminæ; sublimes at 100°; fuses at 120.6°; and boils at 224°. Moderately soluble in water, more readily in alcohol and ether. Heated with phosphoric anhydride, it is decomposed into conine and water. Treated with sodium, it is also converted into conine.

2. *Nicotine.*
$C^{10}H^{14}N^{2}$.

Occurrence. In the leaves and seed of the tobacco species in varying quantity; in poor qualities of tobacco as much as 7 and 8 per cent., in Havana tobacco only 2 per cent.

Preparation. Tobacco leaves are digested repeatedly with water containing sulphuric acid, pressed, and the liquid evaporated half down. It is then distilled with caustic potassa, and the nicotine exhausted from the distillate by ether. The ether is removed from the ethereal solution by evaporating, finally elevating the temperature to 140°. The nicotine, which is still impure, of a brown color, is distilled at 180° in a current of dry hydrogen over quicklime.

Properties. Colorless liquid of a weak odor; when heated, of a suffocating tobacco-odor; specific gravity, 1.048; soluble in water, alcohol, and ether. Boils at 250° with partial decomposition; can, however, be slowly distilled over, even at 146°. Has an alkaline reaction; turns brown, and is decomposed in contact with the air. Exceedingly poisonous. Diatomic base.

The salts are easily soluble, and crystallize with difficulty. The free base as well as its salts give crystallizing compounds with iodine, bromine, and metallic salts.

Nicotine hydrochloro-chloromercurate, $C^{10}H^{14}N^{2}.HCl + 4HgCl^{2}$, is obtained by adding an excess of a solution of corrosive sublimate to a solution of nicotine, neutralized with hydrochloric acid. Crystallizes from water in colorless, radiating groups of needles.—*Nicotine chloromercurate,* $C^{10}H^{14}N^{2} + 3HgCl^{2}$, crystallizes in large prisms, when sufficient of a solution of corrosive sublimate is added to a dilute solution of nicotine hydrochlorate to just cause it to remain turbid.

Bromonicotine, $C^{10}H^{12}Br^{2}N^{2}$. When an ethereal solution of nicotine is poured into an ethereal solution of bromine, shiny, bright-red prisms, $C^{10}H^{12}Br^{2}N^{2}.Br^{2}.HBr$, are deposited, which lose hydrobromic acid in contact with the air, and when boiled with water or alcohol, or when their solution is allowed to stand for a long time, are converted into *bromonicotine hydrobromate*, $C^{10}H^{12}Br^{2}N^{2}.HBr$, bromine being given up. Potassa or ammonia separates free bromonicotine from the cold solutions of these salts.—Crystallizes from water in long colorless needles, permanent in the air. Difficultly soluble in cold water, easily soluble in hot water and in alcohol. A weaker base than nicotine. By boiling with caustic potassa it is reconverted into nicotine.

Nicotine combines with the iodides of alcohol radicles, forming crystalline iodine-compounds, from which silver oxide separates strongly alkaline ammonium bases, which are not volatile.

3. *Sparteine.*
$C^{15}H^{26}N^{2}$.

Occurrence. In *Spartium scoparium.*

Preparation. The plant is exhausted with water, which is slightly acidified with sulphuric acid; the extract evaporated down to a small volume, and distilled with caustic soda. The distillate is evaporated to dryness with hydrochloric acid; and the residue distilled with solid potassium hydroxide.

Properties. Colorless, thick oil, of a bitter taste; sparingly soluble in water; boils at 288°. Strongly alkaline. Has a narcotic action. Diatomic base.

Conducts itself towards alcoholic iodides in the same manner as nicotine.

4. *Opium Bases.*

In opium, the dried juice of the capsules of the poppy (*Papaver somniferum*), are contained, in addition to meconic acid (p. 430) and meconin (p. 382), six well-investigated alkaloids:—

Morphine, $C^{17}H^{19}NO^{3}$,
Codeine, $C^{18}H^{21}NO^{3}$,
Thebaine, $C^{19}H^{21}NO^{3}$,
Papaverine, $C^{21}H^{21}NO^{4}$,
Narcotine, $C^{22}H^{23}NO^{7}$,
Narceine, $C^{23}H^{29}NO^{9}$.

In all varieties of opium, morphine and narcotine are contained in the largest quantity.

Preparation. Opium is broken up and exhausted with a small quantity of water of 65°; the extract mixed with calcium chloride; filtered from precipitated calcium meconate; the filtrate concentrated by evaporation, and allowed to stand undisturbed for a long time. Morphine and codeine hydrochlorate crystallize out, and are separated from the black, treacly mother-liquor by pressing. To separate the two from each other, ammonia is added to their solution, which precipitates only the morphine, the codeine remaining in the liquid. This is concentrated by evaporation, when more morphine is deposited, and the codeine now precipitated by an excess of concentrated caustic potassa, in which any morphine, which may still be present, remains dissolved.

The mother-liquor, from morphine hydrochlorate and codeine, is diluted with water; strained through a cloth; and thoroughly precipitated with ammonia. The precipitate, collected on a cloth filter, and purified by repeated pressing and moistening with water, consists essentially of narcotine with a little papaverine and thebaine, and a great deal of resin. It is stirred with concentrated potassa-ley, forming a paste; after a time water is added, and the deposited narcotine, after being washed with water repeatedly, crystallized from boiling alcohol. Papaverine and thebaine remain in the mother-liquor. After distilling off the alcohol, the residue is exhausted with hot dilute acetic acid, and from the solution, narcotine, papaverine, and the resin precipitated with basic lead acetate. Thebaine remains in solution, and after removing the lead with sulphuric acid, it is precipitated with ammonia. For the purpose of separating the papaverine from narcotine and the resin, the precipitate is boiled with alcohol, the solution evaporated, and the residue extracted with hydrochloric acid. After evaporating again, and allowing to stand for several days, papaverine hydrochlorate, which is difficultly soluble, separates, while narcotine remains dissolved.

The ammoniacal liquid, filtered off from narcotine, papaverine, and thebaine, which contains the narceine, is mixed with lead acetate; filtered; the lead removed from the filtrate by sulphuric acid; then supersaturated with ammonia; and evaporated at a gentle heat, until a thin crust shows itself upon the surface. In a few days narceine separates in a crystalline form, and is purified by recrystallizing from water and alcohol.

Preparation of morphine and narcotine. The separation of all the bases can only be accomplished when large quantities of opium are employed. If the object is only to obtain morphine and narcotine, the opium is exhausted by digesting with dilute alcohol, and the filtrate then allowed to stand for a long time mixed with an excess of ammonia. The separated bases are treated with caustic potassa. The morphine is dissolved by this, while the narcotine remains undissolved. The latter is purified by recrystallization from alcohol. From the alkaline solution the morphine is reprecipitated by ammonium carbonate; and by dissolving in hydrochloric acid, recrystallizing the hydrochlorate, and decomposing it with ammonia, and recrystallizing the precipitate from alcohol, it is purified.

1. Morphine, $C^{17}H^{19}NO^3 + H^2O$. Crystallized from alcohol it forms small, colorless shiny prisms; precipitated by ammonia, a white powdery mass. Has a slightly bitter taste; an alkaline reaction. Soluble in 500 parts of boiling water, but very slightly in cold water, much more easily soluble in alcohol, insoluble in ether, chloroform, and benzene; easily soluble in caustic potassa, but very slightly in ammonia. Fusible, with loss of water of crystallization, congealing in a crystalline form. Narcotic poison; in small quantity causes sleep.

Monatomic base. *Morphine hydrochlorate*, $C^{17}H^{19}NO^3.HCl + 3H^2O$, forms fine prisms, of a silky lustre; easily soluble in alcohol and hot water, less soluble in cold water (in 16–20 parts). *Morphine sulphate*, $2(C^{17}H^{19}NO^3)H^2SO^4 + 5H^2O$, is similar to the hydrochlorate.

A solution of pure neutral morphine salts and also the free base are colored a beautiful dark blue by iron chloride. Heated with concentrated sulphuric acid, morphine is dissolved; the solution has a dirty, grayish-red color; and is turned a beautiful, bright blood-red by the addition of a drop of nitric acid. Heated with potassa to 200°, morphine evolves methylamine.

Oxymorphine (Pseudomorphine), $C^{17}H^{19}NO^4$. Is occasionally contained in opium. Is produced by heating a solution of one molecule of morphine hydrochlorate with one molecule of silver nitrate to 60°. By treating the precipitate with hydrochloric acid, oxymorphine hydrochlorate is obtained. It is sparingly soluble in cold water, more easily in hot water. When treated with ammonia the free base is thrown down from its solution.—Shiny powder, consisting of fine needles. Insoluble in water, alcohol, ether, and chloroform even at the boiling temperature. Fuses at 245°, at the same time undergoing decomposition; soluble in caustic potassa and soda, and in a large excess of ammonia. Gives the same reaction with iron chloride as morphine.—Monatomic base. The salts are nearly all difficultly soluble in water.

Apomorphine, $C^{17}H^{17}NO^2$. The salts of this base are produced by heating morphine or codeine with concentrated hydrochloric acid at 140–150°; by treating morphine with concentrated sulphuric acid, and by heating morphine hydrochlorate with a concentrated solution of zinc chloride to 120°. Sodium bicarbonate precipitates the free base from it.—White amorphous powder; somewhat soluble in water, soluble in alcohol, ether, and chloroform. Turns green rapidly in the air, and then forms an emerald-green solution in water.

Apomorphine hydrochlorate, $C^{17}H^{17}NO^2.HCl$. Forms colorless crystals, which, when heated or when exposed to the air in a moist condition, also become green.

2. Narcotine, $C^{22}H^{23}NO^7$. Colorless, shiny prisms, without taste; fuses at 176°, and is decomposed when heated a few degrees higher. Insoluble in cold water and caustic potassa, soluble in boiling water and in alcohol and ether. Less poisonous than morphine.

Monatomic base. The salts crystallize either badly or not at all. From their solutions alkalies precipitate narcotine in an amorphous condition.

Narcotine dissolves in concentrated hydrochloric acid, the solution having a yellow color; and this solution becomes blood-red when gently heated, and dark-violet when the heat is increased.

Narcotine, when heated with water to 250°, yields trimethylamine, together with other products; heated with concentrated hydrochloric or hydriodic acid, three methyl groups are successively eliminated, and in this way there are formed three new bases: $C^{21}H^{21}NO^7$, $C^{20}H^{19}NO^7$, and $C^{19}H^{17}NO^7$, which as yet have not been further investigated.

Heated with dilute sulphuric acid and manganese peroxide, narcotine yields *opianic acid* (p. 382) and *cotarnine;* boiled for a long time with water it is resolved into *meconin* (p. 382) and *cotarnine;* warmed with dilute nitric acid, there are formed *opianic acid, cotarnine, meconin, hemipinic acid* (p. 382), and other bodies.

Cotarnine, $C^{12}H^{13}NO^3 + H^2O$, is most readily obtained by heating narcotine with diluted (with ten times its weight of water) nitric acid at 49° until solution results. From the solution, filtered after being allowed to cool, it is precipitated by means of potassa.—Colorless, stellate prisms; soluble in boiling water, in alcohol and ammonia, but not in potassa; fuses at 100°. Monatomic base. When boiled with very dilute nitric acid, it is converted into *cotarnic acid*, $C^{11}H^{12}O^{5}$, and a substance forming good crystals, *apophyllic acid*, $C^{8}H^{7}NO^{4}$. At the same time methylamine nitrate is formed.

3. Codeine, $C^{18}H^{21}NO^{3}$. Crystallizes, anhydrous in octahedrons, or, with one molecule of water of crystallization, in rhombic crystals. Fuses at 150°. Easily soluble in hot water, alcohol, and ether; less soluble in cold water (80 parts), insoluble in potassa, soluble in ammonia. When heated from twelve to fifteen hours with concentrated hydrochloric acid, under a layer of paraffin over a water bath, it yields *chlorocodide hydrochlorate*, $C^{18}H^{20}ClNO^{2}.HCl$, from which, by means of sodium bicarbonate, the chlorinated base may be precipitated in the form of a white powder; easily soluble in alcohol and ether. When the hydrochlorate is heated with water at 130–140°, it is resolved into codeine hydrochlorate. If, on the other hand, codeine or chlorocodide hydrochlorate be heated with concentrated hydrochloric acid at 140–150°, they are both broken up, yielding methyl chloride and apomorphine hydrochlorate (p. 439).

4. Thebaine, $C^{19}H^{21}NO^{3}$. Quadratic plates, of a silvery lustre; insoluble in water, potassa, and ammonia. Easily soluble in alcohol and ether. Soluble in concentrated sulphuric acid, the solution being deep-red. Its salts can only with difficulty be obtained in a crystalline form from water, as they decompose, when their solutions are evaporated. Exceedingly poisonous. When boiled with hydrochloric acid, it is converted into an isomeric base, *thebenine*, which is amorphous, absorbs oxygen from the air, especially in the presence of alkalies, and yields salts that crystallize well.

5. **Papaverine,** $C^{21}H^{21}NO^4$. Colorless prisms; fusing point, 147°; insoluble in water, difficultly soluble in cold alcohol and ether, more easily in the hot liquids.

6. **Narceine,** $C^{23}H^{29}NO^9$. White, fine needles, of a silky lustre; fusing point, 145°. But slightly soluble in cold water and cold alcohol, more easily in the hot liquids, insoluble in ether. Is colored blue by iodine, like starch, if care be taken to avoid an excess of iodine. Taken in small quantity it causes a very sound and quiet sleep.

In addition to these, in some varieties of opium, there occur, in exceedingly small quantity, other alkaloids: *meconidine*, $C^{21}H^{23}NO^4$; *laudanine*, $C^{20}H^{25}NO^3$; *codamine*, $C^{19}H^{23}NO^3$; *cryptopine*, $C^{21}H^{23}NO^5$; *protopine*, $C^{10}H^{19}NO^5$; *laudanosine*, $C^{21}H^{27}NO^4$; *hydrocotarnine*, $C^{12}H^{15}NO^3$; *lanthopine*, $C^{23}H^{25}NO^4$; *opianine*, *metamorphine*, and *rhœadine*, $C^{21}H^{21}NO^6$. The latter base is also contained in *Papaver rhœas*.

5. *Bases of Cinchona-bark.*

In true cinchona-barks there occur principally two alkaloids:—

Quinine, $C^{20}H^{24}N^2O^2$, and
Cinchonine, $C^{20}H^{24}N^2O$,

in varying quantities. Calisaya bark (*China regia*) contains the most quinine (between 2 and 3 per cent. quinine, and 0.2–0.3 cinchonine); Huanaco bark contains the most cinchonine* (2.24 per cent. cinchonine, and 0.85 per cent. quinine). A few other alkaloids, as, for instance, *aricine*, $C^{23}H^{26}N^2O^4$, *paytine*, $C^{21}H^{24}N^2O + H^2O$, occur in only a few cinchona-barks; other bases, isomeric with quinine and cinchonine, as *quinidine*, *cinchonine*, do not appear to occur in the plants originally, but to be formed from quinine and cinchonine by a process of transformation.

* The base, *huanocine*, which has been prepared from this bark, is probably identical with cinchonine.

Preparation. Coarse cinchona powder is repeatedly macerated with water containing hydrochloric acid; the filtered solution mixed with sodium carbonate; the precipitate washed, pressed, and dried. It contains quinine and cinchonine, calcium tannate, and other substances. Both bases are extracted with boiling alcohol; the filtered, strongly colored solution neutralized with dilute sulphuric acid; and the alcohol distilled off. On cooling, quinine sulphate crystallizes out, which is obtained colorless by treatment with animal charcoal, and recrystallization. From the colored mother-liquor cinchonine sulphate is obtained. To isolate the bases their salts are dissolved in water, and precipitated with ammonia.

1. Quinine, $C^{20}H^{24}N^2O^2 + H^2O$. Precipitated by ammonia, it forms a white, earthy mass; and it is difficult to obtain it in a crystalline form even from alcohol. It is fusible, with loss of water, forming a resinous mass; tastes very bitter; reacts alkaline; soluble in 1667 parts water of 20°, in 900 parts boiling water, moderately soluble in ether, very easily soluble in alcohol.

Combines with one and with two molecules of a monobasic acid, forming salts. Most of these are crystallizable, have a very bitter taste, and are precipitated by oxalic acid; also by alkalies, platinum chloride, and tannic acid.

Quinine hydrochlorate. The salt, with one molecule of hydrochloric acid, $C^{20}H^{24}N^2O^2HCl + 1\frac{1}{2}H^2O$, forms long prisms, of a silky lustre; the salt, with two molecules of the acid, is converted into the first salt by the action of water. Platinum chloride gives a bright-yellow precipitate in the hydrochloric acid solution, which, after a time, becomes crystalline and orange-red ($C^{20}H^{24}N^2O^2.2HCl.PtCl^4 + H^2O$); mercury chloride gives a white precipitate.

Quinine sulphate, $2(C^{20}H^{24}N^2O^2)H^2SO^4 + 7\frac{1}{2}H^2O$ (the principal form in which quinine is employed as a medicament), crystallizes out of a hot saturated solution of quinine, in dilute sulphuric acid, in long, shiny prisms; as prepared in manufacturing establishments, it usually forms a white, porous, light mass, consisting of very fine and short needles, which have partially lost their water of crystallization. It tastes exceedingly bitter; is very difficultly soluble in water (in 780 parts at the ordinary temperature), more easily soluble in alcohol, easily soluble in water containing sulphuric acid, forming a blue, fluorescent liquid. Fuses like wax, and, when more strongly heated, turns a beautiful red, and is then carbonized.—If an alcoholic solution of iodine is added to a solution of this salt in acetic acid, after a time large, thin plates, consisting of a compound of quinine sulphate with iodine (herapathite) separate. These crystals are almost colorless in transmitted light; in reflected light they have a beautiful, green color and a metallic lustre, and polarize light like tourmaline plates.

The *biacid salt*, $C^{20}H^{24}N^{2}O^{2}.H^{2}SO^{4} + 7H^{2}O$, crystallizes in transparent, four-sided prisms; is more easily soluble; and has an acid reaction.

If chlorine water is added to a salt of quinine, and then ammonia, it turns an intensely emerald-green color. If, after the addition of chlorine water, a little potassium ferrocyanide and then ammonia are added, a deep red color makes its appearance.

2. Cinchonine, $C^{20}H^{24}N^{2}O$. Precipitated with ammonia, it forms a white, earthy mass. Crystallizes easily from alcohol in shiny prisms. Insoluble in water and ether, soluble in hot alcohol, less easily than quinine.

The salts resemble the salts of quinine, but are more easily soluble. They give no green color with chlorine water and ammonia, but a yellowish-white precipitate.

When heated with bromine or chlorine, substitution-products of cinchonine are formed. *Dibromicinchonine*, $C^{20}H^{22}Br^{2}N^{2}O$, is formed by heating cinchonine hydrochlorate with an excess of bromine, and, on dissolving the product in hot water, and adding alcohol and ammonia, separates on cooling in colorless crystalline laminæ. These are decomposed by boiling with alcoholic potassa, forming potassium bromide and a crystallizing base, *oxycinchonine*, which is isomeric with quinine, but essentially different from it; insoluble in water and ether.

The *chinoidine* of commerce, which in the manufacture of quinine is obtained from the last mother-liquors, contains principally two bases, isomeric with quinine and cinchonine, viz.: *quinidine* and *cinchonidine*.

3. Quinidine (Conquinine), $C^{20}H^{24}N^2O^2 + 2H^2O$. Is contained in all cinchona-barks, but more especially in the Pitaya bark; is obtained from chinoidine by extracting with a little ether, adding alcohol to the filtered solution, and evaporating it slowly.—Crystallizes from alcohol in large prisms, which are sparingly soluble in water, and effloresce readily. Fusing point, 168°. Gives the quinine reaction with chlorine water and ammonia.

4. Cinchonidine and **β-Cinchonine** are the names which have been given to two bases very similar to, but not identical with, cinchonine. A base isomeric with cinchonine is contained in quinoïdine and in commercial quinidine; another, of the composition $C^{18}H^{22}N^2O$, has been found in a few varieties of cinchona. When the sulphates of these four cinchona bases are moistened with water and sulphuric acid, and carefully kept fusing at 130° for a few hours, they are converted into the sulphates of two new resinous bases, *quinicine* and *cinchonicine*. These are isomeric with quinine and cinchonine, but entirely different from them in all their properties. Quinicine is formed from quinine and quinidine, cinchonicine from cinchonine and cinchonidine (that prepared from quinoïdine).

Heated with caustic potassa, the cinchona bases yield volatile bases, *chinoline*, and homologous substances (see end of this section).

6. *Bases of the Strychnos Species.*

In various species of *Strychnos*, particularly in nux-vomica (the seed of *Strychnos nux vomica*), and in the bean of St. Ignatius (seed of *Strychnos Ignatii*), are contained two alkaloids:—

Strychnine $C^{21}H^{22}N^{2}O^{2}$,
and Brucine, $C^{23}H^{26}N^{2}O^{4}$,

which are distinguished by their extraordinary, poisonous properties, and the power of causing tetanus when taken even in very small quantities.

Preparation. The nuts boiled with alcohol, and then dried and powdered, are exhausted by boiling with dilute alcohol. The extracts are freed of alcohol by distillation, and foreign substances precipitated by means of lead acetate; the filtrate, after the removal of lead by sulphuretted hydrogen, evaporated; and the bases precipitated by magnesia. In a week the precipitate is filtered off, dried and boiled with alcohol. On evaporating, strychnine crystallizes at first: in the mother-liquor remains brucine together with strychnine. By neutralizing with very dilute nitric acid, and allowing the strychnine nitrate to crystallize out, the two are separated, as the brucine salt remains in the mother-liquor, and crystallizes out afterwards. The salts decolorized by means of animal charcoal are now dissolved in water, and the bases precipitated by means of ammonia.

1. Strychnine, $C^{21}H^{22}N^{2}O^{2}$. Small colorless prisms of an exceedingly bitter taste; reacts alkaline. Scarcely soluble in water, insoluble in ether and anhydrous alcohol, most easily soluble in dilute alcohol, in benzene, and in chloroform.

Most salts of strychnine are crystallizable, possess an exceedingly bitter taste, and act like strychnine itself as deadly poisons.—Its solution is precipitated in a crystalline form by potassium sulphocyanide.

Strychnine nitrate, $C^{21}H^{22}N^{2}O^{2}.HNO^{3}$. Colorless fascicular needles. But slightly soluble in cold water and alcohol, more easily soluble in hot water.

Strychnine dissolves in concentrated sulphuric acid, forming a colorless liquid, which becomes a beautiful violet, when a few small pieces of potassium bichromate are added.

2. Brucine, $C^{23}H^{26}N^2O^4 + 4H^2O$. Crystalline laminæ or large colorless prisms, which effloresce in the air. Very similar to strychnine, but more easily soluble in water, and particularly in alcohol; and less poisonous. Concentrated nitric acid colors it red; on heating, yellow: and if tin chloride or ammonium sulphide is added, the yellow color is converted into a very intense violet. Concentrated sulphuric acid dissolves it, the solution having a pale red color, which soon passes into yellowish-green.

7. *Bases of the Veratrum Species.*

In the different species of *Veratrum* are contained two alkaloids:—

Veratrine, $C^{32}H^{52}N^2O^8$,
and Jervine, $C^{30}H^{46}N^2O^3$.

Veratrine occurs chiefly in sabadilla seeds (of *Veratrum sabadilla*), together with veratric acid (p. 359); and in the root of *Veratrum album;* jervine occurs only in the latter.

Preparation. In a manner similar to that described in connection with the preceding bases. They can be easily separated from each other by treatment with dilute sulphuric acid, which readily dissolves the veratrine, but converts the jervine into a very difficultly soluble sulphate.

1. Veratrine, $C^{32}H^{52}N^2O^8$. White powder or colorless prisms, becoming untransparent in the air; fuses at 115°, and solidifies, forming a resin-like mass. Scarcely soluble in water, easily soluble in alcohol and ether. Very poisonous; it causes violent sneezing, when introduced into the nose in the form of powder or in solution in small quantity. It dissolves in concentrated sulphuric acid, forming a yellow liquid, which soon becomes reddish-yellow, and finally intense blood-red. It is dissolved by concentrated hydrochloric acid, forming a colorless liquid, which, boiled for a long time, becomes colored an intense violet.

2. Jervine, $C^{30}H^{46}N^2O^3 + 2H^2O$. Colorless prisms, insoluble in water, soluble in alcohol; fuses when heated. Its salts are for the greater part very difficultly soluble in water.

8. *Bases of Berberis Vulgaris.*

In the root of these plants are contained two alkaloids:—

Berberine, $C^{20}H^{17}NO^4$,
and Oxyacanthine, $C^{32}H^{36}N^2O^{11}$ (?).

Berberine occurs besides in a great many other plants; in colombo-root (of *Cocculus palmatus*), in several *Menispermaceæ* and *Ranunculaceæ* (in large quantity, for example, in the wood of *Coscinium fenestratum*, and in the root of *Hydrastis Canadensis*, which is officinal in North America.) The preparation of berberine takes place in the same manner as that of the other bases. For the purpose of purification, the difficult solubility of the nitrate in nitric acid is made use of.

1. Berberine, $C^{20}H^{17}NO^4$. Fine yellow prisms, of a strong bitter taste, easily soluble in hot water and alcohol, insoluble in ether; loses five molecules of water of crystallization at 100°, becoming brown; fuses at 120°. Its salts are yellow and crystallizable, most of them insoluble in an excess of acid. If a dilute solution of iodine in potassium iodide be added, not in excess, to a hot alcoholic solution of a salt of berberine, green crystalline laminæ, of a metallic lustre very similar to herapathite (p. 443), separate from the solution on cooling.

Nascent hydrogen (zinc and dilute sulphuric acid or acetic acid) convert it into another base, *hydroberberine*,

$C^{20}H^{21}NO^4$, which crystallizes from alcohol in small, colorless, granular crystals of a diamond lustre, or long, flat needles, and is reconverted into berberine by nitric acid.

2. Oxyacanthine. White amorphous powder; becomes yellow in direct sunlight. Crystallizes from alcohol and ether in fine colorless prisms. Insoluble in water, soluble in alcohol and ether, particularly in the boiling liquid.

9. *Theobromine.*
$C^7H^8N^4O^2$.

Occurrence. In the cacao-bean.

Preparation. The watery extract of the broken-up beans is precipitated by lead acetate; filtered; the lead removed from the filtrate by sulphuretted hydrogen; then evaporated; and the base extracted from the residue with absolute alcohol.

Properties. White crystalline powder of a weak, bitter taste; but slightly soluble in water, alcohol, and ether, more easily in ammonia; sublimable. Weak base.

The *hydrochlorate*, $C^7H^8N^4O^2.HCl$, crystallizes from a solution in hydrochloric acid.

The solution of the free base in ammonia gives a granular crystalline precipitate of *theobromine-silver*, $C^7H^7AgN^4O^2$, when boiled for a length of time with silver nitrate.

10. *Caffeïne, Theïne (Methyl-Theobromine).*
$C^8H^{10}N^4O^2 + H^2O$.

Occurrence. Contained in coffee, tea, Paraguay tea (of *Ilex Paraguayensis*), in cola-beans and in guarana (a mass prepared from the fruit of *Paullinia sorbilis*); and is obtained from them by the same method as that described for theobromine.

Formation. By heating theobromine-silver with methyl-iodide in sealed tubes for twenty-four hours.

Properties. Colorless, long and very thin prisms of a silky lustre; of a weakly bitter taste; difficultly soluble in cold water and alcohol, more easily in hot water.

Loses its water of crystallization completely at 100°; fuses at 234–235°, and sublimes undecomposed. Weak base.

If a trace of caffeïne is dissolved in chlorine-water and the colorless liquid evaporated, there remains behind a brownish-red spot, which dissolves in ammonia, forming a beautiful violet solution.

By the action of chlorine or nitric acid on caffeïne suspended in water, it is converted into *amalic acid*, $C^{12}H^{12}N^4O^7$ (tetramethylalloxantine, $C^8(CH^3)^4N^4O^7$, see Uric acid, p. 237), methylamine and cyanogen chloride being formed at the same time. It forms colorless, difficultly soluble crystals, which become purple in contact with alkalies, and color the skin red. Further action of chlorine causes the formation of *cholestrophan*, $C^5H^6N^2O^3$ (= dimethyl-parabanic acid, $C^3(CH^3)^2N^2O^3$, p. 235).

Boiled with alcoholic potassa or with barium hydroxide, caffeïne assimilates water and gives up carbonic acid, and is converted into an uncrystalline base, *caffeïdine*, $C^7H^{12}N^4O$, easily soluble in water and alcohol. This is a stronger base than caffeïne. Its sulphate crystallizes in colorless long needles. When boiled continuously with barium hydroxide, there are formed ammonia, methylamine, carbonic acid, formic acid, and sarcosine (p. 85).

11. *Piperine.*
$C^{17}H^{19}NO^3$.

Occurrence. In the various kinds of pepper.

Preparation. Powdered white pepper is exhausted with alcohol; the solution distilled off until it forms an extract; this is then washed with water, mixed with potassa, and again dissolved in alcohol. On evaporating, piperine separates, which is purified by repeatedly dissolving in alcohol, and crystallizing.

Properties. Colorless, four-sided prisms, without taste or odor, fusing at 100°, not volatile. Scarcely soluble in water, easily soluble in alcohol. The solution tastes sharp, like pepper, and is neutral. Soluble in cold concentrated sulphuric acid, giving a dark red colored solution. Very weak base.

Decompositions. Heated with soda-lime, it yields piperidine; by boiling with an alcoholic solution of potassa, it is resolved into piperidine and piperic acid (p. 383), one molecule of water being taken up.

Piperidine, $C^5H^{11}N = C^5H^{10}.NH$. Colorless fluid; mixes with water and alcohol; boils at 106°; strongly alkaline; gives well crystallizing salts with acids.

It conducts itself towards the iodides of the alcoholic radicles exactly like conine.

Methylpiperidine, $C^5H^{10}.N.CH^3$, and **Ethylpiperidine,** $C^5H^{10}.N.C^2H^5$, are colorless fluids, boiling at 118° and 128°, respectively. Piperine is decomposed by benzoyl chloride, forming piperidine hydrochlorate and crystalline *benzoylpiperidine*, $C^5H^{10}.N.C^7H^5O$. Other acid chlorides conduct themselves in an analogous manner. Piperine is a compound of this kind.

12. *Sinapine.*

$C^{16}H^{23}NO^5$.

Occurrence. In the seeds of *Sinapis alba* as sinapine sulphocyanate.

Preparation. Yellow mustard is freed of most of its fatty oil by pressure; first exhausted with cold alcohol, and then with hot 85 per cent. alcohol; most of the alcohol distilled off; and the lighter layer of liquid, which separates on cooling, removed. Sinapine sulphocyanate crystallizes from the residue, which is purified by pressing and recrystallizing from alcohol.

The free base cannot be prepared on account of the ease with which it undergoes decomposition.

Sinapine sulphocyanate, $C^{16}H^{23}NO^5.HCNS$. Colorless very voluminous crystalline mass, consisting of fine needles, difficultly soluble in cold water and alcohol, easily in hot; fuses at 130°.

Sinapine sulphate, $C^{16}H^{23}NO^5.H^2SO^4 + 2H^2O$, crystallizes from a hot alcoholic solution of the sulphocyanate on the addition of sulphuric acid. From this salt the base can be set free by means of baryta, but it remains dissolved, imparting to the solution a deep yellow color; and on evaporating, it is decomposed.

On boiling its salts with potassium or barium hydroxide, sinapine is decomposed, yielding *choline* (p. 140) and *sinapic acid* (p. 381).

13. *Harmaline.*
$C^{13}H^{14}N^2O$.

Occurrence. In the seeds of *Peganum harmala* (a plant growing on the steppes of Russia).

Preparation. The powdered seeds are exhausted with water containing a little sulphuric or acetic acid. The alkaloid is precipitated from the extract with a concentrated solution of sodium chloride, in the form of the hydrochlorate; and this, after being purified by recrystallization, decomposed with ammonia.

Properties. Colorless, rhombic octahedrons; sparingly soluble in water and cold alcohol, more easily in hot alcohol; fuses when heated. Combines with acids, forming yellow salts, which are for the greater part easily soluble. Monatomic base.

Harmine, $C^{13}H^{12}N^2O$, occurs together with harmaline in the seeds of *Peganum harmala*, and can be separated from this by subjecting a warm hydrochloric acid solution to partial precipitation with ammonia. It is formed from harmaline by oxidation, when its nitrate is warmed with alcohol and hydrochloric acid; or from its bichromate, when heated to 120°.—Colorless shiny prisms, but slightly soluble in water, more easily soluble in alcohol.

14. *Cocaïne.*
$C^{17}H^{21}NO^4$.

Occurrence. In coca leaves (from *Erythroxylon coca*).

Preparation. The leaves are repeatedly extracted with water of 60–80°; the extract precipitated with lead acetate; the lead removed from the filtrate by means of sodium sulphate; after concentrating by evaporation, and adding sodium carbonate until the liquid shows a weak alkaline reaction, the cocaïne is extracted by shaking with ether.

Properties. Colorless and tasteless, four- or six-sided monoclinic prisms. Fuses at 98°. But slightly soluble in cold water, more easily in alcohol, very easily in ether; reacts alkaline, and has a weak, bitter taste.

On heating with hydrochloric acid it is decomposed with assimilation of water, yielding benzoic acid, methyl alcohol, and *ecgonine*, $C^9H^{15}NO^3 + H^2O$, a base, easily soluble in water; less soluble in absolute alcohol, in ether insoluble; crystallizing in colorless prisms, of a vitreous lustre, which melt at 198°.

There is also contained in coca leaves, together with cocaïne, a liquid, volatile alkaloid, *hygrine.*

15. *Atropine.*
$C^{17}H^{23}NO^3$.

Occurrence. In all parts of *Atropa belladonna* and *Datura stramonium.*

Preparation. Fresh belladonna leaves, gathered at the commencement of the period of flowering, are pressed; the juice heated to 80–90°; filtered; and after the addition of potassa, the atropine extracted by shaking with chloroform. It is extracted from the roots of the belladonna and from the seeds of the thorn-apple in a manner similar to that described in connection with the other alkaloids.

Properties. Crystallizes in fine, white prisms; fusible at 90°; tastes very disagreeably bitter and sharp. Soluble in thirty parts of boiling water, less in cold water, easily soluble in alcohol. Easily decomposable in solution, even when combined with acids, forming ammonia. *Atropine sulphate* and *hydrochlorate* crystallize in fine needles, are permanent in the air, easily soluble in water.

It is very poisonous, and the smallest quantity causes dilatation of the pupils.

When heated with barium hydroxide or hydrochloric acid, it is resolved into tropic acid (p. 354) and the acids resulting from this, atropic and isatropic acids (p. 376); and into *tropine*, $C^8H^{15}NO$, a base easily soluble in water and alcohol, which crystallizes from ether in colorless plates, fusing at 61°. Water is assimilated in this decomposition.

---

16. *Physostigmine* (Eserine), $C^{15}H^{21}N^3O^2$. In the Calabar bean (the seed of *Physostigma venenosum*, a plant growing in Upper Guinea).—Yellow, amorphous mass, fusing at 45°; sparingly soluble in water, easily soluble in alcohol, ether, benzene, and chloroform. Strongly alkaline; tasteless; exceedingly poisonous; causes a decided contraction of the pupil. The free base as well as its salts are decomposed in aqueous solutions in the air.

17. *Hyoscyamine*, $C^{15}H^{23}NO^3$. In the leaves and seeds of *Hyoscyamus niger* and *albus*.—Fine prisms, of a silky lustre; inodorous when pure, when moist or impure of a very repulsive, suffocating odor, and sharp, disagreeable taste; easily fusible. Moderately soluble in water; alkaline; very decomposable in contact with alkalies. Very poisonous; causes, like atropine, dilatation of the pupil. When heated with barium hydroxide, it is resolved into *hyoscinic acid*, $C^9H^{10}O^3$ (identical or isomeric with tropic acid), and a crystalline base *hyoscine*, $C^5H^{13}N$.

18. *Emetine*. In ipecacuanha (the root of *Cephaëlis ipecacuanha*).—White powder; fusing point, 70°; sparingly soluble in cold water, very easily soluble in alcohol; of a weak, bitter taste.—Even in very small doses it causes violent vomiting.

19. *Aconitine*, $C^{27}H^{39}NO^{10}$ (?). In the leaves and seeds of *Aconitum napellus*, in company with aconitic acid (p. 179).—Colorless, rhombic plates; almost insoluble in water even at the boiling temperature; a drop of acid causes instantaneous solution; soluble in alcohol,

ether, benzene, and chloroform. Has a weak alkaline reaction. Very poisonous.

20. *Colchicine*, $C^{17}H^{19}NO^5$. In all parts of *Colchicum autumnale.*—Colorless, amorphous mass, without odor; of a very bitter and sharp taste. Moderately soluble in water; in alcohol very easily soluble; insoluble in ether. Fuses at 140°. Very poisonous; in small quantity causes vomiting and diarrhœa. Hardly possesses basic properties, and when heated with dilute acids is converted into a substance of the same composition, *colchiceïne*, which crystallizes in needles and possesses weak acid properties.

In addition to those already described, numerous other vegetable alkaloids have been prepared, but for the greater part but slightly investigated.

In the distillation of several natural alkaloids (quinine, cinchonine, strychnine), with potassa, there results a number of fluid bases (*chinoline bases*), very similar to each other, which are distillable without decomposition. These do not occur ready formed in nature, but bases of the same composition, and perhaps identical with them, are produced in the distillation of several other bodies, and are contained in coal tar. They form an homologous series, the better known members of which are *chinoline*, C^9H^7N (boiling point, 238°), *lepidine*, $C^{10}H^9N$ (boiling point, 266–271°), and *cryptidine*, $C^{11}H^{11}N$. They are colorless liquids, sparingly soluble in water, easily soluble in alcohol and ether, and yield with acids easily soluble, crystallizing salts. They contain no hydrogen capable of replacement by alcoholic radicles, but, on the contrary, unite directly with the alcoholic iodides, forming well crystallizing iodides, from which, by treatment with silver oxide, are obtained bases analogous to *tetrethylammonium hydroxide.*

Chinoline, heated with amyl iodide, yields *amyl-*

chinoline iodide, $C^{14}H^{18}NI = C^9H^7.C^5H^{11}NI$, which, when heated with potassa, yields a beautiful, but not very permanent blue dye, *cyanide iodide* (the cyanine of commerce), $C^{28}H^{35}N^2I$. This crystallizes in beautiful green plates, of a metallic lustre; is insoluble in water and ether, easily soluble in warm alcohol; fuses at 100°. It combines directly, and without separation of iodine, with two molecules of hydrochloric acid, forming a colorless salt; when heated with silver oxide, however, it gives up its iodine, and yields a bronze-colored, amorphous base.

Lepidine conducts itself like chinoline, and yields a very similar dye, $C^{30}H^{39}N^2I$. The cyanine of commerce is either the derivative of chinoline or of lepidine, or of a mixture of both.

C. Coloring Matters, Bitter Principles, etc.

These names are applied to a large number of peculiar neutral or weakly acid substances, of which only a few have been moderately well investigated. Least known are the uncrystalline, although these often possess interest from the fact that they are frequently constituents of the so-called vegetable extracts. The following, which are mostly crystalline, are among the more remarkable substances of this kind, arranged in alphabetical order.

Aloïn, $C^{17}H^{18}O^7$. Is the purging, active principle of aloes, the juice, dried in the sun, obtained from various species of aloe, either by cutting the leaves, and allowing it to exude spontaneously, or by pressing the separated leaves. The best sort of aloes consists of brown or dark greenish-brown transparent masses, of a lustrous fracture, of a disagreeable odor and a disagreeable, bitter taste.—Aloïn forms small, colorless crystals, of a sweetish-bitter taste; difficultly soluble in cold water and alcohol; becoming brown and resinous when melted, and readily becoming amorphous under all circumstances.—When aloes is heated with nitric acid, an orange-yellow powder, *aloëtic acid*, $C^7H^2(NO^2)^2O$, is at first produced, and afterwards, by further action, *chrysammic acid* (p. 409). When fused with caustic potassa, it yields orcine (p. 307), paraoxybenzoic (p. 347), aloreic (p. 353), and oxalic acids.

Athamantin, $C^{24}H^{30}O^{7}$. In the root and half-ripe seeds of *Athamanta oreoselinum*.—Lustrous, crystalline mass, consisting of fine needles or large, four-sided prisms. Insoluble in water, easily soluble in alcohol and ether. Combines with dry hydrochloric acid and sulphurous anhydride, forming crystalline compounds. The hydrochloric acid compound is decomposed when heated alone or when its alcoholic solution is evaporated, yielding valeric and hydrochloric acids and *oreoselone*, $C^{14}H^{10}O^{3}$, which crystallizes in colorless needles; insoluble in water, difficultly soluble in alcohol and ether; is converted into a crystalline substance, *oreoselin*, $C^{14}H^{12}O^{4}$, when boiled with water containing hydrochloric acid.

Antiarin, $C^{14}H^{20}O^{5}$, a neutral substance, crystallizing in colorless laminæ; difficultly soluble in alcohol; forms compounds with acids, bases, and metallic salts; is the exceedingly poisonous ingredient of a variety of upas, an extract prepared in Java, from the sap of *Antiaris toxicaria*.

Brasilin, $C^{22}H^{20}O^{7}$(?), the coloring matter of Brazil and Pernambuco wood. Small, reddish-yellow prisms, soluble in water and alcohol, forming a red solution. Acids turn it yellow, citric acid causes this change especially beautifully; when now neutralized with an excess of alkali it turns violet or blue, with ammonia deep carmine-red. It is decolorized by sulphuretted hydrogen and sulphurous anhydride.

Cantharidin, $C^{5}H^{6}O^{2}$. Is contained in beetles of the genera *Lytta*, *Meloe*, and *Mylabris*, especially in Spanish flies (*Lytta vesicatoria*), and can be extracted from them with ether.—Colorless, four-sided prisms, or laminæ. Insoluble in water, sparingly in alcohol, easily soluble in ether; melts at 250°, and sublimes at a lower temperature without decomposition. Raises blisters on the skin. Dissolves when heated for a length of time with aqueous alkalies, assimilating water and forming salts of *cantharidic acid*, $C^5H^8O^3$. These crystallize well, but on the addition of acids to the solutions, cantharidin separates, but no cantharidic acid.

Carotin, $C^{18}H^{24}O$, together with *hydrocarotin*, $C^{18}H^{30}O$, in carrots (*Daucus carota*), deposited in the cells in microscopical crystals, the cause of the color of the carrots.—Small, reddish-brown, cubical crystals; fusible at 168°; insoluble in water, difficultly soluble in alcohol.

Carthamin, $C^{14}H^{16}O^7$, the red coloring principle of safflower, the dried flowers of *Carthamus tinctorius*. After exhausting the yellow coloring principle from pure safflower by means of cold water, the carthamin is extracted by treating with a dilute solution of sodium carbonate. The red liquid is then neutralized with acetic acid, and pure cotton immersed in it, on which all the carthamin is deposited. After washing with water the carthamin is again extracted with a dilute solution of sodium carbonate, precipitated with citric acid, and the beautifully crimson-colored precipitate washed by decantation.—Amorphous, deep-red powder, with greenish iridescence; in thin layers it has a beautiful green metallic lustre. Sparingly soluble in water, more soluble in alcohol. Soluble in alkalies, yielding a deep yellowish-red solution. Very unstable in these solutions. Melted with potassa it yields paraoxybenzoic acid and oxalic acid.

Chlorophyl. The green color of plants is occasioned by the presence of microscopical, green globules, which float in the cells. These so-called *chlorophyl-globules* consist of several substances, which inclose a green coloring principle. The composition, as well as the nature of this coloring principle, is as yet unknown; it appears to contain no nitrogen, but iron, as an essential ingredient. It dissolves in hydrochloric acid, forming a green liquid, from which it can be thrown down with boiling water. It is also soluble in alcohol and ether.

Columbin, $C^{21}H^{22}O^{7}$, in columbo root (of *Cocculus palmatus*), together with berberine and a pale-yellow, almost insoluble substance, columbic acid.—Colorless prisms, having a bitter taste.

Curcumin, $C^{10}H^{10}O^{3}$. The coloring matter of turmeric root. Can be most readily extracted by means of boiling benzene, in which but little of the remaining constituents of the root is soluble.—Orange-yellow prisms, of a weak, vanilla-like odor; fusing point, 165°; almost insoluble in water, difficultly soluble in carbon bisulphide and benzene at the boiling temperature, easily soluble in ether and alcohol; soluble in alkalies and alkaline carbonates, the solutions having a brownish-red color. Acids precipitate the curcumin from the solutions in the form of yellow powder. Paper colored with curcumin turns a brownish-red when brought in contact with liquids that have an alkaline reaction; on drying, this color changes to violet; acids restore the original yellow color; when moistened with a solution of borax, and then dried, it turns orange-yellow, and this color is not changed by dilute acids, but is converted into blue by alkalies.

Gentianin (Gentianic acid), $C^{14}H^{10}O^{5}$, in the root of *Gentiana lutea*, which owes its bitter taste, however, not to this, but to another, uninvestigated substance.—Fine, bright-yellow prisms, without taste; scarcely soluble in water, soluble in alcohol; partially sublimable; soluble in alkalies, forming bright-yellow solutions. Yields, with alkalies, salts which crystallize well, and are decomposed even by carbonic acid.

Hæmatoxylin, $C^{16}H^{14}O^{6}$, in logwood (*Hæmatoxylin campechianum*), from which it can be extracted with water or, better, ether.—Yellow, transparent prisms, which, when heated, give up water of crystallization. It possesses a sweetish taste, and is sparingly soluble in cold water, easily soluble in boiling water, in alcohol and ether, the solution being of a yellow color. Soluble in very large quantity in a saturated solution of borax. Ammonia dissolves it, forming a purple solution, which, in contact with the air, becomes dark-red, and, when evaporated, leaves behind dark-violet crystals of *hæmateïn-ammonia*, $C^{16}H^{9}(NH^{4})O^{5} + 2H^{2}O$(?). From a solution of the latter body, acetic acid throws down a brownish-red, voluminous precipitate of *hæmateïn*, $C^{16}H^{10}O^{5}$(?).

Helenin, $C^{21}H^{28}O^{3}$, in the root of *Inula Helenium*, from which it can be extracted by means of alcohol.—Colorless, four-sided prisms; insoluble in water, easily soluble in alcohol and ether; fuses at 72°. Is decomposed, by heating with phosphoric anhydride, into water, carbonic oxide, and a liquid hydrocarbon, *helenene*, $C^{19}H^{26}$.

Laserpitin, $C^{24}H^{36}O^{7}$, in the root of *Laserpitium latifolium.*—Colorless, rhombic prisms; insoluble in water, easily soluble in alcohol and ether. Fuses at 114°. Sublimes undecomposed. Is resolved into angelic acid (p. 124) and an amorphous substance, *laserol*, $C^{14}H^{22}O^{4}$, when heated with caustic potassa.

Peucedanin (Imperatorin), $C^{12}H^{12}O^{3}$, in the root of *Peucedanum officinale* and *Imperatoria obstruthium.*—Colorless prisms, of bright lustre. Insoluble in water, soluble in alcohol and ether; fuses at 75°; not sublimable. Is decomposed by boiling with an alcoholic solution of potassa into angelic acid and oreosilin (compare Athamantin, p. 456).

Picrotoxin, $C^{12}H^{14}O^{5}$, in the seeds of Cocculus indicus (*Menispermum cocculus*). The powdered seeds are extracted with boiling alcohol, the alcohol distilled off from the extract, and the residue boiled with a large quantity of water. Foreign bodies are precipitated from the aqueous solution by means of lead acetate; the filtrate evaporated after treatment with sulphuretted hydrogen; and the picrotoxin, which now separates, purified by repeated crystallization from water.—Stellate groups of colorless needles, of an intensely bitter taste. Difficultly soluble in cold water, more easily soluble in hot water and in alcohol. Very poisonous. Combines with alkalies, baryta, and lime, forming gummy compounds, which are obtained pure only with great difficulty. When boiled with weak acids, it is converted into non-crystallizing, weak acids. By boiling it with nitric acid, oxalic acid is produced.

Porrisic acid (Euxanthic acid), $C^{19}H^{16}O^{10}$, in purree, a yellowish coloring matter, imported from the East Indies, probably the juice of a plant evaporated with magnesia. Purree consists essentially of magnesium euxanthate.—The acid forms yellow, shiny prisms, sparingly soluble in cold water, easily soluble in alcohol and ether. Its salts, with the alkaline metals, are yellow, crystallizable. The magnesium salt crystallizes particularly beautifully. With chlorine and bromine, it forms yellow-colored crystallizing acids, containing chlorine and bromine ($C^{19}H^{14}Cl^2O^{10}$ and $C^{19}H^{14}Br^2O^{10}$). When heated to 180°, it is decomposed into carbonic anhydride, water, and *euxanthon*, $C^{13}H^8O^4$, which is also formed when the acid is dissolved in concentrated nitric acid; this substance crystallizes in yellow prisms; when melted with potassium hydroxide it yields, first *euxanthonic acid*, $C^{13}H^{10}O^5$, at a higher temperature *hydroquinone* (p. 303).

Quassin, $C^{10}H^{12}O^3$, the bitter ingredient in the wood of *Quassia amara* and *excelsa* from South America.—Fine, colorless, crystalline laminæ, of an exceedingly bitter taste; but slightly soluble in water, easily soluble in alcohol; fusible, solidifying in a resinous state.

Santalic acid (Santalin), $C^{15}H^{14}O^5$, in sandal wood (from *Pterocarpus santalinus*), from which it is extracted with alcohol. It is precipitated from the solution with lead acetate, and the precipitate decomposed with dilute sulphuric acid and alcohol.—Microscopical crystals, of a beautiful red; insoluble in water, soluble in alcohol and ether. Soluble in alkalies, with a violet color.

Santonin (Santonic acid), $C^{15}H^{18}O^{3}$, in the seeds of *Artemisia santonica*, of which it forms the active principle. The seeds are mixed with about half their weight of caustic lime, and extracted with dilute alcohol. The extract, after being distilled, is filtered and boiled with acetic acid. The santonin, which crystallizes out on cooling, is purified by recrystallizing from alcohol, and treating with animal charcoal.—Very shiny, colorless prisms, of a slightly bitter taste; scarcely soluble in water, easily soluble in hot alcohol; fusible at 170°, solidifying in a crystalline form, but when cooled suddenly becoming amorphous; only partially sublimable. Weak acid. The colorless crystals become a bright yellow in direct sunlight, frequently cracking; their composition, however, does not appear to be changed. Exposed to direct sunlight for a long time in an alcoholic solution, it is converted into *photosantonin*, $C^{23}H^{34}O^{6}$(?), formic acid and other products being formed at the same time. Photosantonin crystallizes in colorless laminæ, fusing at 64–65°.

Scoparin, $C^{21}H^{22}O^{10}$, in *Spartium scoparium*, together with sparteine (p. 435).—When its alcoholic solution is allowed to evaporate spontaneously, it is obtained in small, stellate crystals. Slightly soluble in cold water and cold alcohol, easily soluble in the hot liquids. It is dissolved by the alkalies with a yellowish-green color, and from these solutions it is thrown down by acids as a white, amorphous precipitate. Fused with potassa it yields phloroglucin and protocatechuic acid.

Smilacin, $C^{18}H^{30}O^{6}$ (?), in sarsaparilla (the root of various species of *Smilax*), from which it can be obtained by boiling with alcohol.—Fine, colorless prisms. Insoluble in cold water, slightly soluble in hot water, forming a disagreeably tasting and strongly foaming liquid. Easily soluble in ether and hot alcohol.

D. Ethereal Oils.

The name *ethereal* or *volatile oils* has been applied to all those compounds, which pass over with the vapor on heating certain plants or parts of plants with water, and form the odorous constituent of these plants. Most of them are mixtures of compounds containing oxygen and hydrocarbons. The oxygenized bodies are of very various character, and belong to entirely different chemical groups. They are in some cases acids (valeric acid in oil of valerian, pelargonic acid in the oil of *Pelargonium roseum*); in some, aldehydes (cuminol in oil of cumin, cinnamic aldehyde in oil of cinnamon); in others, ethers (methyl salicylate in gaultheria oil); in others still, phenols (thymol in oils of thymian and monarda), etc. They have already been described, as far as they are well known, in connection with these compounds, to which they bear a close chemical relation. The hydrocarbons called *terpenes*, contained in the various ethereal oils, have nearly all the same composition in percentages. Their formula is a multiple of the simple formula, C^5H^8.

By far the greater number boils without decomposition at 160–170°, and these have the molecular formula, $C^{10}H^{16}$. A smaller number boils at 250–260°, and has the molecular formula, $C^{15}H^{24}$; and a still smaller number, which boils above 300°, has the formula, $C^{20}H^{32}$.

The hydrocarbons of the formula $C^{10}H^{16}$ show the greatest similarity in their chemical and physical properties, and with many the observed difference between them is confined to the smell and the action upon polarized light. Most of them are imperfectly investigated, and a more careful investigation will probably show a thorough chemical identity of many of them.

The best known is

Turpentine-oil.
$C^{10}H^{16}$.

Occurrence. In all parts of all coniferous trees. When fir, pine, larch trees, etc., are accidentally bruised or intentionally incised, there flows from them a clear, thick, viscid liquid, turpentine. This is a solution of a resin in oil of turpentine. As it occurs in commerce, it is yellow, sometimes clear, sometimes turbid, of a bitter taste and slight odor. Distilled with water, oil of turpentine passes over and the resin remains behind.

Properties. Colorless, thin oil of a peculiar, unpleasant odor; specific gravity, 0.89; boiling point, 160°. Vapor density, 4.698. Almost insoluble in water, miscible with alcohol and ether in all proportions. It dissolves sulphur, phosphorus, and a great many other substances that are insoluble in water. It absorbs oxygen from the air, and converts it partially into ozone. Towards polarized light it conducts itself differently, according to its origin: that obtained from the turpentine of *Pinus maritima* (French oil of turpentine), of *Pinus Mughus*, *Abies pectinata* (templin oil), and *Laryx europæa*, rotates the plane of polarization towards the left; that from the turpentine of *Pinus australis* (English oil of turpentine), however, towards the right.

Under the influence of heat, acids, etc., it is converted into other varieties with other properties, but without a change in the percentage composition. The oil, which is originally produced in the trees, too, appears to be different from that prepared from turpentine. Pine branches distilled with water give an entirely different, almost agreeably smelling oil, which, when distilled over potassa, becomes ordinary oil of turpentine.

Transformations. Oil of turpentine, left for months in contact with acidified water,* is partially converted into a colorless and inodorous body, *terpine* (hydrate of oil of turpentine), $C^{10}H^{20}O^2 + H^2O$, which crystallizes very regularly; fuses at 100°, losing its water of crystallization; sublimes at a higher temperature undecomposed; is sparingly soluble in cold water, easily soluble in hot water and in alcohol and ether. When its so-

* A well-shaken mixture of eight parts of oil of turpentine, two parts of weak nitric acid, and one part of alcohol, is the best.

lution is heated with a trace of some acid, it is converted into a volatile oil, *terpinol* $C^{20}H^{34}O$, of an odor like hyacinthes; specific gravity, 0.852; boiling point, 168°. Both compounds, terpine and terpinole, form, with hydrochloric acid gas, *bihydrochlorate of oil of turpentine*, $C^{10}H^{16}.2HCl$, a crystallizing substance.—Oil of turpentine also absorbs this gas in large quantity, and forms with it a liquid and a solid compound. Both have the composition $C^{10}H^{17}Cl = C^{10}H^{16}.HCl$. The solid one crystallizes from alcohol, or when carefully sublimed, forming clear shiny prisms; has an odor like camphor, and fuses at 115°. The fluid compound is a neutral, colorless oil, that floats on water. Both, when heated with caustic lime, yield oils of the composition C^5H^8, but differing from oil of turpentine in odor and other physical properties (camphilene, terpilene, terebilene).

The action of fuming hydrochloric acid on oil of turpentine, continued for months, causes the formation of $C^{10}H^{16}.2HCl$, which is identical with the compound resulting from terpine and terpinole.

Chlorine converts oil of turpentine into two isomeric chlorine compounds, $C^{10}H^{12}Cl^4$, one of which is crystalline and fuses at 110–115°, the other a colorless, viscid liquid.

When heated with phosphonium iodide, oil of turpentine is converted into a hydrocarbon, $C^{10}H^{20}$, boiling at 160°.

Boiled continuously with dilute nitric acid, oil of turpentine yields acetic, propionic, butyric, oxalic, toluic, terephtalic, camphresinic acids (see Camphor p. 468), and

Terebic acid, $C^7H^{10}O^4$, a body that crystallizes in colorless prisms, which fuse at 168°. Difficultly soluble in cold water, easily soluble in hot water. Is resolved into carbonic anhydride and *pyroterebic acid*, $C^6H^{10}O^2$ (p. 125), when subjected to distillation.

Terebentilic acid, $C^8H^{10}O^2$, results when terpine in the form of vapor is conducted over heated (to 400°) soda-lime, and the product decomposed with hydrochloric acid.—Small, white needles; fuses at 90°, and boils at 250°; almost insoluble in cold water, more easily soluble in hot water.

The following oils consist entirely of hydrocarbons, which are isomeric with oil of turpentine, and very similar to it:—

Oils of *lemon*, *orange*, *apricot*, and *bergamot*, in the shells of the various species of *Citrus;*

oils of *lavender* and *spike*, in the blossoms and leaves of *Lavandula angustifolia* and *Lavandula latifolia;*

oils of *juniper* and *sabine*, in the berries of *Juniperus communis* and *Juniperus sabina;*

oil of *camphor trees*, the oil in *elemi*, in *balsam of copaiva*, in *black pepper*, in *cubebs*, etc.

The following are mixtures of several compounds, which are partially but little known:—

Anise oil, from the seed of *Pimpinella Anisum* (p. 380).*

Apricot-blossom oil, from the blossoms of *Citrus Aurantium.*

Cajeput oil, from the leaves of species of *Melaleuca.*

Calamus oil, from the root of *Acorus Calamus.*

Caraway oil, from the seeds of *Carum carvi.*

Cascarilla oil, from the bark of *Croton Eluteria.*

Chamomile oil, from the flowers of *Matricaria Chamomilla.* Deep blue.

Cinnamon oil, from the barks of *Persea Cinnamomum* and *Persea cassia* (p. 373).

Clove oil, from cloves (blossom-buds of *Caryophyllus aromaticus*) (p. 381).

Coriander oil, from the leaves of *Coriandrum sativum.*

Curled-mint oil, from the green portions of *Mentha crispa.*

Fennel oil, from the seeds of *Fœniculum officinale.**

Lozenge oil, from *Ruta graveolens* (p. 112).

* The oils marked with a star solidify even above 0°, depositing oxygenized compounds (stearoptenes).

Peppermint oil, from the green portions of *Mentha piperita.**

Roman-caraway oil, from the seeds of *Cuminum Cyminum* (pp. 289 and 385).

Roman-chamomile oil, from *Anthemis nobilis* (p. 124).

Rose oil, from the petals of *Rosa centifolia.**

Rosemary oil, from the green portions of *Rosmarinus officinalis.*

Sage oil, from the green portions of *Salvia officinalis.*

Sassafras oil, from the roots of *Laurus Sassafras.*

Tansy oil, from all parts of *Tanacetum vulgare.*

Taragon oil, from the leaves of *Artemisia Dracunculus* (p. 380).

Thyme oil, from the green portions of *Thymus vulgaris.*

Wormseed oil, from the seeds of *Artemisia santonica.*

Wormwood oil, from the green portions of *Artemisia Absinthium.*

E. Camphor.

1. *Japan Camphor* (*Ordinary Camphor*).
$C^{10}H^{16}O$.

Is obtained in Japan and China by distilling all portions of *Laurus camphora* with water. Is prepared artificially by heating the oil of sage or valerian with nitric acid.—Colorless, translucent, tough mass of peculiar odor and taste. Crystallizes readily, either from its solution in alcohol or by sublimation, in shiny crystals, which refract light very strongly. Floats on water, rotating when in small pieces; fuses at 175°; boils at 204°. In an alcoholic solution, it turns the plane of polarization to the right. Volatilizes, even at the ordinary temperature, and sublimes in crystals. Easily inflammable. Sparingly soluble in water, easily in alcohol, ether, and oils.

When heated with an alcoholic solution of potassa, camphor is resolved into *camphic acid*, $C^{10}H^{16}O^2$, an acid insoluble in water, but little known, and *borneol*, $C^{10}H^{18}O$. Oxidizing substances convert it into *cam-*

phoric and *camphoronic acids.* Distilled with phosphorus pentasulphide, it is resolved into water and cymene (p. 289); the same decomposition takes place when it is distilled over phosphoric anhydride or zinc chloride, but in the two latter cases, toluene, xylene, pseudocumene, and other hydrocarbons are produced in considerable quantity at the same time. When heated with phosphorus chloride, two crystalline chlorine compounds, $C^{10}H^{15}Cl$ and $C^{10}H^{16}Cl^2$, are produced, which lose hydrochloric acid easily, and are then converted into cymene. When heated with hydriodic acid, it yields a mixture of hydrocarbons.

Monochlorcamphor, $C^{10}H^{15}ClO$, is produced by adding camphor to an aqueous solution of hypochlorous acid.—Colorless, crystalline mass; but slightly soluble in water, easily soluble in alcohol and ether. Fuses at 95°, and decomposes at 200°, hydrochloric acid being given off. Heated with alcoholic potassa, it yields *oxycamphor*, $C^{10}H^{16}O^2$, together with other substances as yet unknown. Colorless needles; fuse at 137°; sublime without decomposition; insoluble in water, easily soluble in alcohol.

Monobromcamphor, $C^{10}H^{15}BrO$, and *Dibromcamphor*, $C^{10}H^{14}Br^2O$, are produced by the action of bromine on camphor at 100–120°.—Both compounds crystallize in colorless prisms. Monobromcamphor melts at 76°, and boils without decomposition at 274°; dibromcamphor melts at 114.5°, and boils at 285°, undergoing material decomposition.—When bromine is added to a saturated solution of camphor in chloroform, crystalline *camphor bromide*, $C^{10}H^{16}OBr^2$, is deposited, which, when kept, especially in sunlight, is converted into monobromcamphor.

Campholic acid, $C^{10}H^{18}O^2$, is produced when camphor, in the form of vapor, is passed through a heated mixture of calcium hydroxide and potassium hydroxide; and by the action of potassium on a solution of camphor in petroleum.—Crystallizes from alcohol in colorless prisms; insoluble in water; fuses at 80°; sublimes.

Camphoric acid, $C^{10}H^{16}O^{4} = C^{8}H^{14}(CO.OH)^{2}$, is produced by digesting, for a long time, and repeatedly distilling camphor with 10 parts concentrated nitric acid.—Crystallizes from water in thin, colorless laminæ, of a weak acid taste, without odor. Fuses at 175–178°, and emits a pungent odor. Difficultly soluble in cold water, more easily soluble in hot water and in alcohol. When heated it is decomposed into water and *camphoric anhydride*, $C^{10}H^{14}O^{3}$, which sublimes in long, shiny prisms, and fuses at 217°.

Bibasic acid. The *calcium salt*, $C^{10}H^{14}O^{4}Ca + 8H^{2}O$, forms easily soluble crystals; when heated is resolved into carbonic anhydride, water, and phoron (p. 109).

Camphoronic acid, $C^{9}H^{12}O^{5}$. Is formed, together with the preceding compound, when camphor is heated with nitric acid. Also by direct oxidation of camphoric acid.—Brilliant, white, microscopic needles; very easily soluble in water, alcohol, and ether. When fused with potassium hydroxide, it yields butyric acid.

The substance, formerly described as *camphresinic acid*, $C^{10}H^{14}O^{7}$, is a mixture of camphoric and camphoronic acids.

Oxycamphoronic acid, $C^{9}H^{12}O^{6} + H^{2}O$, is obtained by heating camphoronic acid with two atoms of bromine in sealed tubes.—Crystallizes in the monoclinate system from water; easily soluble in alcohol, ether, and water. Loses its water of crystallization at 100°; begins to melt at 210°; distillable. Appears to be triatomic, bibasic.—Is decomposed by potassium hydroxide like camphoric acid.

Camphocarbonic acid, $C^{11}H^{16}O^{3}$. Sodium acts very violently, but without an evolution of hydrogen, on a solution of camphor in toluene, and a yellowish, amorphous mass is deposited, consisting of sodium-camphor and sodium-borneol. If carbonic anhydride is now conducted into the solution, and, after saturating with this, water is added, and the aqueous solution separated from the toluene, borneol is deposited from the solution in a short time. This is filtered off, and to the filtrate hydrochloric acid added, when camphocarbonic acid is precipitated.—Small, colorless crystals. Difficultly soluble in water, easily soluble in ether. Begins to melt at 118–119°, but at this temperature it is resolved into camphor and carbonic anhydride. Monobasic acid.

A camphor, very similar to the ordinary variety just described, separates from the oil of *Matricaria parthenium*, when that portion of the oil, which boils between 200–220°, is cooled down to —5°. It differs from ordinary camphor only in the fact of its turning the plane of polarization to the left. When oxidized it also yields camphoric acid, but while ordinary camphoric acid is dextro-rotatory, the acid obtained from Matricaria-camphor rotates the plane of polarization just as far to the left, and by mixing equal weights of both acids, an optically inactive camphor is produced. Here exists hence exactly the same relation as between dextro- and laevo-tartaric acids, and racemic acid (p. 184).

2. *Camphor of Borneo* (*Borneol, Camphol*).
$C^{10}H^{18}O = C^{10}H^{17}.OH$.

Obtained in Borneo and Sumatra from *Dryobalanops camphora*. It occurs partially in a solid, crystalline state, in cavities in the trunks of old trees, in company with a volatile oil, which is contained in larger quantity in the younger trees, and flows out of incisions made in them. It is produced from camphor when this is heated with alcoholic potassa, or heated successively with sodium and water (see Camphocarbonic acid, p. 468).—Very similar to ordinary camphor, but more friable, having an odor like that of camphor and of pepper; fusing point, 198°; boiling point, 212°. Heated with phosphoric anhydride, it is resolved into water and *borneene*, $C^{10}H^{16}$, which appears to be identical with the oil of camphor (from *Laurus camphora*), occurring in nature, as well as with the non-oxygenized portion of oil of valerian. When the latter is allowed to stand in contact with caustic potassa, and then subjected to distillation, it is converted into borneol.

Borneol is an alcohol. Heated to 200° with acids, it yields ethers, water being eliminated. *Stearic ether*, $C^{10}H^{17}.O.C^{18}H^{35}O$, is a colorless, thick, volatile oil, solidifying after a time.—When borneol is heated with concentrated hydrochloric acid, a crystallizing chloride, $C^{10}H^{17}Cl$, is produced, which is isomeric with hydrochlorate of oil of turpentine, and very similar to the solid variety of the latter.

Bodies isomeric with borneol are contained in oil of hops, cajeput oil, coriander oil, in the oil from *Osmitopsis asteriscoïdes*, and in oil of Indian geranium.

Patchouli-camphor, $C^{15}H^{28}O$, in oil of patchouli, is homologous with borneol.—Crystalline mass. Fusing point, 54–55°; boiling point, 296°.

3. *Mentha-Camphor* (*Menthol*).

$$C^{10}H^{20}O = C^{10}H^{19}.OH.$$

Is obtained by distilling *Mentha piperita* with water, and separates from the oil (oil of peppermint) that passes over when subjected to a very low temperature.—Colorless, transparent prisms, of a strong odor and taste of peppermint; fuses at 36°, and boils at 210°. Combines with acids, forming ethers, like borneol; when heated with hydrochloric acid or phosphorus chloride, yields a liquid chloride, $C^{10}H^{19}Cl$; and with phosphoric anhydride a liquid hydrocarbon, *menthene*, $C^{10}H^{18}$; boiling at 163°.

F. Resins.

This name is applied to a group of bodies but little known, which occur, very widely distributed, in the most various portions of plants, mostly in company with volatile oils, dissolved in which they frequently flow from trees from accidental or intentional cuts. The crude resins are never crystallized; they have the form of drops, like gum; are colored mostly yellow or brown; translucent, brittle, with a shiny, conchoidal fracture; often possessing a weak smell and taste. In a pure state they are colorless, inodorous, and tasteless; several are then crystallizable. They are fusible, inflammable, not volatile, non-conductors of electricity; insoluble in water, soluble in alcohol, in ether and volatile oils.

Most resins, occurring in nature, consist of several simple compounds, which, however, as a rule, are exceedingly difficult to separate and prepare in a pure condition.

Most resins are weak acids or anhydrides of acids.

The number of resins is very large. Only a few of them, which are of importance on account of technical or pharmaceutical employment, are investigated.

The conduct of a great many resins, when heated with fusing potassa (to 1 part of resin 3 parts potassa), is of interest. They then yield, as a rule, together with fatty acids: protocatechuic acid (p. 356), paraoxybenzoic acid (p. 347), phloroglucin (p. 311), and resorcin (p. 306.)

1. *Resins Proper.*

1. Colophony (Pine-resin). The turpentine, which flows from the pines, firs, larches, and other species of *Pinus*, solidifies gradually on the trees, forming a resin, partially by evaporation, partially by oxidation of the oil. Distilled with water, oil of turpentine passes over, the resin remains behind; it is known under the name of colophony.

Colophony is brownish-yellow, translucent, brittle, fusible; easily soluble in alcohol, ether, fatty and volatile oils. When it is digested for several days with ordinary alcohol (at the strongest 70 per cent.), filtered hot, and water added to the filtrate until a slight turbidness remains, crystals separate in a few hours, consisting of

Sylvic acid (Abietic acid), $C^{20}H^{30}O^{2}$. Crystallizes from alcohol in pointed, oval laminæ. Insoluble in water, soluble in alcohol, ether, benzene, and chloroform; fuses at 120°; monobasic acid. The alkaline salts are yellowish, brittle masses; easily soluble in water and alcohol. The *magnesium*, *calcium*, and *barium* salts are white, flocculent precipitates; difficultly soluble in water, more easily in alcohol.

An acid, isomeric with sylvic acid,

Pimaric acid, $C^{20}H^{30}O^{2}$, forms the principal ingredient of the resin from *Pinus maritima* (Galipot). It is deposited from its alcoholic solution in hard crusts. Fusing point, 149°; perfectly insoluble in water, difficultly soluble in cold alcohol and ether, easily in the hot liquids. Monobasic acid. Yields crystallizable salts. Boils above 320°, and, when distilled, is converted into sylvic acid.

2. Copaiba-resin. From species of *Copaifera*, indigenous in Brazil, is obtained, by means of incisions, *balsam of copaiba*, a bright-yellow, clear, thick liquid, resembling oil of turpentine, which consists of resin and a turpene.

The resin, freed of oil by distillation with water, is an acid, *copaivic acid*, $C^{20}H^{30}O^{2}$(?), isomeric with sylvic and pimaric acids; it can be obtained in exceedingly regular, clear, colorless crystals by dissolving the resin in alcohol and allowing it to evaporate spontaneously; or by shaking the balsam for a long time with a concentrated solution of ammonium carbonate, and then acidifying the lower aqueous solution of the ammonium salt with acetic acid.—On the other hand, the different varieties of balsam appear to contain somewhat different or altered resins, and hence the resin cannot always be obtained in a crystalline form. In Maricaibo balsam there is contained an acid, *metacopaivic acid*, $C^{22}H^{34}O^{4}$, very similar to copaivic acid; it crystallizes in laminæ, and fuses at 205-206.°

3. Elemi, from several species of *Amyris* in East and West Indies.—Yellow, translucent, soft, smelling somewhat of volatile oil. It contains a non-crystallizable resin, easily soluble in cold alcohol, and a crystallizable resin, soluble only in boiling alcohol. The latter can be obtained only in fine needles, and does not combine with bases; takes up water from the air and from alcohol, and becomes amorphous. It is also contained in *anime-resin* and in *euphorbium.*

4. Betulin, in birch-bark. Appears as a fleecy vegetation in the bark when gradually heated. Obtained most readily by boiling the outer bark with water, drying, and boiling with alcohol, from which it crystallizes in nodules. Colorless; fuses at 200°, emitting an odor like that of the bark; is sublimable in a current of air.

5. Lactucone, in the juice of *Lactuca virosa.*—Fine, colorless prisms, solidifying after fusion in an amorphous form; very similar to betulin.

6. Copal, from Africa, East Indies, etc., of various origin.—Large (externally opaque, on a fractured surface clear), slightly yellowish or yellow pieces, frequently inclosing insects; hard, brittle, heavier than water. Fusible, but undergoing a change at the same time. Insoluble in alcohol; soluble in ether; soluble in caustic potassa. There are different varieties of copal; they consist of several difficultly separable resins.

7. Dammara resin, from *Pinus Dammara,* in the Moluccas.—Very similar to copal; fusible, however, without decomposition, and soluble in hot alcohol.

8. Mastic, from *Pistacia Lentiscus* in Greece.—Small, yellowish, translucent, round grains, of a slight aromatic odor and taste. Consists of several amorphous compounds, of different solubility in aqueous alcohol.

9. Olibanum (Incense), from a species of *Boswellia*, a tree in Abyssinia.—Subglobular, pale-yellow, translucent grains; for the greater part soluble in alcohol; fusible with decomposition, and emitting a balsamic odor.

10. Sandarac, from *Thuja articulata*, in Barbary.—Small, pale-yellow, translucent, brittle grains; easily fusible; soluble in alcohol.

11. Gum-lac is produced in consequence of the sting of an insect (*Coccus lacca*) in the branches of certain trees in the East Indies. When still on the branches it is called in commerce *stick-lac*, separated from them *seed-lac*, and in a purified, melted condition *shell-lac*, in which state it forms thin, brittle, brown, translucent pieces. Gum-lac contains several other products, originated by the insects, especially a coloring principle and fats.

12. Benzoin-gum, from *Styrax benzoin*, a tree growing in Sumatra.—Large, brittle lumps, which, on the fractured surface, appear to be conglomerated of smaller white and brownish pieces. It has a pleasant vanilla-like odor, evolves vapors of benzoic acid when heated, which forms about 18 per cent. of the gum. Some varieties contain cinnamic acid in addition to benzoic.

13. Guaiacum, from *Guajacum officinale*, a tree growing in the West Indies.—Large, translucent, brittle lumps, externally bluish-green, on the fractured surface brown. Its powder becomes green in contact with the air, or under the influence of chlorine-water. Its solution in alcohol becomes deep blue when acted upon by ozone, nitrous acid, chromic acid, iron sesquichloride, etc.

The principal ingredient of guaiacum is a weak bibasic acid, crystallizing from acetic acid in concentrically arranged needles, *guaiaretic acid*, $C^{20}H^{26}O^4$, which fuses at 75–80°, and by slow distillation is resolved into *pyroguaiacin*, $C^{19}H^{22}O^3$, a crystalline substance, and into *guaiacol*, $C^7H^8O^2$ (p. 305), a liquid. When the resin is subjected to destructive distillation, there are formed besides these, *creosol*, $C^8H^{10}O^2$ (p. 309), and several other bodies.

14. Acaroid resin, from *Xanthorhœa hastilis*, a tree growing in New Holland.—Yellow colored; yields picric acid abundantly when heated with nitric acid, and phenol when subjected to distillation.

15. Dragon's blood, from a number of trees in the West Indies.—Small, dark-brown, opaque lumps; in the form of powder blood-red; soluble in alcohol with red color. Contains a little benzoic acid. Yields toluene when distilled.

16. Amber, a resinous product of extinct coniferæ, occurring in lignite beds.—Colorless, yellow or brownish yellow, transparent or translucent, hard, often inclosing insects. Fusible, undergoing decomposition, however, the succinic acid contained in it being volatilized. In addition to this acid it contains a volatile oil and two resins, soluble in alcohol and ether. Its principal mass consists of an amorphous substance, insoluble in alcohol, fatty and volatile oils, as well as in alkalies.

2. *Caoutchouc.*

Flows from incisions in various trees growing in South America and the East Indies (especially several varieties of *Siphonia* and *Ficus elastica*), as a juice of creamy consistence which dries up, forming caoutchouc. The juice contains albumen in solution, in which the caoutchouc is suspended in the form of globules. When heated, the albumen coagulates, and the caoutchouc globules adhere together with it in coagulated masses. Pure caoutchouc, as it does not occur in commence, is colorless and transparent.

Its characteristic property is elasticity. It loses this property when kneaded for a long time between warm rollers, and is converted into an homogeneous, black, conglomerated mass, which, as long as it is warm, can be moulded at desire. In this condition other substances, especially sulphur, can be intimately mixed with it (vulcanization of caoutchouc), by which means its valuable properties are materially increased. It is not fusible without decomposition. It is insoluble in alcohol, soluble in ether, carbon bisulphide, and a few volatile oils. Insoluble in caustic potassa. It contains no oxygen, and when subjected to dry distillation is resolved almost entirely into a mixture of liquid hydrocarbons.—*Gutta-percha* is a very similar substance, from various species of *Isonandra* in Madras. It is solid at ordinary temperatures, scarcely elastic, becomes soft and elastic, however, when warmed.

3. *Gum-resins.*

Important on account of their employment in medicine; are usually mixtures of peculiar resins, frequently also caoutchouc with protein compounds, gum and volatile oils. They exude from the plants as milky juices or emulsions, which contain the gum or protein compounds in solution, the oils and resins in suspension, and besides these, frequently other substances. Of these latter may be mentioned, assafœtida, euphorbium, galbanum, gamboge, myrrh, opium, etc. Their consideration belongs to the field of pharmacology.

4. *Balsams.*

Under this head are understood exuded or expressed thick, ropy, odorous liquids from certain trees or shrubs. They are either solutions of resins in ethereal oils, or mixtures of substances which bear a close relation to the latter. The following substances are balsams:—

Turpentine (p. 462).
Canada balsam, from *Abies balsamea.*
Balsam of copaiba (p. 472).
Storax balsam (p. 372).
Peru and Tolu balsams (p. 312).

VII. BILIARY COMPOUNDS.*

1. Glycocholic acid, $C^{26}H^{43}NO^{6}$. Fresh ox-bile is evaporated to dryness over a water-bath, the residue exhausted with absolute alcohol, the alcohol separated from the filtered solution by evaporation or distillation, and the residue, which, if necessary, is diluted with water, mixed with milk of lime, and gently warmed, the greater part of the pigment present being by this means thrown down in combination with lime. The mixture is filtered and to the cold filtrate dilute sulphuric acid is added until turbidness remains (an excess to be avoided). In a few hours the whole liquid has become a pulpy mass, consisting of crystals of glycocholic acid, which is purified by pressing, dissolving in a great deal of lime-water, and reprecipitating with sulphuric acid.—Or the bile, evaporated to dryness, is extracted when cold with absolute alcohol, the solution decolorized by digesting with animal charcoal, filtered and treated with a little ether; hereupon, after standing for several hours, a plastery, colored mass is deposited; from this the liquid is poured off, and again treated with fresh ether. After a long time a mixture of sodium glycocholate and taurocholate (crystallized bile) is deposited in fine, colorless needles, which, after the liquid is poured off, is washed with a little ether, and then dissolved in water. This solution is mixed with dilute sulphuric acid until it is decidedly milky, and then allowed to stand. In twenty-four hours the liquid has become filled with

* On the occurrence of these substances in the bile, see the section *Animal Chemistry, Bile.*

crystals of glycocholic acid, which are purified by recrystallizing from boiling water. The taurocholic acid remains in solution. The amorphous mass at first deposited also usually becomes crystalline after a long time.—Or fresh ox-bile, decolorized with animal charcoal, is precipitated with a solution of sugar of lead, the precipitate exhausted with boiling 85 per cent. alcohol, and this solution treated with sulphuretted hydrogen, while still hot. From the filtrate from lead sulphide, glycocholic acid is deposited in crystals, when water is added until turbidness remains.

Glycoholic acid forms very fine, white needles, which pressed together in a mass represent a leaf of a silky lustre. It has a sweetish-bitter taste; is but slightly soluble in water, easily soluble in alcohol. On evaporating its alcoholic solution, it remains behind as a resinous mass. Fusible, but not volatile. Its alkaline salts are easily soluble, and have a very sweet taste. Heated with sulphuric acid, and a solution of sugar, it gives a violet color.

When boiled with alkalies, glycocholic acid, takes up one molecule of water, and is converted into glycocol (p. 84), and

Cholic acid, $C^{24}H^{40}O^5$. This is obtained most readily by boiling crystallized bile for several days with baryta-water or potassa.—Colorless, shiny octahedrons, almost insoluble in water, soluble in alcohol and ether. A solution of its alkali salts has a strong, bitter taste at first, afterward sweetish. It is precipitated from these solutions by acids, as a soft amorphous mass, which however soon becomes crystalline, especially on the addition of ether. With sulphuric acid and a solution of sugar, it shows the same reaction as glycocholic acid.

When boiled with acids, glycocholic acid is also resolved into glycocol and cholic acid, but in this case the latter immediately undergoes a further change, giving up water, and being converted into *dyslisin*, $C^{24}H^{36}O^3$, a grayish-white, amorphous body, not acid, which, when boiled with an alcoholic solution of potassa, is again converted into cholic acid.

2. Taurocholic acid, $C^{26}H^{45}NSO^{7}$. When fresh ox-bile is mixed with neutral lead acetate, a white, plaster-like precipitate is produced, which contains, besides mucus and coloring matter, particularly lead glycocholate. When the filtered liquid is mixed with basic lead acetate, a similar precipitate is formed, which consists of basic lead glycocholate and lead taurocholate and the lead salts of the fatty acids contained in bile. From this precipitate taurocholic acid can be separated with difficulty.—It is more readily obtained from dog-bile, in which no glycocholic acid is contained, or at the most but traces. The alcoholic extract of the dried bile, decolorized with animal charcoal, is evaporated to dryness, the residue dissolved in a small qauntity of alcohol, and the sodium taurocholate precipitated with ether. To an aqueous solution of this salt lead acetate is added together with some ammonia, the precipitate filtered off, dissolved in boiling absolute alcohol, and decomposed by means of sulphuretted hydrogen. The filtrate from lead sulphide is evaporated down to a small volume at a moderate temperature, and then a large excess of ether added to it, which causes the separation of free taurocholic acid as a syrupy mass. This is after a time converted for the greater part into acicular crystals of a silky lustre. It is easily soluble in water and alcohol. When dry it can be heated above 100° without decomposition. When heated with water to 100°, it is resolved into cholic acid and taurin (p. 141). It suffers the same decomposition, when its salts, or the bile, are boiled with alkalies or acids, or by the putrefaction of bile.

Two acids very similar to the two described are *hyoglycocholic acid*, $C^{27}H^{43}NO^{5}$, and *hyotaurocholic acid*, $C^{27}H^{45}NSO^{6}$, which are contained in the bile of the pig. When boiled with alkalies they are resolved into glycocol, taurin, and an acid very similar to cholic acid, *hyocholic acid*, $C^{25}H^{40}O^{4}$. In goose-bile is also contained a distinct acid, *chenotaurocholic acid*, $C^{29}H^{49}NSO^{6}$, very similar to taurochloric acid, which, when boiled with baryta-water, yields taurin and *chenocolic acid*, $C^{27}H^{44}O^{4}$.

3. Lithofellic acid, $C^{20}H^{36}O^{4}$. Is the principal ingredient of a variety of oriental bezoars, and is probably a product of a metamorphosis of the ingredients of the bile, taking place in the living body of a species of goat or antelope. It can be extracted from the bezoars with boiling alcohol.—Crystallizes from alcohol in colorless, short prisms. Insoluble in water, soluble in alcohol. Fuses at 204°. With sulphuric acid and a solution of sugar, it gives the same reaction as glycocholic acid.

4. Cholesterin, $C^{26}H^{44}O + H^{2}O = C^{26}H^{43}.OH$. It is extracted from evaporated bile by means of ether. It is further an ingredient of the brain, the nerves, the yolk of eggs, the yellow bodies in the ovary of the cow, blood, meconium, of feces, and a number of hydropic fluids. It has also been lately found in the vegetable kingdom, and apparently it is here likewise very widely distributed; especially is it contained in vegetable seeds, for example in rye, in barley, in peas, in maize, and in all the young parts of plants. It is collected in largest quantity in biliary calculi, which often consist entirely of it. These concretions are dissolved in boiling alcohol, and then filtered; on cooling the cholesterin crystallizes out.

It crystallizes from alcohol in colorless laminæ, of a pearly lustre, from a mixture of alcohol and ether in regular, tabular prisms. Inodorous and tasteless; fuses at 145°, and solidifies in crystalline form; heated without access of air, it sublimes for the greater part undecomposed. Insoluble in water, but slightly soluble in cold alcohol. Caustic potassa, even by boiling, produces no change in it. In dry chlorine gas, it becomes heated to fusing, hydrochloric acid gas being evolved. When gradually and completely saturated with chlorine gas, it forms a white, amorphous mass,

$C^{26}H^{37}Cl^{7}O$, insoluble in water, fusing at 60°.—It unites directly with one molecule of bromine, when the latter is added to its solution in carbonbisulphide, as long as the color of the bromine disappears. The resulting compound, *cholesterindibromide*, $C^{26}H^{44}Br^{2}O$, crystallizes in small needles, insoluble in water, difficultly soluble in alcohol, easily soluble in ether; fusing point, 147°; reconverted into cholesterin by nascent hydrogen.

Cholesterin is a monatomic alcohol. When treated with hydrochloric acid or phosphorus chloride, it yields *cholesteryl chloride*, $C^{26}H^{43}Cl$, which, when in a pure state, forms colorless, acicular crystals, soluble in alcohol.—Cholesterin combines with acids, forming ethers. The *stearic ether*, $C^{26}H^{43}.O.C^{18}H^{35}O$, is produced by heating cholesterin with stearic acid to 200° in sealed tubes.—Small, white, needles; fusing at 65°. The *benzoic ether*, $C^{26}H^{43}.O.C^{7}H^{5}O$, prepared in the same manner, forms small, crystalline plates, which melt between 125° and 130°.

Dehydrating substances, concentrated sulphuric or phosphoric acid, convert cholesterin into various crystallizing, isomeric, or polymeric hydrocarbons, $C^{26}H^{42}$.

5. Coloring matters of bile. Biliary calculi from the human being, which contain a great deal of pigment, are pulverized, freed of cholesterin and fat by treatment with ether, and then freed of other bodies by successive extraction with hot water and chloroform. The residue, which contains earthy phosphates and carbonates and compounds of the coloring matters with lime and magnesia, is treated with hydrochloric acid, and the coloring matters, which remain after drying, extracted with chloroform. This solution, on being evaporated to dryness, leaves a dark, crystalline residue behind, from which absolute alcohol extracts *bilifuscin*, while *bilirubin* remains behind, which can be purified by repeatedly dissolving in chloroform and precipitating with alcohol. The portion that remains undissolved by chloroform in the first place, still contains a great deal of bilirubin, together with *biliprasin* and a brown, humus-like body, *bilihumin.* It is first treated with alcohol, in which only biliprasin dissolves with a beautiful green color, and the bilirubin is then extracted by means of boiling chloroform.

Bilirubin, $C^{16}H^{18}N^{2}O^{3}$ (or $C^{9}H^{9}NO^{2}$). Dark-red crystals, of the color of chromic acid. In an amorphous state, as obtained by precipitating it from its solution in chloroform by means of alcohol, an orange-red powder. Fuses when heated and is decomposed, swelling up at the same time. Insoluble in water, very slightly in alcohol and ether, more easily in chloroform, benzene, and carbon bisulphide. It dissolves in alkalies very easily, forming a deep orange-red liquid, which, on the addition of a great deal of water, becomes a pure yellow, and, even in very dilute condition, colors the skin yellow. Hydrochloric acid precipitates the bilirubin from this solution. When calcium or barium chloride or lead acetate or other metallic salts are added to a weakly ammoniacal solution of bilirubin, dark-brown colored, flocculent precipitates separate, which are the metallic compounds (salts) of bilirubin. When an alkaline solution of bilirubin is mixed with commercial nitric acid (containing hyponitric acid), the yellow solution becomes first green, then blue, violet, ruby-red, and, finally, a dirty-yellow; especially do these changes of color take place when alcohol is previously added.

Biliverdin, $C^{16}H^{20}N^{2}O^{6}$ (or $C^{8}H^{9}NO^{2}$). Is produced when the solution of bilirubin in caustic soda, is shaken with air or boiled. It then becomes green, and, on the addition of hydrochloric acid, biliverdin is deposited.—Lively green precipitate; insoluble in water, ether, and chloroform, easily soluble in alcohol. With nitric acid it gives the same reaction as bilirubin.

Bilifuscin, $C^{16}H^{20}N^{2}O^{4}$. Is contained in biliary calculi only in very small quantity. In order to obtain it in a pure condition, its alcoholic solution (see above) is evaporated, the residue first freed of fatty acids by treatment with ether, and then of bilirubin, by means of chloroform (bilifuscin purified with ether is insoluble in chloroform; its solubility in chloroform is caused by the presence of fatty acids), then dissolved in alcohol, and this solution evaporated.—Almost black, lustrous, brittle mass. Insoluble in water, ether, and chloroform, easily soluble in alcohol, forming a solution of a deep-brown color; also in alkalies. With nitric acid it gives the same reaction as bilirubin.

Biliprasin, $C^{46}H^{22}N^{2}O^{6}$. Is obtained in a pure condition when its alcoholic solution (see above) is evaporated, foreign substances removed from the residue with ether and chloroform, the residue redissolved in a little alcohol, and this solution evaporated.—Lustrous, almost black, in pulverized condition greenish-black mass. Fuses when heated, and decomposes, at the same time increasing in volume. Insoluble in water, ether, chloroform; easily soluble in alcohol, forming a clear green solution. If ammonia be added to this solution, it turns brown (difference between it and biliverdin); hydrochloric acid turns it green again.

Bilihumin. Is contained in considerable quantities in biliary calculi, and is produced from all the other biliary coloring-matters when their solutions in soda-ley are exposed to the air for a long time.—Blackish-brown, powdery substance.

In addition to the coloring-matters described, others occasionally occur in bile. These are, however, uninvestigated up to the present.

VIII. PROTEIN COMPOUNDS.

The name, protein compounds, is applied to certain nitrogenized substances, very similar to each other, which are widely distributed in the animal and vegetable kingdom.

Formation. Only in plants. The animal organism receives these most important ingredients ready formed in the food, and it has only power to assimilate them, and to cause multitudinous metamorphoses in them.

Composition. This is for all protein compounds so similar that one might be led to suspect that it is the same, and that the variations found are merely caused by the presence of other substances, which they contain to a certain extent in organized intertexture, and from which they have not as yet been separated.—They all contain carbon, hydrogen, nitrogen, oxygen, and sulphur, but the latter in such small quantity that it is impossible to express its presence by means of a probable formula. The following composition of albumen gives a representation of the composition of these bodies:—

Carbon	53.5	per cent.
Hydrogen	7.0	"
Nitrogen	15.5	"
Oxygen	22.4	"
Sulphur	1.6	"

Properties. Most protein compounds can apparently exist in two conditions: a soluble condition, in which they usually occur in nature, and an insoluble or coagulated condition, into which they are converted either spontaneously or by the action of heat or acids. In the soluble form they are contained in plants and animal fluids, and can, for the greater part, be obtained by evaporating at a temperature below 50°. It is, however, exceedingly difficult and scarcely possible to thoroughly purify them of all foreign substances. In this condition they form translucent masses, similar to gum Arabic; inodorous and tasteless; soluble in water, insoluble in alcohol and ether.—In the insoluble, coagulated condition they are white, amorphous, principally flocculent or clotted masses; insoluble in ordinary solvents. A few of them are soluble in dilute mineral acids. Concentrated acetic acid dissolves them all with the aid of heat, some rapidly, others slowly. Dilute potassa also dissolves them all after a time, when kept at a temperature of 60°, forming potassium sulphydrate and other decomposition-products. In a weak solution in acetic acid, potassium ferrocyanide or ferricyanide, and potassium platinocyanide give white amorphous precipitates. When gently heated with a solution of mercury nitrate (containing nitrous acid),* they turn beautiful red with a slight tinge of violet. Hydrochloric acid dissolves them all with the aid of heat, and this solution, when boiled for a long time, becomes a beautiful and very deep violet. When a solution of sugar and concentrated sulphuric acid is carefully poured upon them, they turn first red, then dark violet, the colors being the more beautiful, the more freely the air has access.

Decompositions. At a high temperature they are decomposed, yielding ammonium carbonate and numerous other products. When kept in a moist condition they easily undergo putrefaction and yield ammonia, ammonium sulphide, acids of the acetic acid series, leucine (p. 98), tyrosin (p. 350), and other bodies, with which we are only very imperfectly acquainted. Leucine and tyrosin are also produced from them when

* *Millon's reagent.* Prepared by dissolving 1 part mercury in 1 part concentrated nitric acid, and diluting the solution with double the volume of water.

they are heated for a long time with dilute sulphuric acid, and when they are carefully fused with potassium hydroxide. In the first case aspartic acid (p. 160) and glutamic acid (p. 163) are formed at the same time from many protein compounds. When oxidized with dilute sulphuric acid and manganese peroxide, or potassium bichromate, they yield numerous products: formic, acetic, and other acids of the same series, benzoic acid, oil of bitter almonds, and aldehydes of the fatty acids, prussic acid, acetonitrile, and homologous nitriles, etc.

The most important varieties of protein compounds are:—

1. Albumen. Three modifications of albumen, differing slightly from each other in their properties, have been distinguished: *Vegetable albumen*, *albumen of serum*, and *albumen of eggs*. Vegetable albumen is contained in nearly all vegetable juices; albumen of serum in blood-serum of vertebrate animals, in lymph, chyle, in the transudates and pathological cystic fluids, in urine in diseases of the kidneys, abundantly in colostrum, and in small quantity in milk. Albumen of eggs is contained only in birds' eggs.—*Vegetable albumen* cannot be obtained in a pure, uncoagulated condition from vegetable juices.—*Serum-albumen* is obtained most readily from blood-serum or hydrocelic fluid by diluting with twenty volumes of water, and precipitating the protein compounds, which accompany the albumen, by the careful addition of acetic acid or continued passage of carbonic anhydride into the solution. The liquid, filtered off after twenty-four hours, is evaporated at 40°, and separated by dialysis from the salts; or precipitated with lead acetate, and the precipitate decomposed with carbonic anhydride.—Pure serum-albumen forms a clear, not very tenacious liquid, from which it can be precipitated with alcohol. Directly after being thrown down this precipitate is soluble in water, but, in a few minutes, it is converted into the coagulated condition. It is not precipitated by carbonic anhydride, dilute mineral acids, and tartaric acid, but is gradually changed by them, and the change takes place the more rapidly, the higher the temperature and the stronger the acid is. Concentrated hydrochloric acid gives a precipitate in the solution, soluble in an excess. Perfectly neutral solutions of serum-albumen coagulate at 72–73°. Acids or salts elevate, alkalies lower the temperature required for coagulation. It is not coagulated by shaking with ether.—For the preparation of *egg-albumen*, white of egg is passed through linen, filtered without access of air, and then further purified in the same manner as serum-albumen. In most of its properties it shows a perfect resemblance to serum-albumen; it, however, rotates the plane of polarization somewhat less strongly to the left. With hydrochloric acid it gives a precipitate, which is very difficultly soluble in water and in an excess of hydrochloric acid; is thoroughly and instantaneously coagulated by alcohol and also when shaken with ether.—Concentrated potassa-ley, added to a concentrated solution of either modification of albumen, causes the formation of a transparent, solid jelly of potassium albuminate.

2. Casein. In milk and the yolk of eggs. In order to prepare it, skimmed milk is mixed with dilute sulphuric acid, the white precipitate, after being filtered off and washed, while still wet, is digested with lead carbonate; the filtered solution, containing the casein, then evaporated, after the removal of the lead with carbonic anhydride or sulphuretted hydrogen.—Or the milk is diluted and precipitated with acetic acid; the precipitate washed with water, alcohol, and ether; dissolved in very dilute caustic soda; again precipitated with acetic acid; and again washed as before.—The nature of casein is not yet sufficiently well known. Its solubility in water appears to be dependent upon the presence of alkalies. Casein, which is free of alkali, is insoluble in water, and in a solution of common salt, but easily soluble in water containing a very little hydrochloric acid or alkali. From this solution it is precipitated (in the absence of alkaline phosphates), when exactly neutralized, in the form of a flocculent, fibrous, non-gelatinous mass. In an alkaline solution and in the milk, it is not coagulated by boiling; the solution only forms a skin of coagulated casein on the surface, which, when removed, is reformed. When a slightly alkaline solution of casein is poured into an excess of an acid, a flocculent precipitate is formed, which is soluble in pure water. This is a compound of casein with the acid employed. —The real coagulation of casein is brought about in a peculiar manner, as yet not satisfactorily explained: *i. e.*, by contact with the internal mucous membrane of the stomach of the calf. Skimmed milk, warmed with a small piece of such a stomach (rennet) at 50–60°, coagulates so thoroughly, that only very small quantities of casein remain in a state of solution in the whey. The coagulum formed in this way, mixed with fat, forms *cheese*, when dried.

In a coagulated condition casein resembles coagulated albumen in nearly all its properties.

3. Legumin. In leguminous and many other seeds, a protein compound, very similar to casein, is contained. In order to prepare it, beans or lentils are softened with warm water and triturated to a paste. This paste is then diluted with water and the skins sieved off. Legumin, in a state of solution, is contained in the liquid that passes through the sieve; starch, in a state of suspension, is also contained in it, but the latter is deposited if the liquid is allowed to stand quietly. By adding a very little acetic acid, the legumin is thrown down as a gelatinous mass; to purify it, it is washed out with water, alcohol, and ether. The crude solution soon becomes acid, if left alone, on account of the formation of lactic acid; and thus coagulates spontaneously. It does not coagulate when boiled, but, as in the case of milk, a skin is formed on the surface, which is always reformed when removed. The dissolved condition of legumin appears, as in the case of casein, to be caused by the presence of alkalies. When oily seeds (*e. g.*, shelled and broken-up sweet almonds), are freed of most of their oil by pressure, and then boiled for a short time with water, most of the legumin, in addition to sugar and gum, is dissolved and can be reprecipitated by acetic acid; the albumen, however, remains behind coagulated. Or, if the last particles of fatty oil are extracted from the pressed sweet almonds by means of ether, and they are then treated with cold water, legumin and albumen are dissolved. If the solution is now heated to boiling, the albumen is thrown down in a coagulated condition, and the legumin can afterward or also previously be precipitated with acetic acid. In addition to this, another protein compound, *emulsin*, is contained in sweet almonds. This compound appears to be different from those described, and is characterized by its peculiar action on amygdalin and salicin (pp. 412 and 414).

4. Fibrin. Only known in the insoluble condition. Separates spontaneously from the blood, a short time after the latter has left the living organism, and forms the principal part of the blood-clot. It is not contained in circulating blood, but is formed after this has left the body, by the union of two albuminoid substances contained in blood and other animal fluids, viz: *fibrinogenous* and *fibrino-plastic substance*. The fibrino-plastic substance (*paraglobulin*) is obtained from blood-serum, by carefully adding acetic acid to serum diluted to twenty times its volume, or better by conducting carbonic anhydride into the diluted solution, and then washing the precipitate with water. The fibrinogenous substance can be prepared in the same manner from the pericardium-fluid of the cow, or the fluid of hydrocele. Both of these protein compounds are insoluble in water, and in a saturated solution of common salt; soluble in a dilute solution of common salt, and in very dilute hydrochloric acid. When one of these substances is dissolved in a dilute solution of sodium chloride and an equal amount of the other substance is added in a moist condition, the whole mass coagulates after a time, forming fibrin.—Fibrin is a grayish-white mass, which, in a moist condition, is tough and elastic, in a dry condition, hard and brittle. It is insoluble in water, dilute hydrochloric acid, and in a solution of sodium chloride, and swells up in the latter, as well as in a solution of saltpetre.

If the blood is allowed to flow from the vein directly into a concentrated solution of sodium sulphate, the precipitation of the fibrin is prevented; if, however, the solution is now poured or filtered off from the blood-globules, and saturated with sodium chloride, there is produced a flocculent precipitate, the aqueous solution of which coagulates in a short time, as fibrin.

A body, very similar to the fibrinogenous and fibrinoplastic substances, is *globulin*, in the crystalline lens. It resembles these substances in nearly all its properties, but with neither of them does it form fibrin. A neutral solution of globulin begins to grow turbid at 73°, but is not coagulated below 93°, when it also exhibits an acid reaction.

5. Vegetable fibrin. A protein compound is contained in the different varieties of grain in a coagulated condition; it very strongly resembles animal fibrin. It is obtained from flour, in largest quantity from wheat flour, by mixing it with water, so as to form a stiff dough, tying this up in a cloth, and then kneading it for a long time in cold water, thus thoroughly washing out the soluble ingredients, starch and albumen. It remains behind as a grayish-yellow, tough, pasty mass, capable of being drawn out in thin layers (glutin). Boiling alcohol extracts from it a sticky substance (vegetable gelatin, glutin), likewise containing nitrogen, which, when dried, is brown and viscid; ether extracts a fatty oil.—When seeds sprout, this protein compound is converted into a soluble substance, *diastase*, which has as yet not been prepared in a pure condition. It is remarkable on account of its property of converting large quantities of starch into glucose, when dissolved in water, and heated to 50–70° (compare p. 194).

6. Myosin. Forms the principal mass of the muscle-clot, coagulated after death during the stiffening of the body. Can be obtained most readily by washing out cut-up muscular substance with water, treating the pressed residue with a mixture of one volume of a saturated solution of sodium chloride and two volumes of water, and precipitating the slimy liquid thus obtained with water, or by the addition of sodium chloride.—A mass, insoluble in water, also insoluble in a concentrated solution of sodium chloride, but soluble in a solution which does not contain more than 10 per cent. of sodium chloride. It dissolves easily in very dilute hydrochloric acid (4 cc. fuming acid to 1 litre of water), and can be precipitated unchanged from this solution immediately afterward by means of sodium carbonate, but undergoes changes when left in contact with hydrochloric acid. It dissolves in dilute alkalies, forming alkaline compounds, the solutions of which coagulate at a higher or lower temperature according as they are more or less alkaline. In the yolk of egg, in the crystalline lens, and a few cystic liquids, there occur protein compounds, which are very similar to myosin.

7. Syntonin (Parapeptone). Is formed from myosin by dissolving it in very dilute hydrochloric acid, and from all other protein compounds by dissolving them in concentrated hydrochloric acid. Water precipitates a compound of syntonin with hydrochloric acid from these solutions. It also occurs in the gastric juice, being probably the first product of transformation of the protein compounds.—Is obtained most readily by dissolving coagulated white of egg or pure fibrin in fuming hydrochloric acid, precipitating the filtered solution by the addition of water, redissolving the precipitate in pure water, and carefully precipitating with sodium carbonate. Or chopped meat is washed with water until it is colorless; and then treated with very dilute hydrochloric acid (0.1 per cent.), which converts the myosin into syntonin, and dissolves it. It is precipitated from the filtered liquid by neutralization.—When thrown down it forms a gelatinous, flocculent precipitate; insoluble in water and in a solution of sodium chloride, easily soluble in dilute hydrochloric acid and very dilute alkaline carbonates. The solutions are not coagulated by boiling; the syntonin, however, separates, when they are mixed cold with sodium chloride, ammonium chloride, and various other salts.

ANIMAL CHEMISTRY.

1. *The Blood.*

As long as it flows in the veins, the blood consists of a clear liquid and numberless so-called *blood-corpuscles*, which are suspended in the liquid. The blood-corpuscles are only recognizable with the aid of the microscope; they are disciform, circular, or elliptical in shape, and of a yellowish-red color in all vertebrate animals. The clear fluid of the blood contains, as its principal ingredients, three dissolved protein compounds: *albumen* (serum-albumen), *fibrinogenous*, and *fibrino-plastic* substances. (In regard to these see pp. 486 and 489.)

When drawn from the veins blood coagulates very soon, the fibrinogenous and fibrino-plastic substances uniting with each other to form insoluble *fibrin* (p. 489), which incloses the blood-corpuscles, and forms with them an adherent, gelatinous mass, the coagulum, *placenta sanguinis*. From this, on further shrinking, the remaining solution of albumen separates as a yellowish, almost clear, alkaline fluid, the serum, *serum sanguinis*. Only in the case of cold-blooded animals does the blood coagulate so slowly, and are the blood-corpuscles so large that they can be separated from the dissolved fibrin by means of filtration before the coagulation. The reason why the blood remains fluid in the living organism, but coagulates when no longer under the influence of life, is as yet not known with certainty. We only know that it is the walls of the bloodvessels which prevent the coagulation in the organism; and that, outside of the animal body, the coagulation may be accelerated by elevated temperature, violent motion, and by access of oxygen; retarded by saturation with carbonic anhydride, by the addition of a small quantity of free potassa or ammonia, by a number of alkaline salts, and by slight acidification with acetic acid; and, finally, entirely prevented by the neutralization of the previously acidified blood with ammonia; or, better, by allowing the blood to flow directly from the vein into a concentrated solution of sodium sulphate.—When blood is beaten while flowing from the vein, the fibrin separates in stringy masses, without inclosing a large amount of corpuscles, which, for the greater part, remain unchanged, suspended in the serum. On account of the slimy character of the latter, however, they cannot be separated from it. If ten times its volume of a mixture of one volume concentrated solution of sodium chloride and from nine to ten volumes water is added, and the whole allowed to stand, the separation becomes possible. They then sink, the supernatant liquid can be poured off, and the blood-corpuscles washed with a solution of sodium chloride of the same strength as that employed in the mixture.

The red blood-corpusles of man and most mammalia consist almost exclusively of a peculiar body, *hæmatoglobulin* or *hæmoglobin*, while in the blood-corpuscles of birds and several mammalia, considerable quantities of albuminous substances occur together with this. When free of albuminous substances, or when these are previously removed, the corpuscles crystallize, on the addition of water at a low temperature, in rhombic crystals, only a very small quantity remaining dissolved in the water. They can be purified by recrystallization from water at a low temperature, if a little alcohol is added. After being dried over sulphuric acid at a temperature below 0°, they form a brick-red powder, still containing 3–4 per cent. of water. When this is dissolved in alcohol, and cooled, crystals are again deposited. It decomposes very readily in the presence of water. If an aqueous solution of pure hæmoglobin is allowed to stand for some time at the ordinary temperature; and if hæmoglobin is dried at a temperature above 100°, it becomes dirty-brown, and decomposes, yielding a brown coloring-matter, two protein compounds similar to fibrin and albumen, and several acids (formic, butyric).

Light that has passed through an aqueous solution of hæmoglobin, or through blood, yields a spectrum which shows two very characteristic absorption bands, lying in yellow and green (between Frauenhofer's lines D and E). If the blood is saturated with carbonic acid, or heated to 40–50° after the addition of a drop of ammonia, or if mixed with a drop of ammonium bisulphide, both of these bands disappear, and, instead of them, there appears a single band (between the lines C and D, nearer C.). The original bands reappear, however, immediately if the blood, thus treated, is shaken with atmospheric air.

The most remarkable property of hæmoglobin is its capability of uniting with oxygen and other gases, to form peculiar unstable compounds, which also crystallize, and give up these gases very readily, even in a vacuum, without losing the capability of reuniting with the gases. Hæmoglobin containing oxygen is bright red, that which contains no oxygen is darker. This is the cause of the different color of arterial (with hæmoglobin containing a great deal of oxygen) and venous blood (with hæmoglobin containing little or no oxygen); the optical phenomena above mentioned also find their explanation in this fact. Only the hæmoglobin containing oxygen (oxyhæmoglobin) gives the characteristic absorption-bands.

Hæmoglobin decomposes hydrogen peroxide, and sets oxygen free.

In contact with alkalies and acids, oxyhæmoglobin is resolved into protein compounds, small quantites of fatty acids, and a coloring-matter, *hæmatin*, which, when dried, has a grayish-brown color, and contains 9 per cent. of iron. Its composition can perhaps be expressed by the formula, $C^{34}H^{34}FeN^{4}O^{5}$. Hæmoglobin containing no oxygen gives another very unstable coloring matter, hæmochromogene, which takes up oxygen with great avidity, and is converted into hæmatin.—If pure hæmoglobin, or the blood-corpuscles, or even the blood itself, be heated with an excess of concentrated acetic acid, with an addition of sodium chloride or other chlorine compounds, *hæmin* is formed. This crystallizes in microscopical, well-developed, rhombic plates, insoluble in water, alcohol, and ether, of a yellowish-red color; it yields hæmatin and metallic chlorides when heated with alkalies, and is hence, probably, a compound of hæmatin with hydrochloric acid.—The formation of these crystals of hæmin is principally made use of for the detection of blood.

A coloring matter differing from those described is *hæmatoïdin*, which occurs as a decomposition product of a constituent of blood, probably hæmoglobin, particularly blood which has been in a stagnated condition for some time outside of the vessels, in extravasations of blood from ruptured Graafian vesicles, in extravasations in the brain, in suppurating cavities, etc. It can be prepared most readily from the yellow bodies of the ovaries of the cow. These are triturated with glass-powder; allowed to stand in contact with chloroform for several days; the filtered, yellow solution evaporated at the ordinary temperature; and the residue treated with a little ether for the purpose of removing fat. It crystallizes in small, transparent prisms of the color of chromic acid, is insoluble in water and alcohol, difficultly soluble in ether, easily soluble in chloroform, forming a yellow solution, and easily soluble in carbon bisulphide, forming a red solution. In many respects, it resembles bilirubin (p. 482), and has frequently been mistaken for this; it differs from it, however, very materially in its insolubility in alkalies. The yellow coloring matter of the yolk of eggs is probably identical with hæmatoïdin.

As, during the circulation of the blood in the body, in its passage through the capillary vessels and the organs of secretion and excretion, transformations of its principal ingredients are incessantly taking place;

and as the material for the formation of its principal ingredients, prepared or formed by digestion from the food, is constantly added in the form of chyle and lymph, which are emptied into it; it must contain many other substances besides the principal ingredients. The discovery and recognition of these have, however, been but very imperfectly successful.

When blood is subjected to microscopical investigation, two other kinds of spherical bodies are seen besides the blood-corpuscles. These are colorless, present in less abundance, and some of them smaller than the corpuscles. The smaller ones are drops of fat, the larger (so-called colorless blood-corpuscles) are the lymph- or chyle-corpuscles. In the chemical analysis of blood, as difficult and imperfect as it is as yet, several other substances besides the principal ingredients are found.

Different kinds of fat are found in it, but in small quantity, partially suspended as minute drops, partially in solution in saponaceous combination; and also cholesterin (p. 480).

The liquid, which remains over after the coagulation of the blood by heating, leaves behind a yellow, extract-like mass when evaporated, consisting of a mixture of organic substances and salts. Urea and succinic acid belong to the first, the latter are principally sodium chloride and salts of potassium and sodium with fatty acids, phosphoric acid, and sulphuric acid. In carnivorous animals sodium phosphate is principally found; in graminivorous, sodium carbonate at the same time. Analyzed as a whole, blood has nearly the same elementary composition as the organic muscular substance, as a whole, of the same animal, and contains also the same amount of inorganic ingredients.

1000 parts by weight of blood-corpuscles contain 688 parts of water and 312 of solid ingredients. Of the latter 8-9 parts are inorganic salts, not reckoning the iron of hæmoglobin.

1000 parts by weight of serum contain 903 of water and 97 of solid ingredients; of the latter 8.5 parts are inorganic salts.

The *Crusta inflammatoria* or *buffy coat*, a yellowish-white, semi-solid, membranous mass, which is sometimes formed on blood let from the vein, is produced by the sinking of the blood-corpuscles to a certain extent before the coagulation of the fibrin, the upper layer of the solution thus coagulating without inclosing blood-corpuscles. It is produced under the most varied conditions, particularly when the specific gravity of the serum of the blood is lowered, so that the corpuscles can sink more rapidly, as, for example, after frequent letting of blood. It is almost always formed in the blood of certain animals, as, for instance, the horse, in which the corpuscles possess the property of sinking readily. It was formerly incorrectly considered as a sign of inflammation.

Many variations in the composition of the blood have been observed in diseased conditions of the body. In diabetes for instance, it contains sugar, which moreover is said to be contained in normal blood, though in exceedingly small quantity.

Respiration. The dark venous blood, mixed with the chyle of the thoracic duct, is poured into the right auricle of the heart, through the two grand trunks of the venous system, the venæ cavæ; from the auricle it passes into the corresponding ventricle, and from this is projected into the lungs. It is returned from the latter to the left auricle as bright-red arterial blood; passes into the left ventricle from which it is thrown into the whole body by means of the principal artery, the aorta. The lungs consist of the fine, terminal, vesicular branches of the bronchial tubes, on the walls of which exceedingly fine networks of capillary blood-vessels are spread out. The inspired air is brought in contact with the venous blood, through the fine walls of these air cells, which are impregnated with water; 4–5 per cent. of the volume of the air being absorbed as oxygen, and a volume of carbonic acid, together with some nitrogen, almost equal to that of the absorbed air, being given off, and in the expiration removed from the body, together with a large amount of water vapor. This carbonic acid is formed in the blood during its circulation in the body, probably in the finest capillary networks and in the tissues of the organs themselves; the oxygen collected in the lungs being at the same time absorbed in its place. Carbonic acid, together with small quantities of oyygen and nitrogen, is found in blood from all parts of the body, more oxygen being present in arterial blood, however, than in venous blood (*cf.* p. 495). Venous blood, on the other hand, contains relatively more carbonic acid than arterial blood, the carbonic acid amounting to about one-fifth the volume of the blood.

The quantity of gases given off at each normal expiration is in the case of man about 500 cc.

The amount of water given off from the lungs in twenty-four hours is about 320 grm. or about 236 pounds per year.

The amount of carbonic acid expired in twenty-four hours is on an average 867 grm., containing 236.5 grm. of carbon. Hence in a year over 172 pounds of carbon are given off from the body, through the lungs, in the form of carbonic acid.

The amount of oxygen consumed in twenty-four hours by the act of respiration is 746 grm., or over 544 pounds per year.

2. *Chyle.*

The chyle contained in the lacteals and in the thoracic duct during digestion in the small intestines, is generally a turbid, milky, yellowish-white liquid, in which, with the aid of the microscope, various kinds of minute bodies may be detected, *chyle-corpuscles.* When removed from the vessels it coagulates in a short time. The clot becomes red in the air, and contains fibrin as the coagulated ingredient. The serum separated from the clot shows a weak alkaline reaction, and contains, in addition to the usual undetermined animal substances and the salts, principally albumen and fat; the latter collects on the surface, and undoubtedly forms one variety of the corpuscles, which are apparently surrounded by a protein compound.

3. *Lymph.*

The lymph in the lymphatics is a clear, pale yellow liquid, in which drops of fat and colorless globules of about the size of the blood-corpuscles may be detected by means of the microscope. It contains for the greater part fibrino-plastic protein compounds, but in very varying quantities, and sometimes they are entirely wanting. When they are present, the lymph coagulates rapidly when removed from the vessels, forming a clear gelatinous mass, which incloses the lymph-corpuscles. The liquid which separates from the fibrin contains albumen and the salts of the blood.

During fasting, only lymph is contained in the chyle-vessels of the intestinal canal; during digestion, however, albuminates, fats, etc., from the food enter this lymph, and it becomes what is called chyle, which is then carried into the blood through the thoracic duct.

4. *Saliva.*

The saliva is secreted by six salivary glands, and emptied into the cavity of the mouth through the excretory ducts during chewing or in consequence of irritation. Mixed with the mucus of the mouth, it shows very small, clear corpuscles under the microscope; it is generally slightly alkaline. When dried, it leaves behind about 1 per cent. of solid ingredients. These consist of mucus, several salts, traces of albumen and organic substance (ptyaline), that has not been separated nor analyzed. It is difficultly soluble in water, insoluble in alcohol; the solution does not become turbid by boiling, and the ptyaline is not precipitated by acids nor metallic salts. At 70° it converts starch into dextrin and sugar.—The most remarkable ingredient of the solid residue is a small quantity of potassium sulphocyanide, which can be extracted by means of alcohol.

The so-called tartar of the teeth, which is deposited from the saliva, consists of bone-earth, held together by the organic ingredients of the saliva. The saliva stones of the horse and ass consist principally of calcium carbonate with a little phosphate.

5. *Gastric Juice.*

The gastric juice, secreted by the small glands of the mucous membrane of the stomach during digestion, is a strongly acid, watery liquid, acid from the presence of free lactic acid, and sometimes hydrochloric, butyric, and acetic acids. At the most it contains 1.5 per cent. of solid ingredients. It contains a great deal of common salt, small quantities of other salts, and an organic matter (pepsin) of unknown nature, which, in the presence of an acid, appears to be the cause of the solvent action which the gastric juice exercises upon articles of food otherwise insoluble, as, for example, coagulated fibrin and albumen. Water slightly acidified with hydrochloric acid, and digested with a small piece of the mucous membrane of the stomach, attains the property of dissolving (digesting) coagulated fibrin and albumen, meat, etc., transforming them into amorphous, white bodies (peptone, parapeptone (p. 491), metapeptone), some of which are soluble in water, and others in acids and alkalies. Boiling temperature destroys this action.

The mucous, alkaline, intestinal fluid has also the property of causing the solution of protein compounds, as well as converting starch into sugar, and sugar into lactic and butyric acids.

6. *Bile.*

Bile is separated from the venous blood of the portal vein in the liver. The liver consists of small cells, which are arranged in net-like, adherent rows. In the interstices between these cells are distributed the finest beginnings of the biliary ducts, which conduct away the secreted bile; the finest branches of the portal vein, from the blood of which the bile is secreted; the finest terminals of the hepatic artery, which convey the blood for the support of the liver; and, finally, the delicate veins, which conduct the blood, already employed in the preparation of the bile, into the hepatic veins, from which it is conveyed back to the lungs through the vena cava, and right auricle of the heart.

The finest biliary ducts convey the secreted bile into branches, which grow larger and larger, and finally unite, forming a single canal, the hepatic duct. This conducts the bile during digestion into the duodenum, or at other times through a particular duct into the gall-bladder, in which it remains collected until digestion commences.

When freshly chopped liver is extracted with water, there is obtained a solution of albumen, which coagulates by heating. This solution further contains *glycogen* (p. 206), *urea*, and the other ordinary constituents of animal fluids. During life the liver contains no sugar. This is, however, rapidly formed after death from the glycogen.

Bile is a mucous, yellowish-green, bitter tasting and disagreeably smelling liquid, differing however in color and odor in different classes of animals. It generally reacts slightly alkaline, never acid. It contains between 10 and 14 per cent. solid ingredients, dissolved in water.

Bile contains, as characteristic, principal ingredients, the potassium or sodium salts of *glycocholic* and *taurocholic acids* (p. 479). In ox-bile both acids are contained in nearly equal quantity; in human bile, principally taurocholic and but little glycocholic acid; in the bile of the dog and several other animals, almost exclusively taurocholic acid. In the bile of mammalia these acids are contained as the sodium salts; in the bile of fishes, especially sea fish, however, the potassium salts also occur.

Bile contains besides, in smaller quantity, *cholesterin* (p. 480), *mucus*, and coloring matters (p. 481). These probably result from the coloring matter of the blood, hæmoglobin; are formed in larger quantity in certain diseases, particularly in icterus; and then occur widely distributed in other portions of the organism. Further, bile contains *fatty acids;* an organic base *cholin;* and undetermined extract-like organic substances.

Thoroughly dried bile leaves behind after combustion, about 12 per cent. of ashes, consisting of the sodium, potassium, calcium, and iron salts of sulphuric, phosphoric, and carbonic acids, and of chlorine.

7. *The Skin and its Secretions.*

Horny Tissue.

The general covering of the body consists of the scarf-skin (*cutis, epidermis*) and the corium (*cutis vera*).

The *epidermis* is a horny layer without bloodvessels. It consists of microscopical flat cells, closely joined together. Under this on the corium lies a softer layer of spherical cells (*rete Malpighii*), without doubt unhardened epidermis substance.

The *corium* is a solid, elastic skin supplied with bloodvessels, composed of strong, interlacing, fibrous bands. Under it lies the subcutaneous areolar tissue, in which are contained the two kinds of small cutaneous glands, which secrete the fluid perspiration, and the sebaceous matter of the skin. The excretory ducts of the first kind open into the pores of the epidermis, those of the other into the hair-follicles. In addition to these excretions a quantity of water, with some carbonic acid, is given off through the skin in gaseous form according to purely physical laws.

When boiled for a long time with water, the corium is converted into gelatin, and is dissolved (see p. 508, Gelatinous Tissues). On cooling, this solution congeals, forming a jelly. This transformation is also brought about more rapidly by acids.—Immersed in a solution of basic iron sulphate or of mercury chloride, the skin combines with these salts, and then does not decay. It possesses the greatest affinity for tannic acid, which it takes up from vegetable infusions containing the acid, and with which it forms a compound (apparently merely mechanical) insoluble in water, and not undergoing decay. Upon this depends the process of tanning, or the conversion of skins into leather (*cf.* p. 424).

The horny tissues, viz., the *epidermis*, *the nails*, *claws*, *talons*, *hoofs*, *horns*, *whalebone*, *wool*, *feathers*, *tortoise-shell*, and similar continuations and coverings of the skin, are formations composed of various substances, the principal mass of which, however, appears to consist of one and the same body (keratin), a substance containing sulphur and nitrogen, and closely allied to the protein compounds. All three formations are soluble in caustic potassa with the aid of heat, evolving at the same time a great deal of ammonia, and forming potassium sulphide. Acids precipitate from the solution a gelatinous, nitrogenous substance. Nitric acid turns them yellow and destroys them; when boiled with dilute sulphuric acid, they form leucine (p. 98), and tyrosin (p. 350); subjected to dry distillation, they yield a large quantity of nitrogenous products. The epidermis contains 0.74 per cent., the nails 2.8 per cent., the horse's hoof 4.2 per cent., whalebone 3.6 per cent. of sulphur. They also contain small quantities of calcium phosphate, iron, and silicic acid, which latter is contained as a constant ingredient in larger quantity in the vane of bird-feathers.*

Human hair contains as principal ingredient a protein-like body, that contains over 5 per cent. of sulphur. The presence of this large amount of sulphur is the cause of the turning black of light hair by means of metallic salts. In addition to some calcium phosphate, and small quantities of other salts, hair also contains iron oxide and silicic acid. The cause of the different colors of hair is unknown; it appears, however, that according as the color of the hair differs, the composition also varies.

The *sebaceous matter* of the human skin contains a

* *Chitin*, a substance, that forms the real skeleton, the testa and coverings of the wings of all insects, is entirely different in composition and chemical properties from these formations. Its composition is probably $C^9H^{15}NO^6$. It is not dissolved even by the most concentrated potassa, and carbonizes without fusing, when heated. When boiled with sulphuric acid, it yields grape-sugar and ammonia.

liquid and a solid fat (olein and palmitin). It is acid from the presence of lactic acid (?), and contains, further, salts from the aqueous secretion. In sheep it consists of several kinds of fat and a saponaceous compound of potassium and calcium with a fatty acid.

The *perspiration* is acid, and contains free acetic, butyric, formic, and carbonic acids. It contains only ½ to 2 per cent. of solid ingredients, consisting of urea, undetermined animal matters, potassium and sodium chlorides, and small quantities of sulphates and phosphates. Strongly smelling perspiration appears to contain also caproic acid, and a volatile organic sulphur compound. In certain diseases, as in cholera and kidney complaints, a large increase of the normal, small quantity of urea, contained in perspiration, takes place. In other diseases sugar and uric acid, and under certain conditions also succinic acid, have been detected in perspiration.

8. *Muscles.*

The finest recognizable parts of voluntary muscles are microscopical, reddish, transversely striated fibres, which are united in bundles. The finest bundles are inclosed in sheaths of cellular tissue, and are united by cellular tissue, forming larger bundles. A large number of such larger bundles, bound together by a sheath of cellular tissue, forms a single muscle. In the sheaths is distributed a network of fine bloodvessels and nerves.

The principal ingredient of muscular tissue, congealed after death, is a protein compound, *myosin* (p. 491). It is not yet decided whether this substance is, as such, contained in a state of solution in the living muscle or, similar to blood-fibrin, is formed after the cessation of life. The peculiar phenomenon of *rigor mortis* is, however, undoubtedly caused by the coagulation of the myosin, and this occurs quite independently of the acid, which makes its appearance in muscular tissue after death, and generally after the rigidity.

After thorough drying, flesh leaves behind only about 23 per cent. of solid substance, the remaining 77 per cent. are water. Of the solid residue about 6 per cent. are soluble in water; chopped meat, after being extracted with water, leaving behind only 17 per cent. of solid substance.

The reddish fluid expressed from fresh meat has an acid reaction from the presence of free lactic acid and acid phosphates of the alkalies; it coagulates when heated. The clot is albumen, colored by a brownish-red coloring matter, very similar to, and probably identical with, hæmoglobin (p. 494). Acetic acid and rennet also show the presence of casein.

It contains, further, creatine (p. 248), sarcine (p. 246), *xanthine* (p. 246), inosite (p. 197), dextrin (p. 207), sugar (flesh-sugar, probably identical with grape-sugar), an acid, *inosic acid*, as yet but little known; and salts, particularly potassium paralactate and phosphate, the potassium salts of volatile acids (butyric, acetic, formic(?)), potassium chloride, and magnesium phosphate. Sodium chloride and calcium phosphate are only present in small quantity, and sulphates not at all.

Creatine, *xanthine*, *sarcine* are intermediary products of the process of waste in muscular tissue.

9. *Bones.*

Bones excel all other organs in the large amount of inorganic matter (earthy matter) contained in them. Thin lamellæ of compact bony tissue appear under the microscope as an homogeneous, structureless, transparent mass, which is traversed by small canals (the *Haversian canals*), containing fat and vessels; and, in the interstices between these, by minute, regularly arranged cavities, with numerous fine tubes issuing from all parts of their circumference.

If a bone is placed in very dilute hydrochloric acid, the earthy matter is extracted, and the organic portion, *cartilage*, interwoven with all the fine vessels and membranes, contained in the bone, remains behind as a flexible, soft, translucent mass, having the form of a bone. When dried it shrinks together somewhat, becomes hard and brittle, but remains translucent. By boiling with water it is dissolved, forming glutin.

Water, heated above 100°, *i. e.* under high pressure, extracts all the cartilage from bone, dissolved as gelatin, leaving the pure earthy matter behind.

When bones are burned with access of air, the organic ingredients are destroyed, and the earthy matter remains behind as a white substance, having the form of the bone. It consists of neutral calcium phosphate mixed with calcium carbonate, in varying quantities in different animals; and small quantities of magnesium phosphate and calcium fluoride. Calcium carbonate is contained, as such, in the living bone. Whether bony substance is a chemical compound of cartilage with calcium phosphate, or is merely a mixture, is undetermined. The facility with which the two constituents can be separated, however, without necessitating a change in the form of the bone, speaks for the latter view.

The amount of organic and earthy matter contained in bones, estimated by calcining the bones, is found to vary somewhat in bones of different parts of the body, of different age, and in the bones of different classes of animals. In the parietal bone of man, for example, 68.3 per cent., in the sternum 64.7 per cent., in the tibia 65.5 per cent. of earthy matter have been found.—Human bones, thoroughly dried, contain over 8 per cent. of calcium carbonate. The average amount of calcium phosphate is 57 per cent., that of earthy matter 33 per cent. The bones of all mammalia are very similar in their composition to those of man; those of birds, however, are much richer in inorganic ingredients. In the femur of the pigeon, for example, 89 per cent. of earthy matter was found, of which 82 per cent. consisted of calcium phosphate. In the bones of amphibious animals and fish, on the contrary, the amount of organic matter is decidedly greater.

Fish scales have a composition similar to that of bones, only containing more organic matter. This does not, however, differ in its chemical composition from cartilage; and by boiling with water is likewise converted into gelatin.

The *teeth* also contain the same ingredients as bones, but less organic matter. The tooth-bone, dentine, in man, contains over 64 per cent. of calcium phosphate, over 6 per cent. of calcium and magnesium carbonates, and 28 per cent. of organic matter, affording gelatin. The *enamel* of the teeth, on the other hand, which consists of perpendicular, closely arranged, microscopical fibres or rods, contains no organic matter similar to that of bone; it contains 84–90 per cent. of calcium phosphate (with magnesium phosphate and some calcium fluoride), 4–9 per cent. of calcium carbonate, and 3–6 per cent. of organic substance.

The antlers of the deer-family have the same composition as bones.

10. *Tissues yielding Gelatin.*

These belong to the principal ingredients of the animal body, and do not occur in plants. In an organized form they constitute the *cartilages*, the *tendons*, the *ligaments*, *cellular tissue*, *serous membranes*, the *corium*, etc. All of these substances, entirely insoluble themselves in water, possess the property of becoming converted, by continued boiling with water, into an apparently isomeric substance, and of being dissolved for the greater part as gelatin, the solution of which, on cooling, forms a jelly-like mass.

The *elastic tissue*, which forms the yellow bands of the vertebral column, the ligamentum nuchæ, the exterior covering of the arteries, etc., does not suffer this change in the slightest degree.

The gelatin, which results from all these tissues, differing in composition as well as structure, is of two kinds, ordinary gelatin (*glutin*) and *chondrin;* and based upon this, the fundamental substance of the tissues has been divided into *collagen* and *chondrigen*.

Glutin is produced from bone-cartilage, deer-horns, fish-bones and fish-scales, the skin (corium), tendons, serous membranes, isinglass. The solution, obtained from these substances by boiling them with water, coagulates on cooling, forming a thick jelly, which, when dried, constitutes ordinary carpenter's glue. Pure gelatin is obtained most readily by boiling rasped deer-horns, isinglass, or pure bone-cartilage freed of earthy matter by means of hydrochloric acid, with water, and filtering the solution at about 50°.—Glutin is colorless, transparent, hard, tasteless, and inodorous; softens when heated, and is then destroyed. In cold water it swells up, and when heated dissolves. The solution forms, on cooling, a clear jelly, even when it contains but one per cent. of gelatin; this however varies in the gelatin from different tissues. It is insoluble in alcohol and ether, and is precipitated by alcohol from its aqueous solution as a flocculent mass. When subjected to combustion it always leaves behind some earthy matter.

A solution of this gelatin is *not* precipitated by alum, neutral iron sulphate, neutral and basic lead acetate.

Tannic acid precipitates it completely from its solution. The precipitate, which is at first white and flocculent, generally contracts, forming a thick, tough, sticky mass. Tissues, which have the power to yield gelatin, and are not yet converted into it, take up tannic acid completely from its aqueous solution; upon this property is founded the process of tanning (converting hides into leather).—Acetic acid readily dissolves gelatin; the solution possesses the properties of glue but does not gelatinize.

Glutin contains about 18 per cent. of nitrogen and a very small quantity ($\frac{1}{2}$ per cent.) of sulphur. Its composition cannot be expressed by a probable formula.

When boiled for a long time and particularly at a temperature above 100°, its solution loses the property of gelatinizing. On evaporation it then dries up, forming a yellowish, gummy mass, which is easily soluble in cold water. The change that thus takes place is not understood.—Subjected to dry distillation, it yields a large number of products, among which the most remarkable are ammonium carbonate and the volatile bases: methylamine, di- and trimethylamine, pyridin, etc.*—When distilled with manganese peroxide, or potassium bichromate and sulphuric acid, gelatin yields the same numerous products as the protein compounds under similar treatment (p. 486).

When a solution of gelatin is boiled with sulphuric acid or potassa, there are produced, besides ammonia and some not well known products, *glycocol* (p. 85), and *leucine* (p. 98).

Chondrin is produced from permanent (non-ossifying) cartilages, as from the cartilages of the ribs, the joints, bronchi, nose, from the cornea, from bone-cartilage before ossification, by boiling with water.—Its solution congeals on cooling, like that of ordinary gelatin; in a dried condition it looks like the latter, but its solution is not only precipitated by tannic acid, but by acetic and hydrochloric acids, dilute sulphuric acid, alum, lead acetate, and iron sulphate; all of which do not precipitate glutin. The precipitate with alum forms large, compact, white flocks, soluble in an excess of alum and several other salt solutions. The precipitates with hydrochloric and sulphuric acids, but not that with acetic acid, are easily redissolved in an excess of the precipitating substance.—On combustion, chondrin likewise leaves behind earthy matter.—It contains between 14 and 15 per cent. of nitrogen and a small quantity of sulphur.

Its decomposition-products are the same as those of glutin; by boiling with sulphuric acid, however, only leucine, but no glycocol, is formed. When boiled with hydrochloric acid, it yields a fermentable sugar.

The *gelatin from the bones of placoidians* differs from the two other varieties of gelatin in the fact that its solution does not gelatinize; otherwise it conducts itself like chondrin.

In silk is contained a peculiar body, *fibroïn*, $C^{15}H^{23}N^{5}O^{6}$, which constitutes about 66 per cent. of

* These volatile bases are contained in the substance called *Oleum animale Dippelii*. It is obtained by rectification of fetid animal oil, which is a by-product in the preparation of bone-black on the large scale, from bones free of fat (see Pyridin bases, p 130).

raw silk. It can be obtained pure most readily by repeatedly digesting silk with water at 30°, and treating the bright-yellow, lustrous residue with alcohol and ether. By boiling with dilute sulphuric acid, it yields tyrosin, leucine, and some glycocol.

In addition to fibroïn, silk contains a species of gelatin, in many respects similar to glutin, *silk-gelatin* (seracin), $C^{15}H^{25}N^{5}O^{8}$, which can be extracted by boiling water. It is formed apparently from fibroïn by the assimilation of oxygen and water. In a dried condition, it forms a colorless and inodorous powder, which swells up largely with water, and dissolves in it more readily than glutin. A solution which contains less than 1 per cent. still congeals on cooling, forming a consistent jelly. By long boiling with dilute sulphuric acid, it yields a little leucine, about 5 per cent. of tyrosin, and about 10 per cent. of serine (p. 175).

11. *Fat.*

Fat occurs in a great many forms in the animal organism, partially as minute drops or globules, suspended in fluids, as in the milk, in blood, partially deposited in a free state in the tissues or inclosed in particular fat cells; in the latter manner for instance in the upper portion of the subcutaneous cellular tissue.

In connection with glycerin it has already been mentioned, that the fats which are most widely distributed in the animal kingdom are identical with the vegetable fats of most general occurrence. In the same connection, the details in regard to the occurrence of the various animal fats, their properties and composition, were given.

12. *Mucus.*

In the mucus secreted by mucous membranes are detected microscopical clear granules, and separated cells or particles of the external coat (epithelium) of the mucous membranes.

The characterizing ingredient of mucus is a peculiar

nitrogenized body (mucin). It does not appear to be dissolved in the water of the mucus, but to be swollen up into a colloïd state. The liquid contains, besides this, potassium and sodium chlorides, and small quantities of other salts. Mucus is not coagulated by heating, but precipitated by alcohol and dilute acetic acid.

13. *Transudates of Serous Membranes.*

The fluid, which collects in dropsical affections, contains albumen in varying, frequently in very large quantity; and, in addition to this, the ordinary salts and undetermined substances. It is usually alkaline. Occasionally it contains urea and cholesterin suspended in fine laminæ. The amniotic fluid and the fluid in hydatids contain the same ingredients. When boiled or treated with nitric acid, these fluids become more or less turbid or coagulated.

Pus is a creamy, thick, intransparent liquid, which consists of a clear, colorless, or slightly yellow serum (pus-serum), and, suspended in this, the pus-corpuscles and fat-globules. Pus-serum contains albumen, which coagulates by heat, and further, leucine, sodium chloride, and other inorganic salts. Pus-corpuscles possess the greatest resemblance to the colorless blood-corpuscles.

14. *The Eye.*

The *sclerotic*, formed of very compactly interwoven cartilaginous fibres, can, like the corium, be dissolved as gelatin, by long-continued boiling with water.

The *cornea* is formed of a peculiar tissue, and conducts itself chemically like chondrigenous cartilage, but swells up in acetic acid.

The *black pigment* (melanin), which is deposited in the form of microscopical, brown granules in separate, closed cells in the choroid, is insoluble in water, alcohol, and dilute acids; soluble in potassa, forming a dark-yellow liquid; is reprecipitated by acids. It contains 13–14 per cent of nitrogen. When subjected to combustion it leaves behind an ash containing iron. It is probably a metamorphosis-product of the coloring matter of the blood. Whether the pigment in the *rete mucosum* of the negro, and many pigments deposited in diseases, are identical with it, is not decided.

The *vitreous humor* and *aqueous humor* consist of water with not quite 2 per cent. of solid substances dissolved in it. In the vitreous humor these are albumen, sodium chloride, undetermined organic substances, and urea; the aqueous humor, on the other hand, contains no albumen.

The *crystalline lens* consists of concentric layers or laminæ, which are composed of compactly arranged, clear fibres (probably tubes), and contain a very concentrated liquid. This latter contains about 60 per cent. of fat, cholesterin, and inorganic salts, and 35 per cent. of a protein compound, *globulin* (p. 490), very similar to fibrinogenous substance.

15. *The Nervous System.*

Without entering here into a detailed consideration of the fine structure of the brain, it may be remarked in general, that the hemispheres of the cerebrum and cerebellum consist of two masses differing essentially from each other in construction and, without doubt, also in composition. These are an outer gray layer, the *substantia cinerea*, and a white fibrous mass, covered by the former, the *substantia medullaris*.

The gray matter is very abundantly supplied with bloodvessels and poorly with brain fibres; its principal mass consists of peculiar microscopical globules.

The marrow is less abundantly supplied with bloodvessels and water, and is very fibrous. Examined under the microscope it is found to consist of very delicate, transparent cylinders, formed of a thin membrane. They contain a semi-fluid, oily, clear mass, the nerve-marrow. The white matter contains more fat than the gray. In certain portions of the human brain there have also been discovered microscopical bodies, which conduct themselves towards iodine like cellulose, but are essentially cholesterin (p. 480).

The spinal marrow and the nerves, that have their origin in it and in the brain, have a similar structure.

100 parts of fresh human brain, dried at 100°, leave behind 21.5 parts of solid residue.

The characteristic ingredient of the brain substance is *lecithin* (protagon). To separate it, brain substance, freed as thoroughly as possible from blood and coverings, is reduced to a pulp, and shaken with water and ether. The mixture is allowed to stand at 0°, until the ethereal solution appears at the top; this is then removed, and this process repeated several times, the greater portion of the cholesterin being removed in this way by the ether, while the ingredients, which are easily soluble in water, remain dissolved in this solvent. The ether and water are then filtered off as thoroughly as possible, and the residue digested with 85 per cent. alcohol at 45° over a water bath, and filtered while still warm. This solution, when cooled down to 0°, throws down an abundant precipitate, which is collected and washed with ether until cholesterin can no longer be detected in the filtrate. The residue is dried in a vacuum over sulphuric acid, then moistened with a little water and dissolved in alcohol at 45°. By gradual cooling of the filtered solution, protagon is deposited in crystals, which may be purified by recrystallization.

It forms fine, radiate needles; after drying over sulphuric acid, a light, flocculent powder. Difficultly soluble in cold alcohol and cold ether, more easily in the warm liquids. When heated with absolute alcohol to a temperature higher than 55°, it is dissolved, undergoing at the same time partial decomposition. When treated with water it swells up, forming an opaque, pasty mass, which, with more water, yields a clear but opalescent solution, from which protagon is precipitated as a flocculent mass by boiling with concentrated solutions of calcium and sodium chlorides and other salts. It dissolves in glacial acetic acid, and crystallizes from this solution unchanged, on cooling.

Protagon contains carbon, hydrogen, nitrogen, oxygen and phosphorus.—It decomposes below 100°, the more readily the more anhydrous it is.—When boiled for a long time with baryta-water it yields barium glycerinphosphate, solid fatty acids, and *neurine* (p. 140). Protagon is also contained in blood, in yolk of eggs, and in the vegetable kingdom (*e. g.* in maize).

In addition to the substances mentioned, there are contained in the brain protein compounds (particularly casein), cholesterin, lactic acid, inosite, and very small quantities of creatine, xanthine, sarcine, uric acid, and inorganic salts.

Other bodies prepared from the brain, and but little known, as *cerebrin*, *cerebric acid*, etc., appear to be mixtures.

16. *The Egg.*

A hen's egg when laid consists of the shell, the white, and the yolk.

The *egg-shell*, provided with small pores penetrable by air, is covered on the inside with a solid membrane, consisting of two layers, which separate at the larger end of the egg, and leave a space between, which is filled with air. The shell consists of 97 per cent. of calcium carbonate; 1 per cent. of calcium phosphate with magnesium phosphate; and 2 per cent. of organic substance, which remains undissolved, when the shell is treated with hydrochloric acid.

The *white of the egg* surrounds the yolk in three layers, of which the outermost is the most liquid. It is inclosed in thin, transparent, membranous cells. At 75° it coagulates, forming a solid, white, elastic mass. It contains 12–14 per cent. of albumen, mostly dissolved in water, as sodium albuminate, besides a very small quantity of fat and grape-sugar, and about 0.7 per cent. of inorganic ingredients. These consist of soda, potassium and sodium chlorides, and earthy phosphates.

The *yolk*, inclosed in a thin membrane, appears under the microscope as a pulpy mass closely filled with very fine granules, in which yellowish globules and fat-drops are floating. The globules are bubbles or cells, which contain a yellowish oil.

The analysis of the yolk of egg shows on an average 45 per cent. of water, 30 per cent. of fat, 15 per cent. of protein compounds, and 1 per cent. of inorganic salts.

The fat, which can be obtained from the yolk by shaking with ether or, after the yolk is boiled hard, partially by means of pressure, is reddish-yellow, colored by a coloring principle as yet comparatively unknown, perhaps identical with hæmatoïdin (p. 496). It consists of palmitin and olein.

The protein compound (formerly called *vitellin*) is a mixture of casein and other protein compounds. Besides these, there are contained in the yolk lecithin (p. 514), cholesterin, and apparently also glycogen.

The inorganic ingredients are soda, potassium and sodium chlorides, potassium, calcium and magnesium phosphates, and iron oxide. The potassium salts are more abundantly present than the sodium salts, and earthy phosphates are present in much larger quantity than in the white.

It is very probable, that the eggs of all classes of animals contain the same ingredients. In the yolk of fishes and several amphibious animals are observed under the microscope transparent crystalline plates, which, however, in different species of animals, possess different forms and properties. They appear to be protein compounds, or at least to be very similar to these.

17. *Semen.*

Animal semen, in a pure condition as formed in the testicles, is a whitish, ropy, inodorous liquid of high specific gravity, and neutral or alkaline reaction; when ejaculated, it is more translucent, more strongly alkaline, and of a peculiar odor, on account of the presence of the secretions of the prostate and Cowper's glands. It consists of a watery liquid, which contains, in a state of suspension, as peculiar, morphological elements, the *spermatic filaments* (spermatozoa), microscopic, fibrous bodies, distinguished by the power of motion; and the *seminal cells* (seminal granules), cells very similar to the colorless blood-corpuscles.

Semen contains 10–12 per cent. of solid ingredients, which consist of fat, inorganic salts, particularly calcium phosphate, and a peculiar but slightly known body, *spermatin*, very similar to mucin (p. 512). The latter is the cause of the gelatinous consistence of semen. It is not precipitated from its solution by boiling; by evaporation, however, it is converted into a modification, which is completely insoluble in water.

18. *Milk.*

The characterizing ingredients of milk are fat, casein, (p. 487), and sugar of milk (p. 200). The two latter are present in a state of solution; the fat is suspended in the form of globules. Besides these, milk contains the ordinary undetermined substances and the salts of the animal fluids (particularly alkaline and earthy phosphates), and also some iron oxide.

Milk appears under the microscope as a clear liquid, filled with numberless clear globules of different sizes, mostly however smaller than the blood-corpuscles. They are surrounded by an envelope, which incloses the fat. Hence ether, shaken with milk, takes up hardly any fat. This takes place only when the milk has been treated with alkali or acetic acid, and the envelopes broken up by this means.

The quantity of solid ingredients in milk is varying in different animals, and different individuals. Woman's milk contains 11–13; cow's and goat's milk, 13–14; mare's milk, 16; bitch's milk, 25 per cent. The amount of single ingredients contained in it is just as varying. Fat 3–5, casein 2–8, sugar of milk 2–9, salts 0.25–1.5 per cent.

Milk is, as a rule, slightly alkaline. It does not coagulate by heating, but easily by acids, by voluntary acidification, and in the presence of the mucous membrane of the calves' stomach (rennet). When evaporated, a crust of coagulated casein is formed on its surface.

The *colostrum* (the secretion of the glands of the breast for the first two or three days after parturition), contains a larger supply of solid ingredients than ordinary milk. In addition to the much smaller milk-globules, larger spherical masses, the so-called granular bodies, which are apparently conglomerations of casein and fat-vesicles, are observed in it.

Cream, which separates from milk on standing undisturbed, is formed of the milk-globules, which, being specifically lighter, rise to the surface. Churning breaks up the envelopes of the globules, and their contents then adhere together, forming butter. The yellow color of butter is accidental, and arises from certain constituents of the food. In rancid butter, traces of fatty acids have become free. Butter fuses at about 32°.

Judging from the products that have been obtained by the saponification of cow's milk, the latter is a mixture of several varieties of fat (compare Glycerin p. 168), which, however, have not as yet allowed of separation. The fatty acids obtained from them are palmitic, stearic, and oleic acids, which form the principal quantity; further myristic, butyric, caproic, caprylic, and capric acids.—Whether the same varieties of fat are contained in the milk of all animals is uninvestigated.

19. *Urine.*

The urine is secreted by the kidneys from arterial blood. The kidneys consist of microscopical canals (tubuli uriniferi), which are distributed in the cortical substance and towards their open ends unite forming pyramidal tufts, of which each one is grasped at the apex by a short cylindrical sheath, the calyx. The calyces empty into a common larger sac, the pelvis of the kidney. This is continued by the ureter, which in its turn conducts the urine into the bladder. The tubes of the cortical substance commence partially as small *culs-de-sac*, in each of which lies a plexus of capillary vessels.

Chopped kidney substance, when ground in a mortar, becomes almost liquid. If this is strained, a comparatively very small quantity of solid substance remains behind, consisting of the membranes of the fine bloodvessels and the tubuli uriniferi. The strained, milky and mucous liquid coagulates when heated, forming a gelatinous mass, which consists principally of albumen.

The watery extract of kidneys contains, further, in small quantity xanthine, sarcine, inosite, taurine, and leucine.

Normal human urine is acid, principally owing to the presence of acid sodium phosphate; it has an unpleasant saltish and bitter taste; has a mean specific gravity of 1.020; always deposits a cloudy layer of mucus; and after a time becomes more strongly acid, microscopic crystals of uric acid and sometimes calcium oxalate being thrown down. Later it again becomes neutral, finally alkaline, commencing to undergo decomposition and emitting a foul odor, the formation of ammonium carbonate and crystals of magnesium ammonio-phosphate taking place.

In its ordinary condition urine contains between 7–8 per cent. of solid ingredients; the rest is water. This relative proportion is, however, exceedingly varying, according to the quantity of liquid taken into the body as drink, according to the evaporation from the skin and the condition of health.

The characterizing ingredients of human urine are *urea* (p. 227), and *uric acid* (p. 232).

When urine is evaporated down to the consistence of honey, and allowed to stand for a long time covered up, crystals of urea or of a compound of it with sodium chloride are formed. If in this concentrated condition it is mixed with an excess of nitric acid, it forms a pulp of crystalline scales, which are urea nitrate. Normal human urine contains between 2.5 and

3.2 per cent. of urea; a healthy man secretes 22–36 grms. of urea in twenty-four hours.

When fresh urine is mixed with an acid, the uric acid falls after a time, sometimes immediately, as a brownish or reddish powder. Its amount is about 0.1 per cent.

Human urine contains, further, creatine (about 0.1 per cent.), frequently succinic acid, traces of hippuric acid and of ammonium oxalurate, occasionally xanthine and several other organic substances of undetermined nature, which are obtained as an extractive mass in the analysis.

It contains about 2 per cent. of inorganic salts, potassium and sodium chlorides, potassium and sodium sulphates, acid sodium phosphate, calcium and magnesium phosphates, further, a small quantity of silicic acid and iron. The salts of the alkaline earths can be precipitated from it by means of ammonia.

Urine may, further, contain various foreign substances, which are brought into the body in a soluble condition, and extracted from the blood by the kidneys. A number of salts, for example, saltpetre, potassium ferrocyanide, etc., pass unchanged from the stomach into the urine; also organic acids, tartaric, oxalic acids, etc. Their salts with the alkaline metals, however, are decomposed during digestion, and they are found in the urine in the form of alkaline carbonates, imparting an alkaline reaction to the urine. Further, several organic coloring principles, volatile oils, resins, etc., pass unchanged into the urine, imparting to it color and odor. Benzoic acid, oil of bitter almonds, cinnamic acid and quinic acid, are found in the urine, transformed into hippuric acid.

In diseases the character of the urine is changed in various ways. Occasionally it becomes neutral or even alkaline, and is then turbid from the separation of calcium phosphate, and microscopical crystals of magnesium ammonio-phosphate, and calcium oxalate. Or it becomes too concentrated, and on cooling deposits gray or reddish sediments, consisting of alkaline urates. In fevers this sediment is of a brick-color or rosy-red, and consists principally of sodium urate, colored by a very small quantity of a red substance, as yet uninvestigated. In diseases, substances are frequently found in the urine, which it does not contain in a healthy condition. In many varieties of dropsy, and a few other diseases, it contains albumen; it then becomes turbid on the addition of nitric acid and by heating. In jaundice it contains ingredients of bile; in diabetes, grape-sugar, frequently in very large quantity, and is then secreted to an enormous extent. It is in this condition fermentable, and afterwards, on being subjected to distillation, it yields alcohol. When the origin of the pneumogastric nerve in the brain is injured, sugar occurs in the urine. Further, lactic acid, lactates, indigo, or, rather, a substance capable of producing indigo (*cf.* p. 384), leucine, tyrosin, taurine, etc., are occasionally contained in urine.

In certain diseased conditions of the body, difficultly soluble ingredients of the urine are deposited even in the urinary canals, and form concretions (gravel and urinary calculi), frequently of great size and hardness, and of very varying composition. Most of them consist of *uric acid* with *ammonium urate;* others are mixtures of *calcium phosphate* with *magnesium ammonio-phosphate;* others consist of *calcium-oxalate;* many are formed of alternating layers of all of these substances. Calculi consisting of *cystine* (p. 175) and *xanthine* (p. 246) are the most rare.

Urine is of very varying character according to the class of animals from which it is obtained. That of the higher classes always contains urea in predominant quantities; in that of the lower classes, on the other hand, uric acid is more abundant. The urine of the lion and tiger is so abundantly supplied with urea, that frequently, without previous evaporation, the addition of nitric acid causes the nitric acid compound to crystallize out in laminæ. In the urine of dogs there is frequently obtained a peculiar crystallizing acid, *kynurenic acid*, as yet but little known.—The urine of birds and amphibious animals is a white, pulpy mass (after drying, earthy), which consists almost exclusively of acid ammonium urate.—The urine of herbivorous mammalia, as, for example, that of horses and cattle, is usually alkaline; contains urea, but little uric acid; on the other hand, a large quantity of hippuric acid (p. 336), and frequently phenol (p. 290); further, potassium bicarbonate and lactate, but no alkaline phosphate; it deposits a sediment of calcium and magnesium carbonates. The urine of sucking calves contains allantoïne (p. 243) and no hippuric acid. The urine of insects contains uric acid and guanine (p. 247).

20. *Excrements.*

Normal human excrements contain about 25 per cent. of solid ingredients, and of these, on an average, 6.5 per cent. are inorganic salts; the rest is water. Their nature varies according to the food. The ash of human excrements contains 25–30 per cent. of soluble salts, and about 30 per cent. of phosphoric acid in the form of sodium, potassium, calcium, and magnesium salts. The excrements of herbivorous animals contain all the phosphoric acid which is separated from the organism, as this acid is entirely wanting in the urine. —Human excrements, the organic ingredients of which soon begin to undergo decay, contain mucus, undigested remnants of food, altered ingredients of the bile (taurin, cholesterin), a peculiar, crystallizing compound, *excretin*, containing sulphur, and but little known and undetermined matters.—The excrements of cattle contain a large quantity of undigested cellulose, colored green by chlorophyl.

INDEX.

ERRATA.

Page 106, line 3 from bottom, *read* "Thialdine" *for* "Trialdine"
" 133, line 10 from top, *read* "Propargylic" *for* "Propagylic."
" 150, after paragraph on Lactamide, insert:

Lactimide, $C^2H^4 \left\{ \begin{matrix} NH \\ CO, \end{matrix} \right.$ is formed by heating alanin in a current of hydrochloric acid gas at 180–200°.—Colorless, transparent needles or laminæ. Fusing point, 275°; easily soluble in water and alcohol.

Page 195, line 7 from bottom, *read* "Glucinic" *for* "Glucic."
" 316, " 5 from top, *read* "Stiryl" *for* "Styryl."
" 377, " 18 from bottom, *read* "Hydrocoumarinic" *for* "Hydrocoumaric."
" 453, " 7 from top, *read* "Physostigmine" *for* "Thysostigmine."
" 465, erase the bottom line—"Lozenge oil," etc.
" 466, after Rosemary oil, insert "*Rue-oil* from *Ruta-graveolens* (p. 112)."
" 511, line 8 from top, *read* "Sericin" *for* "Seracin."

CATALOGUE OF BOOKS

PUBLISHED BY

HENRY C. LEA.

(LATE LEA & BLANCHARD.)

The books in the annexed list will be sent by mail, post-paid, to any Post Office in the United States, on receipt of the printed prices. No risks of the mail, however, are assumed, either on money or books. Gentlemen will therefore, in most cases, find it more convenient to deal with the nearest bookseller.

Detailed catalogues furnished or sent free by mail on application. An illustrated catalogue of 64 octavo pages, handsomely printed, mailed on receipt of 10 cents. Address,

HENRY C. LEA,
Nos. 706 and 708 Sansom Street, Philadelphia.

PERIODICALS,

Free of Postage.

AMERICAN JOURNAL OF THE MEDICAL SCIENCES. Edited by Isaac Hays, M.D., published quarterly, about 1100 large 8vo. pages per annum,
MEDICAL NEWS AND LIBRARY, monthly, 384 large 8vo. pages per annum,
} For five Dollars per annum, in advance

OR,

AMERICAN JOURNAL OF THE MEDICAL SCIENCES, Quarterly,
MEDICAL NEWS AND LIBRARY, monthly,
MONTHLY ABSTRACT OF MEDICAL SCIENCE, 48 pages per month, or nearly 600 pages per annum.
In all, about 2100 large 8vo. pages per annum,
} For six Dollars per annum, in advance.

MEDICAL NEWS AND LIBRARY, monthly, in advance, $1 00.

MONTHLY ABSTRACT OF MEDICAL SCIENCE, in advance, $2 50.

OBSTETRICAL JOURNAL. With an American Supplement, edited by J. V. Ingham, M.D. $5 00 per annum, in advance. Single Numbers, 50 cents. Is published monthly, each number containing ninety-six octavo pages.

ASHTON (T. J.) ON THE DISEASES, INJURIES, AND MALFORMATIONS OF THE RECTUM AND ANUS. With remarks on Habitual Constipation. Second American from the fourth London edition, with illustrations. 1 vol. 8vo. of about 300 pp. Cloth, $3 25.

ASHWELL (SAMUEL). A PRACTICAL TREATISE ON THE DISEASES OF WOMEN. Third American from the third London edition. In one 8vo. vol. of 528 pages. Cloth, $3 50.

ASHHURST (JOHN, Jr.) THE PRINCIPLES AND PRACTICE OF SURGERY. FOR THE USE OF STUDENTS AND PRACTITIONERS. In 1 large 8vo. vol. of over 1000 pages, containing 533 wood-cuts. Cloth, $6 50; leather, $7 50.

ATTFIELD (JOHN). CHEMISTRY; GENERAL, MEDICAL, AND PHARMACEUTICAL. Seventh edition, revised by the author. In 1 vol. 12mo. Cloth, $2 75; leather, $3 25.

BROWNE (EDGAR A.) HOW TO USE THE OPHTHALMOSCOPE. Elementary instruction in Ophthalmoscopy for the Use of Students. In one small 12mo. vol, many illust. Cloth, $1. *(Now ready.)*

BLOXAM (C. L.) CHEMISTRY, INORGANIC AND ORGANIC. With Experiments. In one handsome octavo volume of 700 pages, with 300 illustrations. Cloth, $4 00; leather, $5 00.

BRINTON (WILLIAM). LECTURES ON THE DISEASES OF THE STOMACH. From the second London ed. 1 vol. 8vo. Cloth, $3 25.

BIGELOW (HENRY J.) ON DISLOCATION AND FRACTURE OF THE HIP, with the Reduction of the Dislocations by the Flexion Method. In one 8vo. vol. of 150 pp., with illustrations. Cloth, $2 50.

BASHAM (W. R.) RENAL DISEASES; A CLINICAL GUIDE TO THEIR DIAGNOSIS AND TREATMENT. With illustrations. 1 vol. 12mo. Cloth, $2 00.

BUMSTEAD (F. J.) THE PATHOLOGY AND TREATMENT OF VENEREAL DISEASES. Third edition, revised and enlarged, with illustrations. 1 vol. 8vo., of over 700 pages. Cloth, $5; leather, $6.

—— **AND CULLERIER'S ATLAS OF VENEREAL.** See "CULLERIER."

BARLOW (GEORGE H.) A MANUAL OF THE PRACTICE OF MEDICINE. 1 vol. 8vo., of over 600 pages. Cloth, $2 50.

BAIRD (ROBERT). IMPRESSIONS AND EXPERIENCES OF THE WEST INDIES. 1 vol. royal 12mo. Cloth, 75 cents.

BARNES (ROBERT). A PRACTICAL TREATISE ON THE DISEASES OF WOMEN. In one handsome 8vo. vol. of about 800 pages, with 169 illustrations. Cloth, $5; leather, $6.

BRYANT (THOMAS). THE PRACTICE OF SURGERY. In one handsome octavo volume, of over 1000 pages, with many illustrations. Cloth, $6 25; leather, $7 25.

BRISTOWE (JOHN SYER). A MANUAL OF THE PRACTICE OF MEDICINE. A new work, edited with additions by James H. Hutchinson, M.D. In one handsome 8vo. volume of over 1100 pages. Cloth, $5 50; leather, $6 50. *(Just issued)*

BOWMAN (JOHN E.) A PRACTICAL HAND-BOOK OF MEDICAL CHEMISTRY. Sixth American, from the fourth London edition. With numerous illustrations. 1 vol. 12mo. of 350 pp. Cloth, $2 25.

—— INTRODUCTION TO PRACTICAL CHEMISTRY, INCLUDING ANALYSIS. Sixth American, from the sixth London edition, with numerous illustrations. 1 vol. 12mo. of 350 pp. Cloth, $2 25.

BELLAMY'S MANUAL OF SURGICAL ANATOMY. With numerous illustrations. In one royal 12mo. vol. Cloth, $2 25. *(Lately issued.)*

BURNETT (CHARLES H.) THE EAR: ITS ANATOMY, PHYSIOLOGY, AND DISEASES. A Practical Treatise for the Use of Students and Practitioners. In one handsome 8 o vol., with many illustrations. (*In press.*)

BLANDFORD (G. FIELDING). INSANITY AND ITS TREATMENT. With an Appendix of the laws in force in the United States on the Confinement of the Insane, by Dr. Isaac Ray. In one handsome 8vo vol., of 471 pages. Cloth, $3 25.

CARTER (R. BRUDENELL). A PRACTICAL TREATISE ON DISEASES OF THE EYE. With additions and test-types, by John Green, M.D. In one handsome 8vo. vol. of about 500 pages, with 124 illustrations. Cloth, $3 75.

CHAMBERS (T. K.) A MANUAL OF DIET IN HEALTH AND DISEASE. In one handsome octavo volume of 310 pages. Cloth, $2 75. (*Just issued.*)

—— RESTORATIVE MEDICINE. An Harveian Annual Oration delivered at the Royal College of Physicians, London, June 21, 1871. In one small 12mo. volume. Cloth, $1 00.

COOPER (B. B.) LECTURES ON THE PRINCIPLES AND PRACTICE OF SURGERY. In one large 8vo. vol. of 750 pages. Cloth, $2 00.

CARPENTER (WM. B.) PRINCIPLES OF HUMAN PHYSIOLOGY. From the Eighth English Edition. In one large vol. 8vo., of 1083 pages. With 373 illustrations. Cloth, $5 50; leather, raised bands, $6 50. (*Just issued.*)

—— PRIZE ESSAY ON THE USE OF ALCOHOLIC LIQUORS IN HEALTH AND DISEASE. New edition, with a Preface by D. F. Condie, M.D. 1 vol. 12mo. of 178 pages. Cloth, 60 cents.

CLELAND (JOHN) A DIRECTORY FOR THE DISSECTION OF THE HUMAN BODY. In one small royal 12mo. vol. Cloth, $1 25. (*Just issued.*)

CENTURY OF AMERICAN MEDICINE.—A HISTORY OF MEDICINE IN AMERICA, 1776–1876. By E. H. Clarke, M.D., Late Prof. of Materia Medica in Harvard Univ.; Henry J. Bigelow, M.D., Prof. of Surgery in Harvard Univ.; Samuel D. Gross, M.D., D.C.L. Oxon., Prof. of Surgery in Jefferson Med. Coll., Philada.; T. Gaillard Thomas, Prof. of Obstetrics, etc., in Coll. of Phys. and Surgeons, N. Y.; J. S. Billings, M.D., U.S.A., Librarian of National Medical Library, Washington. In one handsome royal 12mo. volume of 366 pages. Cloth, $2 25.

CHRISTISON (ROBERT). DISPENSATORY OR COMMENTARY ON THE PHARMACOPŒIAS OF GREAT BRITAIN AND THE UNITED STATES. With a Supplement by R. E. Griffith. In one 8vo. vol. of over 1000 pages, containing 213 illustrations. Cloth, $4.

CHURCHILL (FLEETWOOD). ON THE THEORY AND PRACTICE OF MIDWIFERY. With notes and additions by D. Francis Condie, M.D. With about 200 illustrations. In one handsome 8vo. vol. of nearly 700 pages. Cloth, $4; leather, $5.

—— ESSAYS ON THE PUERPERAL FEVER, AND OTHER DISEASES PECULIAR TO WOMEN. In one neat octavo vol. of about 450 pages. Cloth, $2 50.

CONDIE (D. FRANCIS). A PRACTICAL TREATISE ON THE DISEASES OF CHILDREN. Sixth edition, revised and enlarged. In one large 8vo. vol. of 800 pages. Cloth, $5 25; leather, $6 25.

CLOWES (FRANK) AN ELEMENTARY TREATISE ON PRACTICAL CHEMISTRY AND QUALITATIVE INORGANIC ANALYSIS. Especially adapted for Lab ratory Use. From the Second English Edition. In one royal 12mo. vol. (*In press.*)

CULLERIER (A.) AN ATLAS OF VENEREAL DISEASES. Translated and edited by FREEMAN J. BUMSTEAD, M.D. A large imperial quarto volume, with 26 plates containing about 150 figures, beautifully colored, many of them the size of life. In one vol., strongly bound in cloth, $17.

—— Same work, in five parts, paper covers, for mailing, $3 per part.

CYCLOPEDIA OF PRACTICAL MEDICINE. By Dunglison, Forbes, Tweedie, and Conolly. In four large super-royal octavo volumes, of 3254 double-columned pages, leather, raised bands, $15. Cloth, $11.

CAMPBELL'S LIVES OF LORDS KENYON, ELLENBOROUGH, AND TENTERDEN. Being the third volume of "Campbell's Lives of the Chief Justices of England." In one crown octavo vol. Cloth, $2.

DALTON (J. C.) A TREATISE ON HUMAN PHYSIOLOGY. Sixth edition, thoroughly revised, and greatly enlarged and improved, with 316 illustrations. In one very handsome 8vo. vol. of 830 pp. Cloth, $5 50; leather, $6 50. (*Just issued.*)

DAVIS (F. H.) LECTURES ON CLINICAL MEDICINE. Second edition, revised and enlarged. In one 12mo. vol. Cloth, $1 75.

DON QUIXOTE DE LA MANCHA. Illustrated edition. In two handsome vols. crown 8vo. Cloth, $2 50; half morocco, $3 70.

DEWEES (W. P.) A TREATISE ON THE DISEASES OF FEMALES. With illustrations. In one 8vo. vol. of 536 pages. Cloth, $3.

DRUITT (ROBERT). THE PRINCIPLES AND PRACTICE OF MODERN SURGERY. A revised American, from the eighth London edition. Illustrated with 432 wood engravings. In one 8vo. vol. of nearly 700 pages. Cloth, $4; leather, $5.

DUNGLISON (ROBLEY) MEDICAL LEXICON; a Dictionary of Medical Science. Containing a concise explanation of the various subjects and terms of Anatomy, Physiology, Pathology, Hygiene, Therapeutics, Pharmacology, Pharmacy, Surgery, Obstetrics, Medical Jurisprudence, and Dentistry. Notices of Climate and of Mineral Waters; Formulæ for Officinal, Empirical, and Dietetic Preparations, with the accentuation and Etymology of the Terms, and the French and other Synonymes. In one very large royal 8vo. vol. New edition. Cloth, $6 50; leather, $7 50. (*Just issued.*)

—— HUMAN PHYSIOLOGY. Eighth edition, thoroughly revised. In two large 8vo. vols. of about 1500 pp., with 532 illus. Cloth, $7.

—— NEW REMEDIES, WITH FORMULÆ FOR THEIR PREPARATION AND ADMINISTRATION. Seventh edition. In one very large 8vo. vol. of 770 pages. Cloth, $4.

DE LA BECHE'S GEOLOGICAL OBSERVER. In one large 8vo. vol. of 700 pages, with 300 illustrations. Cloth, $4.

DANA (JAMES D.) THE STRUCTURE AND CLASSIFICATION OF ZOOPHYTES. With illust. on wood. In one imp. 4to. vol. Cloth, $4.

ELLIS (BENJAMIN). THE MEDICAL FORMULARY. Being a collection of prescriptions derived from the writings and practice of the most eminent physicians of America and Europe. Twelfth edition, carefully revised by A. H. Smith, M. D. In one 8vo. volume of 374 pages. Cloth, $3.

ERICHSEN (JOHN). THE SCIENCE AND ART OF SURGERY. A new and improved American, from the sixth enlarged and revised London edition. Illustrated with 630 engravings on wood. In two large 8vo. vols. Cloth, $9 00; leather, raised bands, $11 00.

ENCYCLOPÆDIA OF GEOGRAPHY. In three large 8vo. vols. Illustrated with 83 maps and about 1100 wood-cuts. Cloth, $5.

FOTHERGILL'S PRACTITIONER'S HANDBOOK OF TREATMENT. In one handsome 8vo. vol. of about 550 pp. Cloth, $4. (*Just issued.*)

FENWICK (SAMUEL). THE STUDENTS' GUIDE TO MEDICAL DIAGNOSIS. From the Third Revised and Enlarged London Edition. In one vol. royal 12mo. Cloth, $2 25.

FLETCHER'S NOTES FROM NINEVEH, AND TRAVELS IN MESOPOTAMIA, ASSYRIA, AND SYRIA. In one 12mo. vol. Cloth, 75 cts.

FOX (TILBURY). EPITOME OF SKIN DISEASES, with Formulæ for Students and Practitioners. In one small 12mo. vol. Cloth, $1.

FLINT (AUSTIN). A TREATISE ON THE PRINCIPLES AND PRACTICE OF MEDICINE. Fourth edition, thoroughly revised and enlarged. In one large 8vo. volume of 1070 pages. Cloth, $6; leather, raised bands, $7. (*Just issued.*)

—— A PRACTICAL TREATISE ON THE PHYSICAL EXPLORATION OF THE CHEST, AND THE DIAGNOSIS OF DISEASES AFFECTING THE RESPIRATORY ORGANS. Second and revised edition. One 8vo. vol. of 595 pages. Cloth, $4 50.

—— A PRACTICAL TREATISE ON THE DIAGNOSIS AND TREATMENT OF DISEASES OF THE HEART. Second edition, enlarged In one neat 8vo. vol. of over 500 pages, $4 00.

—— ON PHTHISIS: ITS MORBID ANATOMY, ETIOLOGY, ETC. in a series of Clinical Lectures. A new work. In one handsome 8vo. volume. Cloth, $3 50. (*Just issued.*)

—— A MANUAL OF PERCUSSION AND AUSCULTATION; of the Physical Diagnosis of Diseases of the Lungs and Heart, and of Thoracic Aneurism. In one handsome royal 12mo. volume. Cloth, $1 75. (*Just issued.*)

—— MEDICAL ESSAYS. In one neat 12mo. volume. Cloth, $1 38.

FOWNES (GEORGE). A MANUAL OF ELEMENTARY CHEMISTRY. From the tenth enlarged English edition. In one royal 12mo. vol. of 857 pages, with 197 illustrations. Cloth, $2 75; leather, $3 25.

FULLER (HENRY). ON DISEASES OF THE LUNGS AND AIR PASSAGES. Their Pathology, Physical Diagnosis, Symptoms, and Treatment. From the second English edition. In one 8vo. vol. of about 500 pages. Cloth, $3 50.

GALLOWAY (ROBERT). A MANUAL OF QUALITATIVE ANALYSIS. From the fifth Eng. ed. In one 12mo. vol. Cloth, $2 50.

GLUGE (GOTTLIEB). ATLAS OF PATHOLOGICAL HISTOLOGY. Translated by Joseph Leidy, M.D., Professor of Anatomy in the University of Pennsylvania, &c. In one vol. imperial quarto, with 320 copperplate figures, plain and colored. Cloth, $4.

GREEN (T. HENRY). AN INTRODUCTION TO PATHOLOGY AND MORBID ANATOMY. Second Amer., from the third Lond. Ed. In one handsome 8vo. vol., with numerous illustrations. Cloth, 2 75. (*Just issued.*)

GRAY (HENRY). ANATOMY, DESCRIPTIVE AND SURGICAL. A new American, from the fifth and enlarged London edition. In one large imperial 8vo. vol. of about 900 pages, with 462 large and elaborate engravings on wood. Cloth, $6; leather, $7. (*Lately issued.*)

GRIFFITH (ROBERT E.) A UNIVERSAL FORMULARY, CONTAINING THE METHODS OF PREPARING AND ADMINISTERING OFFICINAL AND OTHER MEDICINES. Third and Enlarged edition. Edited by John M. Maisch. In one large 8vo. vol. of 800 pages, double columns. Cloth, $4 50; leather, $5 50.

GROSS (SAMUEL D.) A SYSTEM OF SURGERY, PATHOLOGICAL, DIAGNOSTIC, THERAPEUTIC, AND OPERATIVE. Illustrated by 1403 engravings. Fifth edition, revised and improved. In two large imperial 8vo. vols. of over 2200 pages, strongly bound in leather, raised bands, $15.

GROSS (SAMUEL D.) A PRACTICAL TREATISE ON THE DISeases, Injuries, and Malformations of the Urinary Bladder, the Prostate Gland, and the Urethra. Third Edition, thoroughly Revised and Condensed, by Samuel W. Gross, M.D., Surgeon to the Philadelphia Hospital. In one handsome octavo volume, with about two hundred illustrations. Cloth, $4 50. (*Just issued.*)

—— A PRACTICAL TREATISE ON FOREIGN BODIES IN THE AIR PASSAGES. In one 8vo. vol. of 468 pages. Cloth, $2 75.

GIBSON'S INSTITUTES AND PRACTICE OF SURGERY. In two 8vo. vols. of about 1000 pages, leather, $6 50.

GOSSELIN (L.) CLINICAL LECTURES ON SURGERY, Delivered at the Hospital of La Charité. Translated from the French by Lewis A. Stimson, M.D., Surgeon to the Presbyterian Hospital, New York. With illustrations. (*Publishing in the Medical News and Library for* 1876–7.)

HUDSON (A.) LECTURES ON THE STUDY OF FEVER. 1 vol. 8vo., 316 pages. Cloth, $2 50.

HEATH (CHRISTOPHER). PRACTICAL ANATOMY; A MANUAL OF DISSECTIONS. With additions, by W. W. Keen, M.D. In 1 volume; with 247 illustrations. Cloth, $3 50; leather, $4.

HARTSHORNE (HENRY). ESSENTIALS OF THE PRINCIPLES AND PRACTICE OF MEDICINE. Fourth and revised edition. In one 12mo. vol. Cloth, $2 63; half bound, $2 88. (*Lately issued.*)

—— CONSPECTUS OF THE MEDICAL SCIENCES. Comprising Manuals of Anatomy, Physiology, Chemistry, Materia Medica, Practice of Medicine, Surgery, and Obstetrics. Second Edition. In one royal 12mo. volume of over 1000 pages, with 477 illustrations. Strongly bound in leather, $5 00; cloth, $4 25. (*Lately issued.*)

—— A HANDBOOK OF ANATOMY AND PHYSIOLOGY. In one neat royal 12mo. volume, with many illustrations. Cloth, $1 75.

HAMILTON (FRANK H.) A PRACTICAL TREATISE ON FRACTURES AND DISLOCATIONS. Fifth edition, carefully revised. In one handsome 8vo. vol. of 830 pages, with 344 illustrations. Cloth, $5 75; leather, $6 75. (*Just issued.*)

HOLMES (TIMOTHY). SURGERY, ITS PRINCIPLES AND PRACTICE. In one handsome 8vo. volume of 1000 pages, with 411 illustrations. Cloth, $6; leather, with raised bands, $7. (*Just ready.*)

HOBLYN (RICHARD D.) A DICTIONARY OF THE TERMS USED IN MEDICINE AND THE COLLATERAL SCIENCES. In one 12mo. volume, of over 500 double-columned pages. Cloth, $1 50; leather, $2.

HODGE (HUGH L.) ON DISEASES PECULIAR TO WOMEN, INCLUDING DISPLACEMENTS OF THE UTERUS. Second and revised edition. In one 8vo. volume. Cloth, $4 50.

—— THE PRINCIPLES AND PRACTICE OF OBSTETRICS. Illustrated with large lithographic plates containing 159 figures from original photographs, and with numerous wood-cuts. In one large quarto vol. of 550 double-columned pages. Strongly bound in cloth, $14.

HOLLAND (SIR HENRY). MEDICAL NOTES AND REFLECTIONS. From the third English edition. In one 8vo. vol. of about 500 pages. Cloth, $3 50.

HODGES (RICHARD M.) PRACTICAL DISSECTIONS. Second edition. In one neat royal 12mo. vol., half bound, $2.

HUGHES. SCRIPTURE GEOGRAPHY AND HISTORY, with 12 colored maps. In 1 vol. 12mo. Cloth, $1.

HORNER (WILLIAM E.) SPECIAL ANATOMY AND HISTOLOGY. Eighth edition, revised and modified. In two large 8vo. vols. of over 1000 pages, containing 300 wood-cuts. Cloth, $6.

HILL (BERKELEY). SYPHILIS AND LOCAL CONTAGIOUS DISORDERS. In one 8vo. volume of 467 pages. Cloth, $3 25.

HILLIER (THOMAS). HAND-BOOK OF SKIN DISEASES. Second Edition. In one neat royal 12mo. volume of about 300 pp., with two plates. Cloth, $2 25

HALL (MRS. M.) LIVES OF THE QUEENS OF ENGLAND BEFORE THE NORMAN CONQUEST. In one handsome 8vo. vol. Cloth, $2 25; crimson cloth, $2 50; half morocco, $3.

JONES (C. HANDFIELD). CLINICAL OBSERVATIONS ON FUNCTIONAL NERVOUS DISORDERS. Second American Edition. In one 8vo. vol. of 348 pages. Cloth, $3 25.

KIRKES (WILLIAM SENHOUSE). A MANUAL OF PHYSIOLOGY. A new American, from the eighth London edition. One vol., with many illus., 12mo. Cloth, $3 25; leather, $3 75.

KNAPP (F.) TECHNOLOGY; OR CHEMISTRY, APPLIED TO THE ARTS AND TO MANUFACTURES, with American additions, by Prof. Walter R. Johnson. In two 8vo. vols., with 500 ill. Cloth, $6.

KENNEDY'S MEMOIRS OF THE LIFE OF WILLIAM WIRT. In two vols. 12mo. Cloth, $2.

LEA (HENRY C.) SUPERSTITION AND FORCE; ESSAYS ON THE WAGER OF LAW, THE WAGER OF BATTLE, THE ORDEAL, AND TORTURE. Second edition, revised. In one handsome royal 12mo. vol., $2 75.

—— STUDIES IN CHURCH HISTORY. The Rise of the Temporal Power—Benefit of Clergy—Excommunication. In one handsome 12mo. vol. of 515 pp. Cloth, $2 75.

—— AN HISTORICAL SKETCH OF SACERDOTAL CELIBACY IN THE CHRISTIAN CHURCH. In one handsome octavo volume of 602 pages. Cloth, $3 75.

LA ROCHE (R.) YELLOW FEVER. In two 8vo. vols. of nearly 1500 pages. Cloth, $7.

—— PNEUMONIA. In one 8vo. vol. of 500 pages. Cloth, $3.

LINCOLN (D. F.) ELECTRO-THERAPEUTICS. A Condensed Manual of Medical Electricity. In one neat royal 12mo. volume, with illustrations. Cloth, $1 50. (*Just issued.*)

LEISHMAN (WILLIAM). A SYSTEM OF MIDWIFERY. Including the Diseases of Pregnancy and the Puerperal State. Second American, from the Second English Edition. With additions, by J. S. Parry, M.D. In one very handsome 8vo. vol. of 800 pages and 200 illustrations. Cloth, $5; leather, $6. (*Just issued.*)

LAURENCE (J. Z.) AND MOON (ROBERT C.) A HANDY-BOOK OF OPHTHALMIC SURGERY. Second edition, revised by Mr. Laurence. With numerous illus. In one 8vo. vol. Cloth, $2 75.

LEHMANN (C. G.) PHYSIOLOGICAL CHEMISTRY. Translated by George F. Day, M. D. With plates, and nearly 200 illustrations. In two large 8vo. vols., containing 1200 pages. Cloth, $6.

—— A MANUAL OF CHEMICAL PHYSIOLOGY. In one very handsome 8vo. vol. of 336 pages. Cloth, $2 25.

LAWSON (GEORGE). INJURIES OF THE EYE, ORBIT, AND EYELIDS, with about 100 illustrations. From the last English edition. In one handsome 8vo. vol. Cloth, $3 50.

LUDLOW (J. L.) A MANUAL OF EXAMINATIONS UPON ANATOMY, PHYSIOLOGY, SURGERY, PRACTICE OF MEDICINE, OBSTETRICS, MATERIA MEDICA, CHEMISTRY, PHARMACY, AND THERAPEUTICS. To which is added a Medical Formulary. Third edition. In one royal 12mo. vol. of over 800 pages. Cloth, $3 25; leather, $3 75.

LYNCH (W. F.) A NARRATIVE OF THE UNITED STATES EXPEDITION TO THE DEAD SEA AND RIVER JORDAN. In one large octavo vol., with 28 beautiful plates and two maps. Cloth, $3.

—— Same Work, condensed edition. One vol. royal 12mo. Cloth, $1.

LEE (HENRY) ON SYPHILIS. In one 8vo. vol. Cloth, $2 25.

LYONS (ROBERT D.) A TREATISE ON FEVER. In one neat 8vo. vol. of 362 pages. Cloth, $2 25.

MARSHALL (JOHN). OUTLINES OF PHYSIOLOGY, HUMAN AND COMPARATIVE. With Additions by FRANCIS G. SMITH, M. D., Professor of the Institutes of Medicine in the University of Pennsylvania. In one 8vo. volume of 1026 pages, with 122 illustrations. Strongly bound in leather, raised bands, $7 50. Cloth, $6 50.

MACLISE (JOSEPH). SURGICAL ANATOMY. In one large imperial quarto vol., with 68 splendid plates, beautifully colored; containing 190 figures, many of them life size. Cloth, $14.

MEIGS (CHAS. D.). ON THE NATURE, SIGNS, AND TREATMENT OF CHILDBED FEVER. In one 8vo. vol. of 365 pages. Cloth, $2.

MILLER (JAMES). PRINCIPLES OF SURGERY. Fourth American, from the third Edinburgh edition. In one large 8vo. vol. of 700 pages, with 240 illustrations. Cloth, $3 75.

—— THE PRACTICE OF SURGERY. Fourth American, from the last Edinburgh edition. In one large 8vo. vol. of 700 pages, with 364 illustrations. Cloth, $3 75.

MONTGOMERY (W. F.) AN EXPOSITION OF THE SIGNS AND SYMPTOMS OF PREGNANCY. From the second English edition. In one handsome 8vo. vol. of nearly 600 pages. Cloth, $3 75.

MULLER (J.) PRINCIPLES OF PHYSICS AND METEOROLOGY. In one large 8vo. vol. with 550 wood-cuts, and two colored plates. Cloth, $4 50.

MIRABEAU; A LIFE HISTORY. In one 12mo. vol. Cloth, 75 cts.

MACFARLAND'S TURKEY AND ITS DESTINY. In 2 vols. royal 12mo. Cloth, $2.

MARSH (MRS.) A HISTORY OF THE PROTESTANT REFORMATION IN FRANCE. In 2 vols. royal 12mo. Cloth, $2.

NELIGAN (J. MOORE). AN ATLAS OF CUTANEOUS DISEASES. In one quarto volume, with beautifully colored plates, &c. Cloth, $5 50.

NEILL (JOHN) AND SMITH (FRANCIS G.) COMPENDIUM OF THE VARIOUS BRANCHES OF MEDICAL SCIENCE. In one handsome 12mo. vol. of about 1000 pages, with 374 wood-cuts. Cloth, $4; leather, raised bands, $4 75.

NIEBUHR (B. G.) LECTURES ON ANCIENT HISTORY; comprising the history of the Asiatic Nations, the Egyptians, Greeks, Macedonians, and Carthagenians. Translated by Dr. L. Schmitz. In three neat volumes, crown octavo. Cloth, $5 00.

ODLING (WILLIAM). A COURSE OF PRACTICAL CHEMISTRY FOR THE USE OF MEDICAL STUDENTS. In one 12mo. vol. of 261 pp., with 75 illustrations. Cloth, $2.

PLAYFAIR (W. S.) A TREATISE ON THE SCIENCE AND PRACTICE OF MIDWIFERY. In one handsome octavo vol. of 576 pp., with 166 illustrations, and two plates. Cloth, $4; leather, $5. (*Just issued.*)

PAVY (F. W.) A TREATISE ON THE FUNCTION OF DIGESTION, ITS DISORDERS AND THEIR TREATMENT. From the second London ed. In one 8vo. vol. of 246 pp. Cloth, $2.

—— A TREATISE ON FOOD AND DIETETICS, PHYSIOLOGICALLY AND THERAPEUTICALLY CONSIDERED. In one neat octavo volume of about 500 pages. Cloth, $4 75. (*Just issued.*)

PARRISH (EDWARD). A TREATISE ON PHARMACY. With many Formulæ and Prescriptions. Fourth edition. Enlarged and thoroughly revised by Thomas S. Wiegand. In one handsome 8vo. vol. of 977 pages, with 280 illus. Cloth, $5 50; leather, $6 50.

PIRRIE (WILLIAM) THE PRINCIPLES AND PRACTICE OF SURGERY. In one handsome octavo volume of 780 pages, with 316 illustrations. Cloth, $3 75.

PEREIRA (JONATHAN). MATERIA MEDICA AND THERAPEUTICS. An abridged edition. With numerous additions and references to the United States Pharmacopœia. By Horatio C. Wood, M. D. In one large octavo volume, of 1040 pages, with 236 illustrations. Cloth, $7 00; leather, raised bands, $8 00.

PULSZKY'S MEMOIRS OF AN HUNGARIAN LADY. In one neat royal 12mo. vol. Cloth, $1.

PAGET'S HUNGARY AND TRANSYLVANIA. In two royal 12mo. vols. Cloth, $2.

REMSEN (IRA). THE PRINCIPLES OF CHEMISTRY. In one handsome 12mo. vol Cloth, $1 50. ((*Just issued.*)

ROBERTS (WILLIAM). A PRACTICAL TREATISE ON URINARY AND RENAL DISEASES. A second American, from the second London edition. With numerous illustrations and a colored plate. In one very handsome 8vo. vol. of 616 pages. Cloth, $4 50.

RAMSBOTHAM (FRANCIS H.) THE PRINCIPLES AND PRACTICE OF OBSTETRIC MEDICINE AND SURGERY. In one imperial 8vo. vol. of 650 pages, with 64 plates, besides numerous woodcuts in the text. Strongly bound in leather, $7.

RIGBY (EDWARD). A SYSTEM OF MIDWIFERY. Second American edition. In one handsome 8vo. vol. of 422 pages. Cloth, $2 50.

RANKE'S HISTORY OF THE TURKISH AND SPANISH EMPIRES in the 16th and beginning of 17th Century. In one 8vo. volume, paper, 25 cts.

—— HISTORY OF THE REFORMATION IN GERMANY. Parts I., II., III. In one vol. Cloth, $1.

SCHAFER (EDWARD ALBERT) A COURSE OF PRACTICAL HISTOLOGY: A Manual of the Microscope for Medical Students. In one handsome octavo vol. With many illust. Cloth, $2. (*Just issued.*)

SMITH (EUSTACE). ON THE WASTING DISEASES OF CHILDREN. Second American edition, enlarged. In one 8vo. vol. Cloth, $2 50.

SARGENT (F. W.) ON BANDAGING AND OTHER OPERATIONS OF MINOR SURGERY. New edition, with an additional chapter on Military Surgery. In one handsome royal 12mo. vol. of nearly 400 pages, with 184 wood-cuts. Cloth, $1 75.

SMITH (J. LEWIS.) A TREATISE ON THE DISEASES OF INFANCY AND CHILDHOOD. Third Edition, revised and enlarged. In one large 8vo. volume of 724 pages, with illustrations. Cloth, $5; leather, $6. *(Just issued.)*

SHARPEY (WILLIAM) AND QUAIN (JONES AND RICHARD). HUMAN ANATOMY. With notes and additions by Jos. Leidy, M.D., Prof. of Anatomy in the University of Pennsylvania. In two large 8vo. vols. of about 1300 pages, with 511 illustrations. Cloth, $6.

SKEY (FREDERIC C.) OPERATIVE SURGERY. In one 8vo. vol. of over 650 pages, with about 100 wood-cuts. Cloth, $3 25.

SLADE (D. D.) DIPHTHERIA; ITS NATURE AND TREATMENT. Second edition. In one neat royal 12mo. vol. Cloth, $1 25.

SMITH (HENRY H.) AND HORNER (WILLIAM E.) ANATOMICAL ATLAS. Illustrative of the structure of the Human Body. In one large imperial 8vo. vol., with about 650 beautiful figures. Cloth, $4 50.

SMITH (EDWARD). CONSUMPTION; ITS EARLY AND REMEDIABLE STAGES. In one 8vo. vol. of 254 pp. Cloth, $2 25.

STILLE (ALFRED). THERAPEUTICS AND MATERIA MEDICA. Fourth edition, revised and enlarged. In two large and handsome volumes 8vo. Cloth, $10; leather, $12. *(Just issued.)*

STILLE (ALFRED) AND MAISCH (JOHN M.) THE NATIONAL DISPENSATORY: Embracing the Chemistry, Botany, Materia Medica, Pharmacy, Pharmacodynamics, and Therapeut'cs of the Pharmacopœias of the United States and Great Britain. For the Use of Physicians and Pharmaceutis s. In one handsome 8vo. vol, with numerous illustrati ns. *(n press.)*

SCHMITZ AND ZUMPT'S CLASSICAL SERIES. In royal 18mo.

CORNELII NEPOTIS LIBER DE EXCELLENTIBUS DUCIBUS EXTERARUM GENTIUM, CUM VITIS CATONIS ET ATTICI. With notes, &c. Price in cloth, 60 cents; half bound, 70 cts.

C. I. CÆSARIS COMMENTARII DE BELLO GALLICO. With notes, map, and other illustrations. Cloth, 60 cents; half bound, 70 cents.

C. C. SALLUSTII DE BELLO CATILINARIO ET JUGURTHINO. With notes, map, &c. Price in cloth, 60 cents; half bound, 70 cents.

Q. CURTII RUFII DE GESTIS ALEXANDRI MAGNI LIBRI VIII. With notes, map, &c. Price in cloth, 80 cents; half bound, 90 cents.

P. VIRGILII MARONIS CARMINA OMNIA. Price in cloth, 85 cents; half bound, $1.

M. T. CICERONIS ORATIONES SELECTÆ XII. With notes, &c. Price in cloth, 70 cents; half bound, 80 cents.

ECLOGÆ EX Q. HORATII FLACCI POEMATIBUS. With notes, &c. Price in cloth, 70 cents; half bound, 80 cents.

ADVANCED LATIN EXERCISES, WITH SELECTIONS FOR READING. Revised. Cloth, price 60 cents; half bound, 70 cents.

SWAYNE (JOSEPH GRIFFITHS). OBSTETRIC APHORISMS. A new American, from the fifth revised English edition. With additions by E. R. Hutchins, M. D. In one small 12mo. vol. of 177 pp., with illustrations. Cloth, $1 25.

STURGES (OCTAVIUS). AN INTRODUCTION TO THE STUDY OF CLINICAL MEDICINE. In one 12mo. vol. Cloth, $1 25.

SCHOEDLER (FREDERICK) AND MEDLOCK (HENRY). WONDERS OF NATURE. An elementary introduction to the Sciences of Physics, Astronomy, Chemistry, Mineralogy, Geology, Botany, Zoology, and Physiology. Translated from the German by H. Medlock. In one neat 8vo. vol., with 679 illustrations. Cloth, $3.

STOKES (W.) LECTURES ON FEVER. In one 8vo. vol. Cloth, $2.

SMALL BOOKS ON GREAT SUBJECTS. Twelve works; each one 10 cents, sewed, forming a neat and cheap series; or done up in 3 vols., cloth, $1 50.

STRICKLAND (AGNES). LIVES OF THE QUEENS OF HENRY THE VIII. AND OF HIS MOTHER. In one crown octavo vol., extra cloth, $1; black cloth, 90 cents.

——MEMOIRS OF ELIZABETH, SECOND QUEEN REGNANT OF ENGLAND AND IRELAND. In one crown octavo vol., extra cloth, $1 40; black cloth, $1 30.

TANNER (THOMAS HAWKES). A MANUAL OF CLINICAL MEDICINE AND PHYSICAL DIAGNOSIS. Third American from the second revised English edition. Edited by Tilbury Fox, M. D. In one handsome 12mo. volume of 366 pp. Cloth, $1 50.

—— ON THE SIGNS AND DISEASES OF PREGNANCY. From the second English edition. With four colored plates and numerous illustrations on wood. In one vol. 8vo. of about 500 pages. Cloth, $4 25.

TUKE (DANIEL HACK). INFLUENCE OF THE MIND UPON THE BODY. In one handsome 8vo. vol. of 416 pp. Cloth, $3 25.

TAYLOR (ALFRED S.) MEDICAL JURISPRUDENCE. Seventh American edition. Edited by John J. Reese, M.D. In one large 8vo. volume of 879 pages. Cloth, $5; leather, $6. (*Just issued.*)

—— PRINCIPLES AND PRACTICE OF MEDICAL JURISPRUDENCE. From the Second English Edition. In two large 8vo. vols. Cloth, $10; leather, $12. (*Just issued.*)

—— ON POISONS IN RELATION TO MEDICINE AND MEDICAL JURISPRUDENCE. Third American from the Third London Edition. 1 vol. 8vo. of 788 pages, with 104 illustrations. Cloth, $5 50; leather, $6 50. (*Just issued.*)

THOMAS (T. GAILLARD). A PRACTICAL TREATISE ON THE DISEASES OF FEMALES. Fourth edition, thoroughly revised. In one large and handsome octavo volume of 801 pages, with 191 illustrations. Cloth, $5 00; leather, $6 00. (*Just issued.*)

TODD (ROBERT BENTLEY). CLINICAL LECTURES ON CERTAIN ACUTE DISEASES. In one vol. 8vo. of 320 pp., cloth, $2 50.

THOMPSON (SIR HENRY). CLINICAL LECTURES ON DISEASES OF THE URINARY ORGANS. Second and revised edition. In one 8vo. volume, with illustrations. Cloth, $2 25. (*Just issued.*)

—— THE DISEASES OF THE PROSTATE, THEIR PATHOLOGY AND TREATMENT. Fourth edition, revised. In one very handsome 8vo. vol. of 355 pp., with 13 plates. Cloth, $3 75.

THOMPSON (SIR HENRY). THE PATHOLOGY AND TREATMENT OF STRICTURE OF THE URETHRA AND URINARY FISTULÆ. From the third English edition. In one 8vo. vol. of 359 pp., with illustrations. Cloth, $3 50.

WALSHE (W. H.) PRACTICAL TREATISE ON THE DISEASES OF THE HEART AND GREAT VESSELS. Third American from the third revised London edition. In one 8vo. vol. of 420 pages. Cloth, $3.

WATSON (THOMAS). LECTURES ON THE PRINCIPLES AND PRACTICE OF PHYSIC. A new American from the fifth and enlarged English edition, with additions by H. Hartshorne, M.D. In two large and handsome octavo volumes. Cloth, $9; leather, $11.

WÖHLER'S OUTLINES OF ORGANIC CHEMISTRY. Translated from the 8th German edition, by Ira Remsen, M.D. In one neat 12mo. vol Cloth, $3 00. (*Lately issued.*)

WELLS (J. SOELBERG). A TREATISE ON THE DISEASES OF THE EYE. Second American, from the Third English edition, with additions by I. Minis Hays, M.D. In one large and handsome octavo vol., with 6 colored plates and many wood-cuts, also selections from the test-types of Jaeger and Snellen. Cloth, $5 00; leather, $6 00.

WHAT TO OBSERVE AT THE BEDSIDE AND AFTER DEATH IN MEDICAL CASES. In one royal 12mo. vol. Cloth, $1.

WEST (CHARLES). LECTURES ON THE DISEASES PECULIAR TO WOMEN. Third American from the Third English edition. In one octavo volume of 550 pages. Cloth, $3 75; leather, $4 75.

—— LECTURES ON THE DISEASES OF INFANCY AND CHILDHOOD. Fifth American from the sixth revised English edition. In one large 8vo. vol. of 670 closely printed pages. Cloth, $4 50; leather, $5 50. (*Just issued.*)

—— ON SOME DISORDERS OF THE NERVOUS SYSTEM IN CHILDHOOD. From the London Edition. In one small 12mo. volume. Cloth, $1.

WILLIAMS (CHARLES J. B. and C. T.) PULMONARY CONSUMPTION: ITS NATURE, VARIETIES, AND TREATMENT. In one neat octavo volume. Cloth, $2 50.

WILSON (ERASMUS). A SYSTEM OF HUMAN ANATOMY. A new and revised American from the last English edition. Illustrated with 397 engravings on wood. In one handsome 8vo. vol. of over 600 pages. Cloth, $4; leather, $5.

—— ON DISEASES OF THE SKIN. The seventh American from the last English edition. In one large 8vo. vol. of over 800 pages. Cloth, $5.

Also, A SERIES OF PLATES, illustrating "Wilson on Diseases of the Skin," consisting of 20 plates, thirteen of which are beautifully colored, representing about one hundred varieties of Disease. $5 50.

Also, the TEXT AND PLATES, bound in one volume. Cloth, $10.

—— THE STUDENT'S BOOK OF CUTANEOUS MEDICINE. In one handsome royal 12mo. vol. Cloth, $3 50.

WINCKEL ON PATHOLOGY AND TREATMENT OF CHILDBED. With Additions by the Author. Translated by Chadwick. In one handsome octavo volume of 484 pages. Cloth, $4. (*Just issued.*)

ZEISSL ON VENEREAL DISEASES. Translated by Sturgis. (*Preparing.*)

CPSIA information can be obtained
at www.ICGtesting.com
Printed in the USA
LVHW051747120921
697675LV00001B/11